Nonlinear Theory of Generalized Functions

CHAPMAN & HALL/CRC
Research Notes in Mathematics Series

Submission of proposals for consideration
Suggestions for publication, in the form of outlines and representative samples, are invited by the Editorial Board for assessment. Intending authors should approach one of the main editors or another member of the Editorial Board, citing the relevant AMS subject classifications. Alternatively, outlines may be sent directly to the publisher's offices. Refereeing is by members of the board and other mathematical authorities in the topic concerned, throughout the world.

Preparation of accepted manuscripts
On acceptance of a proposal, the publisher will supply full instructions for the preparation of manuscripts in a form suitable for direct photo-lithographic reproduction. Specially printed grid sheets can be provided. Word processor output, subject to the publisher's approval, is also acceptable.

Illustrations should be prepared by the authors, ready for direct reproduction without further improvement. The use of hand-drawn symbols should be avoided wherever possible, in order to obtain maximum clarity of the text.

The publisher will be pleased to give guidance necessary during the preparation of a typescript and will be happy to answer any queries.

Important note
In order to avoid later retyping, intending authors are strongly urged not to begin final preparation of a typescript before receiving the publisher's guidelines. In this way we hope to preserve the uniform appearance of the series.

Michael Grosser, Günther Hörmann,
Michael Kunzinger,
and Michael Oberguggenberger
(Editors)

Nonlinear Theory of Generalized Functions

Proceedings of the Workshop *Nonlinear Theory of Nonlinear Functions*

Erwin-Schrödinger-Institute, Vienna
October – December 1997

CRC Press
Taylor & Francis Group
Boca Raton London New York

CRC Press is an imprint of the
Taylor & Francis Group, an **informa** business
A CHAPMAN & HALL BOOK

Chapman and Hall/CRC
2385 NW Executive Center Drive, Suite 320, Boca Raton FL 33431

and by CRC Press
4 Park Square, Milton Park, Abingdon, Oxon, OX14 4RN

First issued in hardback 2020

CRC Press is an imprint of Taylor & Francis Group, LLC

Library of Congress Cataloging-in-Publication Data

Nonlinear theory of generalized functions : proceedings of the workshop
nonlinear theory of nonlinear functions / edited by Michael Grosser
... [et al.].
p. cm. -- (Chapman & Hall/CRC research notes in mathematics
series)
Includes bibliographical references.
ISBN 0-8493-0649-3 (alk. paper)
1. Theory of distributions (Functional analysis)--Congresses.
2. Nonlinear theories--Congresses. I. Grosser, Michael.
II. Series.
QA324.N66 1999
515'.782—dc21 98-52356
CIP

Library of Congress Card Number 98-52356

ISBN: 978-1-138-41774-8 (hbk)
ISBN: 978-0-8493-0649-5 (pbk)

Visit the Taylor & Francis Web site at
http://www.taylorandfrancis.com

and the CRC Press Web site at
http://www.crcpress.com

Preface

This volume contains the proceedings of the workshop "Nonlinear Theory of Generalized Functions", held at the Erwin-Schrödinger International Institute for Mathematical Physics, Vienna in the period October – December 1997. The project brought together 30 invited visitors as well as about 20 guests and local participants from various fields in mathematical analysis and physics.

The general goal of the project was to investigate the interplay of nonlinearity and creation and propagation of singularities in models involving differential equations and generalized functions. These questions arise in a variety of fields, such as nonlinear partial differential equations, noise-driven stochastic partial differential equations, general relativity and geometry with singularities. In all these cases, non-differentiable functions or distributions arise as candidates for solutions to nonlinear differential equations. To cope with this situation new tools are needed which may go beyond classical concepts of weak solutions. The theory of differential algebras of generalized functions, developed in the past two decades, served as the central theme of the project. One decisive aim, though, was to address applications, and to relate these concepts to and compare them with more classical approaches.

Thus, to our delight, this volume has resulted in a compilation of various approaches to the problem of singularities in nonlinear models. It demonstrates the many facets research on these questions currently exhibits, and also exemplifies the encompassing scope of its applications.

For ease of presentation, we have grouped the papers in five sections. The section on nonlinear partial differential equations includes the investigation of weak solutions in evolution equations, shock formation and delta-waves, boundary layers as well as solutions in generalized function algebras. The section on structure theory deals with various current topics in generalized function algebras. Generalized actions of Lie groups as well as generalized functions methods in the theory of relativity are the subjects of the section on geometry and general relativity. Stochastic nonlinear partial differential equations enter the field due to the strong singularities introduced

by the driving noise. Here white noise calculus and generalized function algebras offer alternative approaches. The final section completes the picture by presenting methods from nonstandard analysis, which is also capable of handling singularities.
It is our pleasure to thank all visitors and participants of the workshop for their scientific contributions, active engagement in joint research and discussions, and for their joyful spirit in making this project such a success. The Erwin-Schrödinger-Institute provides an excellent atmosphere for research, discussions and exchange of ideas. We are grateful to the Advisory Committee, and especially to President J. Yngvason, Directors P. Michor and K. Schmidt and Honorary President W. Thirring for granting and support of the project. Special thanks are due to the administrative director L. Hartyani and secretary A. Ojeda for their efficient handling of the daily organization while at the same time providing such a pleasant and friendly atmosphere.
We thank the staff of CRC, especially M. Lince and M. Williams for their support and patience in publishing this volume.

Vienna and Innsbruck, September 1998,

Michael Grosser
Günther Hörmann
Michael Kunzinger
Michael Oberguggenberger

Contents

PARTIAL DIFFERENTIAL EQUATIONS

On the Structure of Singularities in Solutions of the Nonlinear Schrödinger Equation for the Critical Case, $p = 4/n$

JAIME ANGULO, JERRY L. BONA, FELIPE LINARES, AND MARCIA SCIALOM

1 Introduction

This paper is concerned with nonlinear dispersive wave equations whose solutions sometimes develop singularities in finite time. Here, attention is given principally to the initial-value problem in the *critical case* for the focusing nonlinear Schrödinger equation (NLS-equation henceforth),

$$\begin{cases} iu_t + \Delta u + \tau |u|^p u = 0, & x \in \mathbb{R}^n,\ t \geqq 0, \\ u(x,0) = u_0(x) \end{cases} \tag{1.1}$$

where Δ is the Laplace operator on $\mathbb{R}^n$, $\tau > 0$, and $u : [0,T) \times \mathbb{R}^n \to \mathbb{C}$ for some $T > 0$.

For $p < \frac{4}{n}$, it is well known that the problem (1.1) has global solutions no matter what the size of the initial data $u_0(x)$ in $H^1(\mathbb{R}^n)$ (cf. [C], Chapter 6). Moreover, for any p with $0 < p < \frac{4}{n-2}$ (any $p > 0$ if $n = 1$ or 2), the NLS-equation admits standing-wave solutions of the form

$$u_\lambda(x,t) = G(x)e^{i\lambda t}, \tag{1.2}$$

where $\lambda > 0$ and $G = G_\lambda$ is a real-valued, positive, radially symmetric function (called a ground-state) which is rapidly decreasing to zero at infinity [BL1], [BL2], [BLP], [S]. When $p < \frac{4}{n}$, these standing waves are orbitally stable in the following sense (see [C], [CL], [W4]). For every $\epsilon > 0$, there is a $\delta = \delta(\epsilon) > 0$ such that if $u_0 \in H^1(\mathbb{R}^n)$ and $\|u_0 - G\|_1 \leqq \delta$, then there are maps $\theta : \mathbb{R} \to \mathbb{R}$ and $\sigma : \mathbb{R} \to \mathbb{R}^n$ such that if u is the solution of (1.1) with initial data u_0, then

$$\|u(\cdot,t) - e^{i\theta(t)}G(\cdot - \sigma(t))\|_1 \leqq \epsilon \tag{1.3}$$

for all $t > 0$, where $\|h\|_1^2 = \|h\|^2 + \|\nabla h\|^2$ connotes the standard norm of a function $h \in H^1(\mathbb{R}^n)$ and the unadorned symbol $\|\cdot\|$ is the $L^2(\mathbb{R}^n)$-norm. It is worth remark also that G is unique up to spatial translations and phase shifts.

If the initial data u_0 is small enough in the $H^1(\mathbb{R}^n)$-norm and $p \geqq \frac{4}{n}$, there still exist global solutions for (1.1), but the ground states are known to be unstable in this case (cf. [BC]).

The case $p = \frac{4}{n}$ is called the *critical case* for the NLS-equation because solutions of (1.1) with $\tau > 0$, $p \geqq \frac{4}{n}$ and initial data $u_0 \in H^1(\mathbb{R}^n)$ may blow up in finite time, whereas those for $p < \frac{4}{n}$ do not. This singular behavior is an indication of certain physical phenomena, for instance, the energy transfer to particles via the "Langmuir collapse" in plasma physics [Z], and, in nonlinear optics, the collapse of solutions is the "self-focusing" of light pulses in a nonlinear and dispersive medium [CGT].

The existence of solutions and stability or instability of standing waves for the NLS-equation has been considered in detail by many scientists. A small sample of papers that emphasize especially the aspects of interest here might include [BL1], [BL2], [C], [CL], [GSS], [GV], [LBSK], [S], [W1], [W4] and the references cited in these works.

As mentioned, our focus is on the critical case $p = \frac{4}{n}$ where light will be shed on how the blow-up occurs. (Some of our results are still valid for $p > \frac{4}{n}$.) To understand the blow-up phenomenon, consider a ground state $G = G_\lambda$ which defines a solution of the NLS-equation as in (1.2). The function G necessarily satisfies

$$\Delta G - \lambda G + G^{p+1} = 0, \tag{1.4}$$

where τ has been set equal to 1. (As long as $\tau > 0$, we can always assume $\tau = 1$ by rescaling the dependent variable.) If one assumes in (1.1) with $p = \frac{4}{n}$ that $u_0 \in H^1(\mathbb{R}^n)$ has $\|u_0\| < \|G\|$, then the corresponding solution is globally defined (cf. [W2]). This result is sharp in the sense that for any $R \geqq \|G\|$, there are initial data u_0 with $\|u_0\| = R$ such that there is a $t^* < \infty$ for which the solution of the NLS-equation corresponding to u_0 belongs to the class $C([0, t^*); H^1(\mathbb{R}^n))$ and

$$\lim_{t \to t^{*-}} \|\nabla u(\cdot, t)\| = +\infty. \tag{1.5}$$

It is well understood that for (1.5) to hold, it must be the case that the energy H, which is a conserved quantity under the NLS-flow, satisfies

$$H(u_0) = \|\nabla u_0\|^2 - \frac{2}{p+2} |u_0|_{p+2}^{p+2} < 0.$$

As above and henceforth, the unadorned symbol $\|\cdot\|$ is the $L^2(R^n)$-norm while the symbol $|\cdot|_q$ represents the $L^q(R^n)$-norm, $1 \leqq q \leqq \infty$ and, incidentally, the inner-product in $L^2(R^n)$ will be indicated by $< \cdot, \cdot >$.
The first result describing the asymptotic behavior of blowing-up solutions of the NLS-equation appears to have been that of M. Weinstein in [W3] and [W5]. Later, Laedke *et al.* in [LBSK] studied Weinstein's result and attempted to extend it to a larger class of initial data. Our purpose here is to provide some details related to the last cited work, and to add precision to some of the resulting theory. Roughly speaking, the result in view is that for suitable, negative energy data that are close to an unstable ground state, the corresponding solutions blow up in finite time. Moreover, it will be shown that the blow up near ground-state solutions for the case $p = \frac{4}{n}$ is *stable* in a sense made precise in Theorem 2.2 below. We also give a detailed description of the evolution of the stability parameters θ and σ, analogous to those in (1.3), that appear in the proof of Theorem 2.2. These results are modelled on those provided by Bona and Soyeur (see [BS], Theorem 7) in the case $p < \frac{4}{n}$.
This note is organized as follows. Section 2 is devoted to the proof of the stability result and Section 3 is concerned with the evolution of the stability parameters θ and σ.

2 Main result

Henceforth, it is assumed in (1.1) that $\tau = 1$, $p = \frac{4}{n}$ and $u_0 \in H^1(\mathbb{R}^n)$. If it is further assumed that either

$$(*) \quad \begin{cases} a) & n = 1 \text{ and } H(u_0) < 0, \\ b) & n \geqq 2, \; H(u_0) < 0, \text{ and } u_0 \text{ is radially symmetric, or} \\ c) & n \geqq 1, \; |x|u_0 \in L^2(R^n) \text{ and } H(u_0) \leqq 0, \end{cases}$$

then it is known that (1.5) holds for the corresponding solution u of (1.1), (see [N] and [OT]).
Following [LBSK] and [W2], introduce the functions

$$\begin{cases} \phi(x,t) = \mu(t)^{-\frac{n}{2}} u(\frac{x}{\mu(t)}, t), \text{with} \\ \mu(t) = \frac{\|\nabla u(\cdot,t)\|}{\|\nabla G\|}, \qquad 0 \leqq t < t^*, \text{ and } \mu(0) = 1, \end{cases} \tag{2.1}$$

where t^* is the maximal time of existence of the solution of (1.1) under consideration. Note that, unless u is the zero-solution, $0 < \mu(t) < \infty$ for $0 < t < t^*$. The normalization $\mu(0) = 1$ is a temporary one made to simplify the presentation of the argument. It will be dispensed with later. It is easy to check that the function ϕ verifies the identities

$$i) \qquad \|\phi(\cdot,t)\| = \|u(\cdot,t)\| = \|u_0\|, \tag{2.2}$$

$$ii) \qquad \|\nabla\phi(\cdot,t)\| = \|\nabla G\|, \tag{2.3}$$

$$iii) \quad H(\phi(\cdot,t)) = \|\nabla\phi(\cdot,t)\|^2 - \tfrac{2}{p+2}|\phi(\cdot,t)|_{p+2}^{p+2} = \tfrac{1}{\mu^2(t)}H(u(\cdot,t)). \tag{2.4}$$

The identity $i)$ follows from the fact that the charge $N(u) = \|u\|^2$ is a conserved quantity for the NLS-equation. Our first lemma states that the function ϕ is in the same class as u.

Lemma 2.1 *If $u \in C([0,t^*);H^1(\mathbb{R}^n))$, then $\phi \in C([0,t^*);H^1(\mathbb{R}^n))$.*

Proof. This follows immediately since $\mu \in C([0,t^*);\mathbb{R})$ and $0 < \mu(t) < \infty$ for $0 \leqq t < t^*$. ∎

As the stability considered here is with respect to form, i.e., up to translation in space and phase, it is propitious to introduce the orbit

$$\mathcal{O}(G_\lambda) \equiv \{g | g(x) = G_\lambda(x+\alpha_0)e^{i\alpha_1}, (\alpha_0,\alpha_1) \in \mathbb{R}^n \times [0,2\pi)\}$$

of G_λ. An induced metric on the space $H^1(\mathbb{R}^n)$ factored by the closed subset $\mathcal{O}(G_\lambda)$ provides a pseudo-metric on $H^1(\mathbb{R}^n)$ (see [B], [Bo], [CL] and [W4]), namely

$$\rho_\lambda(\phi(\cdot,t),G_\lambda)^2 \equiv \inf_{\substack{\alpha_0\in\mathbb{R}^n \\ \alpha_1\in[0,2\pi)}} \{\|\nabla\phi(\cdot+\alpha_0,t)e^{i\alpha_1} - \nabla G(\cdot)\|^2 + \lambda\|\phi(\cdot+\alpha_0,t)e^{i\alpha_1} - G(\cdot)\|^2\}. \tag{2.5}$$

Define the set $\mathcal{S}$ to be

$$\mathcal{S} = \{u_0 | u_0 \in H^1(\mathbb{R}^n) \text{ and condition (*) holds for } u_0\}. \tag{2.6}$$

Observe that for initial data $u_0 \in \mathcal{S}$, the conditions (*) implies the corresponding solution u of (1.1) obeys (1.5) for some $t^* > 0$.

Theorem 2.2 *Let $p = \frac{4}{n}$, $\lambda > 0$, and let $G = G_\lambda$ be a ground-state solution of (1.2). For any $\epsilon > 0$, there is a $\delta = \delta(\epsilon)$ such that if $u_0 \in \mathcal{S}$ with $\rho_\lambda(u_0,G) < \delta$, and u is a solution of (1.1) corresponding to u_0 whose blow-up time is t^*, say, then $u \in C([0,t^*);H^1(\mathbb{R}^n))$ and*

$$\inf_{\substack{\alpha_0\in\mathbb{R}^n \\ \alpha_1\in[0,2\pi)}} \Big\{\lambda\|u(\cdot,t) - \mu^{\frac{n}{2}}(t)G(\mu(t)\cdot) + \alpha_0)e^{-i\alpha_1}\|^2 + \mu^{-2}(t)\|\nabla u(\cdot,t) - \mu^{\frac{n}{2}}(t)\nabla G(\mu(t)\cdot) + \alpha_0)e^{-i\alpha_1}\|^2\Big\} < \epsilon \tag{2.7}$$

for all $t \in [0, t^*)$, where $\mu(t)$ is as in (2.1).

Proof. Suppose at the outset that $\mu(0) = 1$. The proof is based on the explicit time-dependent functional (see [LBSK], [LBS] and [LS])

$$L_t[u] = \frac{1}{\mu^2(t)} H(u(\cdot,t)) + \lambda\Big(\frac{\|u(\cdot,t)\|}{\|G\|}\Big)^{2k} (\|u(\cdot,t)\|^2 - \|G\|^2), \tag{2.8a}$$

where $k \in \mathbb{N}$ is a parameter that will be determined later. Observe that if u is a solution of (1.1), then at any time $t \in [0, t^*)$, $L_t[u] = L_t[u_0]$ since both the energy H and the charge $\|\cdot\|$ are preserved by the NLS-flow. The functional $L_t[u]$ may be written as a functional defined on ϕ for which the explicit dependence on μ disappears, viz.

$$\widetilde{L}_t[\phi] \equiv L_t[u] = H(\phi(\cdot,t)) + \lambda\Big(\frac{\|\phi(\cdot,t)\|}{\|G\|}\Big)^{2k} (\|\phi(\cdot,t)\|^2 - \|G\|^2). \tag{2.8b}$$

Suppose we are able to establish the inequalities

$$i)\ \Delta\widetilde{L}_t \leqq c_0\|u_0 - G\|_1 \quad \text{and} \tag{2.9}$$

$$ii)\ \Delta\widetilde{L}_t \geqq c_1\|\phi(\cdot,t) - G\|_1^2 - c_2\sum_{j=1}^{4}\|\phi(\cdot,t) - G\|_1^{j+2} -$$

$$-\sum_{j=1}^{2k} c_{k,j}\|\phi(\cdot,t) - G\|_1^{j+2}, \tag{2.10}$$

for $\Delta\widetilde{L}_t = \widetilde{L}_t[\phi] - \widetilde{L}_t[G]$, modulo translations and phase, where $c_0, c_i, c_{k,j}$ are fixed constants. The result in Theorem 2.2 will then follow. Indeed, if $\|u_0 - G\|_1 < \delta_0$, we have from (2.9) that

$$\Delta\widetilde{L}_t \leqq c_0\delta_0. \tag{2.11}$$

As will be argued in more detail later, (2.10), (2.11) and the identity

$$\|\phi(\cdot,t) - G\|_1^2 = \|u(\cdot,t) - \mu^{\frac{n}{2}}(t)G(\mu(t)\cdot)\|^2 + \mu^{-2}(t)\|\nabla u(\cdot,t) - \mu^{\frac{n}{2}}(t)\nabla G(\mu(t)\cdot)\|^2$$

lead to the desired result.

The upper bound (2.9) is a straightforward consequence of $H(u_0) < 0$ and $H(G) \geqq 0$, where the constant c_0 depends on $\|G\|$.

To prove (2.10), write a renormalized version of ϕ as G plus a remainder w, thusly;

$$\phi(x + \alpha_0, t)e^{i\alpha_1} = G(x) + w(x,t), \qquad w = a + ib, \tag{2.12}$$

where $\alpha_0 = \alpha_0(t)$ and $\alpha_1 = \alpha_1(t)$ will be chosen later and where a and b are real functions. Thus w depends on α_0 and α_1. Using the representation (2.12) of the translated and phase-shifted ϕ, we calculate

$$\begin{aligned}
\Delta\widetilde{L}_t &= \widetilde{L}_t[\phi(\cdot+\alpha_0)e^{i\alpha_1}] - \widetilde{L}_t[G] = \widetilde{L}_t[G+w] - \widetilde{L}_t[G] \\
&= H(G+w) - H(G) + \lambda\Big(\frac{\|G+w\|}{\|G\|}\Big)^{2k}(\|G+w\|^2 - \|G\|^2) \\
&= <-\Delta a, a> + <-\Delta b, b> + 2 <a, \Delta G> - 2 <a, G^{p+1}> \\
&\quad -(p+1) <G^p a, a> - <G^p b, b> + 2\lambda <a, G> + \lambda <a, a> \\
&\quad + \lambda <b, b> + 4\frac{k\lambda}{\|G\|^2} <G, a>^2 + R.
\end{aligned}$$

The remainder R can be bounded below in terms of powers of $\|w\|_1$ of cubic order and higher by using the Sobolev imbedding of $H^1(\mathbb{R}^n)$ into $L^p(\mathbb{R}^n)$, which is valid since $p = \frac{4}{n}$. The upshot of these ruminations is the inequality

$$\begin{aligned}
\Delta\widetilde{L}_t \geqq\; & <\mathcal{L}^+ b, b> + <\mathcal{L}^- a, a> + \frac{4k\lambda}{\|G\|^2}(<a, G>)^2 - c_2(\lambda)\sum_{j=1}^{4}\|w\|_1^{j+2} \\
& - \sum_{j=1}^{2k} c_{k,j}(\lambda)\|w\|_1^{j+2}. \qquad (2.13)
\end{aligned}$$

In (2.13) we have followed [LBSK] in introducing the naturally occurring differential operators

$$\mathcal{L}^+ = -\Delta - G^p + \lambda, \qquad (2.14)$$

$$\mathcal{L}^- = -\Delta - (p+1)G^p + \lambda, \qquad (2.15)$$

which are, respectively, the real and imaginary parts of the operator $\widetilde{L}_t$ linearized about the ground state.

The first step to obtaining a suitable lower bound on the quadratic form comprised of the first three terms on the right-hand side of (2.13) is to show there exist maps $\alpha_0 = \alpha_0(t)$ and $\alpha_1 = \alpha_1(t)$ minimizing the function

$$\Omega_t(\alpha_0, \alpha_1) = \|\nabla\phi(\cdot+\alpha_0, t)e^{i\alpha_1} - \nabla G(\cdot)\|^2 + \lambda\|\phi(\cdot+\alpha_0, t)e^{i\alpha_1} - G(\cdot)\|^2.$$

The following lemma is the analog of [Bo], Lemma 1, in the present context.

Lemma 2.3 *Suppose that for some $t_0 \in [0,t^*)$ and some $(\widetilde{\alpha}_0, \widetilde{\alpha}_1) \in \mathbb{R} \times [0,2\pi)$, it is the case that*

$$\Omega_{t_0}(\widetilde{\alpha}_0, \widetilde{\alpha}_1) < \|(-\Delta+\lambda)^{1/2}G\|^2. \tag{2.16}$$

Then, it follows that

$$\inf \{\Omega_{t_0}(\alpha_0,\alpha_1) |\ (\alpha_0,\alpha_1) \in \mathbb{R}^n \times [0,2\pi)\} \tag{2.17}$$

is attained at least once in $\mathbb{R}^n \times [0,2\pi)$.

Proof. It is immediate that $\Omega_{t_0}(\alpha_0,\alpha_1)$ is a continuous function of (α_0,α_1) on $\mathbb{R}^n \times [0,2\pi)$. Moreover, for any $\alpha_1 \in [0,2\pi)$, we have

$$\begin{aligned}\lim_{|\alpha_0|\to\infty} \Omega_{t_0}(\alpha_0,\alpha_1) &= \|\nabla\phi(\cdot,t_0)\|^2 + \|\nabla G\|^2 + \lambda(\|\phi(\cdot,t_0)\|^2 + \|G\|^2) \\ &= \|(-\Delta+\lambda)^{1/2}G\|^2 + \lambda\|u_0\|^2 + \|\nabla\phi(\cdot,t_0)\|^2. \end{aligned} \tag{2.18}$$

The hypothesis (2.16), the continuity of Ω_{t_0}, and (2.18) imply the result. ■

Next, it is established that the infimum in (2.17) is attained at points (α_0,α_1) at least for t_0 in some interval of the form $[0,T]$. To this end, it is sufficient to obtain condition (2.16) in such an interval. Let $\epsilon > 0$ be such that

$$\epsilon^2 < \frac{1}{2}\max\{1,\lambda\}\|(-\Delta+\lambda)^{\frac{1}{2}}G\|^2.$$

The standing-wave solution $G(x)e^{it\lambda}$ is globally defined, and hence by the continuous dependence theory for (1.1) (see [C], Chapter 4 or [GV]), for the value of ϵ just specified and $T > 0$, there exists a $\delta > 0$ such that if $\|u_0 - G\|_1 < \delta$, then the solution u of (1.1) corresponding to u_0 exists at least for $0 \leqq t \leqq T$ and, in addition,

$$\|u(\cdot,t) - G(\cdot)e^{i\lambda t}\|_1 < \epsilon/2,$$

for all $t \in [0,T]$. Of course, if $u_0 \in S$ as well, it follows that there exists $t^* > T$ such that $u \in C([0,t^*); H^1(\mathbb{R}^n))$ and u blows up at t^* in the sense of (1.5).

Because both ϕ and u are continuous mappings of the time-axis into $H^1(\mathbb{R}^n)$ over the interval $[0,t^*)$, it follows that there is a $T_1 > 0$ such that for all $t \in [0,T_1]$,

$$\|\phi(\cdot,t) - \phi(\cdot,0)\|_1 \leqq \frac{\epsilon}{4} \quad \text{and} \quad \|u(\cdot,t) - u(\cdot,0)\|_1 \leqq \frac{\epsilon}{4}.$$

Then, for $0 \leqq t \leqq \widetilde{T} = min\{T, T_1\}$, we have

$$\begin{aligned}\|\phi(\cdot,t) - Ge^{i\lambda t}\|_1 &\leqq \|\phi(\cdot,t) - u(\cdot,t)\|_1 + \|u(\cdot,t) - Ge^{i\lambda t}\|_1 \\ &\leqq \|\phi(\cdot,t) - \phi(0)\|_1 + \|u(0) - u(\cdot,t)\|_1 + \frac{\epsilon}{2} \leqq \epsilon. \end{aligned} \tag{2.19}$$

Thus, the infimum (2.17) is taken on at values $(\alpha_0(t), \alpha_1(t))$ throughout the time-interval $[0, \widetilde{T}]$. We take such values of α_0 and α_1 as providing a meaning for the definition of w in (2.12), at least for $t \in [0, \widetilde{T}]$.

The result of Lemma 2.3 together with (1.4) provide us with compatibility relations on a and b, namely

$$\int_{\mathbb{R}^n} a(x,t) G^p(x) G_{x_i}(x)\, dx = 0, \qquad i = 1, .., n, \tag{2.20a}$$

and,

$$\int_{\mathbb{R}^n} b(x,t) G^{p+1}(x)\, dx = 0. \tag{2.20b}$$

for all $t \in [0, t^*)$. These relations are obtained by differentiating Ω_t with respect to α_0 and α_1 and evaluating at values that minimize Ω_t.

The issue of obtaining the lower bound advertised in (2.13) is addressed in the next several lemmas.

Lemma 2.4 *If $\mathcal{L}^+$ is as defined in (2.14), then there is a positive value C such that*

$$\inf_{\substack{\|f\|=1 \\ f \perp G^{p+1}}} < f, \mathcal{L}^+ f > \ \geqq \ C. \tag{2.21}$$

Proof. Because of the variational characterization of the ground state G as the minimizer of the action (see [BL1], [BL2]), it follows that $\mathcal{L}^+$ is a nonnegative definite, self-adjoint operator on $L^2(\mathbb{R}^n)$ with null space spanned by G. Hence, the infimum on the left-hand side of (2.21) is non-negative. (cf. [C]).

Supposing that this infimum is zero, let $\{f_j\}$ be a sequence of $H^1(\mathbb{R}^n)$-functions with $\|f_j\| = 1$, $f_j \perp G^{p+1}$ and

$$\lim_{j \to \infty} < f_j, \mathcal{L}^+ f_j > = \ 0.$$

Then, for any $\eta > 0$, there is a J such that for $j > J$,

$$0 < \lambda \leqq \int |\nabla f_j|^2\, dx + \lambda \int |f_j|^2\, dx \leqq \int G^p |f_j|^2\, dx + \eta. \tag{2.22}$$

Since $\|G\|_\infty < \infty$, (2.22) implies $\|f_j\|_1$ to be uniformly bounded as j varies. By standard arguments, it follows that there is a subsequence of the $\{f_j\}$, which we denote again by $\{f_j\}$, and an $f^* \in H^1(\mathbb{R}^n)$ such that $f_j \to f^*$ weakly in $H^1(\mathbb{R}^n)$, pointwise almost everywhere, and in $L^2_{loc}(\mathbb{R}^n)$. By virtue of the weak convergence,

f^* satisfies the condition $f^* \perp G^{p+1}$. It is a straightforward consequence of the just mentioned properties of the sequence f_j and the exponential decay of G to 0 as $|x| \to \infty$ that

$$\int G^p |f_j|^2 \, dx \to \int G^p |f^*|^2 \, dx$$

as $j \to \infty$. Taking the limit in (2.22) as $j \to \infty$ thus yields

$$0 < \lambda \leqq \int G^p |f^*|^2 \, dx + \eta.$$

As $\eta > 0$ is arbitrary, it must be the case that $f^* \neq 0$.

It is now shown that the infimum is achieved. Indeed, weak convergence is lower semicontinuous, so

$$\|\nabla f^*\| \leqq \liminf_{j \to \infty} \|\nabla f_j\|_2.$$

Since, also, $< G^p f_j, f_j > \to < G^p f^*, f^* >$ as $j \to \infty$, it is adduced that

$$< f^*, \mathcal{L}^+ f^* > \leqq \liminf_{j \to \infty} < f_j, \mathcal{L}^+ f_j > = 0.$$

Since $f^* \neq 0$, define $g^* = \frac{f^*}{\|f^*\|}$. Then, we have $\|g^*\| = 1, g^* \perp G^{p+1}$, and $< g^*, \mathcal{L}^+ g^* > = 0$.

A consequence of the last reasoning is that there exist nontrivial critical points (g^*, β, θ) for the Lagrange multiplier problem,

$$\begin{cases} \mathcal{L}^+ f = \beta f + \theta G^{p+1}, & \text{subject to} \\ \|f\| = 1 \quad \text{and} & \\ f \perp G^{p+1} & \end{cases} \tag{2.23}$$

Using (2.23) and the fact that $< \mathcal{L}^+ g^*, g^* > = 0$, it is easily seen that $\beta = 0$. It is thereby concluded that

$$\mathcal{L}^+ f = \theta G^{p+1} \tag{2.24}$$

has nontrivial solutions (g^*, θ) satisfying the side constraints. Taking the inner product of (2.24) with G it is determined that $\theta = 0$, and therefore $g^* = \nu G$ for some $\nu \neq 0$, a contradiction since G is not orthogonal to G^{p+1}. Therefore, the minimum in (2.21) is positive and the proof of the lemma is complete. ■

It follows readily from (2.21) that for some $\widetilde{D}_1 > 0$,

$$< f, \mathcal{L}^+ f > \geqq \widetilde{D}_1 \|f\|_1^2, \qquad \text{if} \qquad f \perp G^{p+1}. \tag{2.25}$$

The next lemma will be used to prove an estimate similar to (2.25) for the operator $\mathcal{L}^-$.

Lemma 2.5 *Let A be a self-adjoint operator on $L^2(\mathbb{R}^n)$ having exactly one negative eigenvalue β with corresponding ground-state eigenfunction $f_\beta \geqq 0$ and let $g \in N^\perp(A)$. Assume $< g, f_\beta > \neq 0$ and that*

$$-\infty < \alpha \equiv \min_{\substack{\|f\|=1 \\ <f,g>=0}} < Af, f > .$$

If $< A^{-1}g, g > \leqq 0$, then it must be the case that $\alpha \geqq 0$.

Proof. See Lemma E.1 in [W2]. ∎

Corollary 2.6 *There exists $\gamma < 0$ such that if $g = G + \gamma \Delta G$, then*

$$\min_{\substack{\|f\|=1 \\ <f,g>=0}} \langle \mathcal{L}^- f, f \rangle = 0. \tag{2.26}$$

Proof. The operator $\mathcal{L}^-$ has a unique negative eigenvalue with eigenfunction $f_\lambda > 0$ from Sturm-Liouville theory in dimension $n = 1$ (see [CLe], [ABH]), and variational methods and the mini-max principle in dimension $n \geqq 2$ (see [W4]). Notice that whatever the value of γ, it follows from the equation (1.4) for G that

$$< f_\lambda, g > = < f_\lambda, G > + \gamma < f_\lambda, \lambda G - G^{p+1} > .$$

Since both f_λ and G are everywhere positive, it follows that for small values of γ, the inner product $< f_\lambda, g >$ is non-zero. On the other hand, for any given γ, define the function f by

$$f(x) = -\frac{n}{4\lambda} G(x) - \frac{\lambda\gamma + 1}{2\lambda} x \cdot \nabla G(x)$$

for $x \in \mathbb{R}^n$. Short calculations show that $\mathcal{L}^- f = G + \gamma \Delta G = g$ and that

$$< f, g > = \left(\frac{n}{4} ||G||^2 + \frac{1}{2\lambda} ||\nabla G||^2 \right) \gamma + \left(\frac{2-n}{4} ||\nabla G||^2 \right) \gamma^2.$$

It thus becomes obvious that for small, negative values of γ, it is possible to have both $< f_\lambda, g > \neq 0$ and $< (\mathcal{L}^-)^{-1} g, g > = < f, g > < 0$.

Since $\mathcal{N}(\mathcal{L}^-) = \text{Span}\{G_{x_i} | i = 1, ..., n\}$ (see [W2], [K] and [M]), it is clear from its definition that $g \in \mathcal{N}^\perp(\mathcal{L}^-)$. It follows immediately from Lemma 2.5 that the minimum in (2.26) is non-negative. On the other hand, G_{x_i}, suitably normalized, is orthogonal to g, has norm 1 and $\mathcal{L}^-$ vanishes there, so the minimum cannot be positive. ∎

Lemma 2.7 *If $g \equiv G + \gamma\Delta G$, with $\gamma < 0$ as in the last Corollary, then*

$$\inf\{<\mathcal{L}^- f, f> \ : \ \|f\| = 1, <f, g>= 0, f \perp G^p G_{x_i}, 1 \leqq i \leqq n\} \ = \ D_2 > 0. \quad (2.27)$$

Proof. Because of the Corollary, we know that $D_2 \geqq 0$. Suppose that $D_2 = 0$. Following the proof in Lemma 2.4, one ascertains that there exists f^* such that $\|f^*\| = 1$, $<\mathcal{L}^- f^*, f^*>= 0$, $<f^*, g>= 0$ and $<f^*, G^p G_{x_i}>= 0$, $i = 1, ..., n$. Thus there exists ξ, θ, χ such that

$$\mathcal{L}^- f^* = \xi f^* + \theta(G + \gamma\Delta G) + \chi G^p G_{x_i}, \qquad i = 1, ..., n. \qquad (2.28)$$

In consequence of the conditions satisfied by f^*, it must be that $<\mathcal{L}^- f^*, f^*>= \xi = 0$ and $<\mathcal{L}^- f^*, G_{x_i}>=<f^*, \mathcal{L}^- G_{x_i}>= 0 = \chi \int G^p (G_{x_i})^2 \, dx$, which implies $\chi = 0$. Therefore, $\mathcal{L}^- f^* - \theta g = 0$. But, if f is the auxiliary function arising in the proof of Corollary 2.6, then $\mathcal{L}^- f = g$, whence $\mathcal{L}^-(f^* - \theta f) = 0$, and therefore $f^* - \theta f \in \mathcal{N}(\mathcal{L}^-)$. From the property $<f, g> \neq 0$ established in Corollary 2.6, it follows from the preceding that $\theta = 0$. Therefore, for some non-zero $l \in \mathbb{R}^n$, it is true that $f^* = l \cdot \nabla G$, which is a contradiction since such a function cannot be orthogonal to $G^p G_{x_i}$ for $1 \leqq i \leqq n$. This completes the proof. ■

Remark 2.8. Comparing the results of Corollary 2.6 and Lemma 2.7 with similar results of Weinstein (the case $\gamma = 0$ in [W4]), it is worth pointing out that in his work, the identity (2.26) is true for $p \leqq \frac{4}{n}$, but (2.27) is only true for $p < \frac{4}{n}$.

Attention is now turned to estimating the term $<\mathcal{L}^- a, a> + \frac{4k\lambda}{\|G\|^2}(<a, G>)^2$ in (2.13), where a satisfies the compatibility relations (2.20a). We continue to carry over the notation from Corollary 2.6 and Lemma 2.7. In particular, γ is chosen so that the conclusions of Corollary 2.6 are valid. Let $a_{\shortparallel} = \frac{<a, g>}{\| g\|^2} g$ and let $a_\perp = a - a_{\shortparallel}$. It follows from the properties of a and $g = G + \gamma\Delta G$ that $<a_\perp, \ g>= 0$ and $\int G^p G_{x_i} a_\perp \, dx = 0$, $i = 1, ..., n$. Without loss of generality, take it that $<a, \ g> < 0$. Thus, from Lemma 2.7, the Cauchy-Schwarz inequality and the properties of $a, a_\perp, a_{\shortparallel}$ and g, it follows that

$$\begin{cases} <a_\perp, \mathcal{L}^- a_\perp> \ \geqq \ D_2 \|a_\perp\|^2 \\ <a_{\shortparallel}, \mathcal{L}^- a_{\shortparallel}>= \frac{\|a_{\shortparallel}\|^2}{\| g\|^2} < \ g, \mathcal{L}^- \ g> \\ <a_\perp, \mathcal{L}^- a_{\shortparallel}>= \frac{<a, \ g>}{\| g\|^2} <a_\perp, \mathcal{L}^- \ g> \geqq -D_3 \|a_\perp\| \|a_{\shortparallel}\|, \end{cases} \qquad (2.29)$$

with $D_i > 0, i = 2, 3$. Identity (2.3) and the elementary properties of Hilbert spaces imply that

$$2 < a, \Delta G >= \|\nabla a\|^2 + \|\nabla b\|^2.$$

Thus, from the Cauchy-Schwartz inequality, we obtain (remember, γ and $< a, g >$ are both negative)

$$\begin{aligned} \frac{4k\lambda}{\|G\|^2}(< a, G >)^2 &\geqq \frac{4k\lambda}{\|G\|^2} < a,\ g >^2 - \gamma\frac{4k\lambda}{\|G\|^2} < a,\ g > (\|\nabla a\|^2 + \|\nabla b\|^2) \\ &\geqq \frac{4k\lambda\|\ g\|^2}{\|G\|^2}\|a_{\shortparallel}\|^2 + \gamma\frac{4k\lambda\|\ g\|}{\|G\|^2}\|a\|(\|\nabla a\|^2 + \|\nabla b\|^2) \\ &\geqq \frac{4k\lambda\|\ g\|^2}{\|G\|^2}\|a_{\shortparallel}\|^2 + 4k\lambda\gamma D_4\|a + ib\|_1^3, \end{aligned} \quad (2.30)$$

with $D_4 > 0$ also.

If $\theta > 0$ is chosen so that $D_2 - 2\theta D_3 = D_5 > 0$ and k is fixed in such a way that

$$\frac{4k\lambda\|\ g\|^2}{\|G\|^2} + \frac{< g, \mathcal{L}^-\ g >}{\|\ g\|^2} - 2\frac{D_3}{\theta} \equiv D_6 > 0,$$

it follows from (2.29), (2.30) and Young's inequality that

$$\begin{aligned} < \mathcal{L}^- a, a > + \frac{4k\lambda}{\|G\|^2}(< a, G >)^2 &\geqq D_6\|a_{\shortparallel}\|^2 + D_5\|a_\perp\|^2 + 4k\lambda\gamma D\|w\|_1^3 \\ &\geqq D'\|a\|^2 - D''\|w\|_1^3, \end{aligned} \quad (2.31)$$

for some positive constants D' and D''. With (2.31) in hand, it follows easily from the specific form of the operator $\mathcal{L}^-$ that,

$$< \mathcal{L}^- a, a > + \frac{4k\lambda}{\|G\|^2}(< a, G >)^2 \geqq \widetilde{D}_2\|a\|_1^2 - \widetilde{D}_3\|w\|_1^3, \quad (2.32)$$

with $\widetilde{D}_2, \widetilde{D}_3 > 0$.

Finally, collecting the results (2.25), (2.32) and substituting them in (2.13), there obtains

$$\begin{aligned} \Delta\widetilde{L}_t &\geqq \widetilde{D}_1\|b\|_1^2 + \widetilde{D}_2\|a\|_1^2 - \widetilde{D}_3\|w\|_1^3 - c_2\sum_{j=1}^{4}\|w\|_1^{j+2} - \sum_{j=1}^{2k} c_{k,j}\|w\|_1^{j+2} \\ &\geqq c_1\|w\|_1^2 - c_2\sum_{j=1}^{4}\|w\|_1^{j+2} - \sum_{j=1}^{2k} c_{k,j}\|w\|_1^{j+2}, \end{aligned} \quad (2.33)$$

where $c_1, c_2, c_{k,j}$ are constants that depend only on λ and the dimension of the space.

Now we are in position to prove Theorem 2.2. Suppose first that $u_0 \in \mathcal{S}$ (defined in (2.5)) and $\|u_0 - G\|_1 = \delta$. Then, for $\widetilde{T}$ as in (2.19) and at least for $t \in [0, \widetilde{T}]$, it follows from (2.9) and (2.10) or (2.33) that

$$q(\rho_\lambda(\phi(\cdot,t),G)) \leqq \Delta\widetilde{L}_t \leqq c_0\delta \tag{2.34}$$

where $q(x) = c_1x^2 - c_2\sum_{j=1}^{4} x^{j+2} - \sum_{j=1}^{2k} c_{k,j}x^{j+2}$ and $\rho_\lambda(\phi(\cdot,t),G)$ is as in (2.5). Since $\|w(t)\|_1^2 = \rho_\lambda(\phi(\cdot,t),G)^2$ is a continuous function of $t \in [0,t^*)$ (see Lemma 2 in [Bo]), it follows from the inequality

$$q(\rho(\phi(\cdot,0),G)) \leqq c_0\delta \tag{2.35}$$

and from (2.34) that for $t \in [0,\widetilde{T}]$ we have

$$\rho(\phi(\cdot,t),G) \leqq \epsilon. \tag{2.36}$$

provided the δ is chosen small enough at the outset.
To finish the proof, we show that the inequality (2.36) is still true for $t \in [0,t^*)$. Following the ideas in [Bo], let

$$\mathcal{A} = \{t : \text{the infimum in (2.17) is attained at finite values of } (\alpha_0,\alpha_1)\}.$$

As shown above, $[0,\widetilde{T}] \subset \mathcal{A}$. Let T_1 be the largest value such that $[0,T_1) \subset \mathcal{A}$ and suppose that $T_1 < t^*$. Then from (2.36) we obtain that

$$\inf\ \Omega_t = \rho_\lambda(\phi(\cdot,t),G)^2 \leqq \epsilon^2 \leqq \frac{\|(-\Delta+\lambda)^{1/2}G\|^2}{2}. \tag{2.37}$$

Since $\inf\ \Omega_t$ is a continuous function of t for all $t \in [0,t^*)$, there is a $T > 0$ such that

$$\inf\ \Omega_t < \|(-\Delta+\lambda)^{1/2}G\|^2$$

for $t \in [T_1, T_1+T]$. But then Lemma 2.3 implies that the infimum in (2.17) is taken at finite values of (α_0,α_1) and this contradicts the choice of T_1. Therefore $T_1 = t^*$ and the stability Theorem 2.2 is established for $\mu(0) = 1$.
Now we discuss the general case where we do not necessarily start with $\mu(0) = 1$. First, remark that if G_λ is a solution of (1.4), then $G_\lambda(x) = \lambda^{1/p}R(\lambda^{1/2}x)$ where R satisfies,

$$\Delta R - R + R^{p+1} = 0.$$

Let $u_0 \in S$ obey the restriction

$$\lambda\|u_0 - G\|^2 + \|\nabla u_0 - \nabla G\|^2 < \delta^2, \tag{2.38}$$

where δ will be determined presently. Corresponding to $\lambda_1 > 0$, a solution G_{λ_1} of (1.4) has $\|\nabla G_{\lambda_1}\|^2 = \lambda_1\|\nabla R\|^2$. It is therefore possible to choose λ_1 such that $\|\nabla G_{\lambda_1}\|^2 = \|\nabla u_0\|^2$. Then if

$$\mu_{\lambda_1}(t) = \frac{\|\nabla u(\cdot,t)\|}{\|\nabla G_{\lambda_1}\|},$$

it is obviously the case that $\mu_{\lambda_1}(0) = 1$.

The idea is to apply the preceding theory to the case $\lambda = \lambda_1$ and then use the triangle inequality to conclude the desired result for the given value of λ and u_0. Thus the program parallels that given first in the context of Korteweg-de Vries-type equations in [B] and [Bo].

An estimate of the quantity

$$I_\lambda(u(\cdot,t),G,\mu) \equiv \mu^{-2}(t)\|\nabla u(\cdot,t) - \mu^{\frac{n}{2}}(t)\nabla G(\mu(t)(\cdot))\|^2 + \lambda\|u(\cdot,t) - \mu^{\frac{n}{2}}(t)G(\mu(t)(\cdot))\|^2$$

will be helpful. Denoting G_{λ_1} and μ_{λ_1} by G_1 and μ_1, respectively, it follows from the definitions of μ and μ_1 that

$$\begin{aligned} I_\lambda(u(\cdot,t),G,\mu) &\leqq 2\Big[\mu^{-2}\,\|\mu^{\frac{n}{2}}\nabla G(\mu(\cdot)) - \mu_1^{\frac{n}{2}}\nabla G_1(\mu_1(\cdot))\|^2 \\ &+\lambda\|\mu^{\frac{n}{2}}\,G(\mu(\cdot)) - \mu_1^{\frac{n}{2}}\,G_1(\mu_1(\cdot))\|^2\Big] + \frac{2\lambda}{\lambda_1}\,I_{\lambda_1}(u(\cdot,t),G_1,\mu_1). \end{aligned} \tag{2.39}$$

Now, we estimate the right-hand side of (2.39). First observe that

$$\lambda\|\mu^{\frac{n}{2}}\,G(\mu(\cdot)) - \mu_1^{\frac{n}{2}}\,G_1(\mu_1(\cdot))\|^2 = 0$$

and

$$\mu^{-2}\,\|\mu^{\frac{n}{2}}\nabla G(\mu(\cdot)) - \mu_1^{\frac{n}{2}}\nabla G_1(\mu_1(\cdot))\|^2 = 0.$$

Thus we only need to estimate the term $I_{\lambda_1}(u(\cdot,t),G_1,\mu_1)$ in (2.39). For this, it suffices to show that $\rho_{\lambda_1}(u_0,G_{\lambda_1}) \leqq \widetilde{C}\delta$, where $\widetilde{C} = \widetilde{C}(\lambda,R) > 0$ and then apply the foregoing theory for the special case $\mu_1(0) = 1$. Because $\rho_{\lambda_1}(u_0,G_{\lambda_1}) \leqq \rho_{\lambda_1}(u_0,G) + \rho_{\lambda_1}(G,G_{\lambda_1})$, we may estimate $\rho_{\lambda_1}(u_0,G)$ and $\rho_{\lambda_1}(G,G_{\lambda_1})$ separately and still reach the desired inequality.

First, estimate the term $\rho_{\lambda_1}(G, G_{\lambda_1})$, viz.

$$
\begin{aligned}
[\rho_{\lambda_1}(G,G_{\lambda_1})]^2 &\leqq \lambda_1\|G-G_{\lambda_1}\|^2+\|\nabla G-\nabla G_{\lambda_1}\|^2\\
&= \lambda_1\int_{\mathbf{R}^n}\Big|R(x)-(\frac{\lambda_1}{\lambda})^{\frac{n}{4}}R((\frac{\lambda_1}{\lambda})^{1/2}x)\Big|^2\,dx\\
&+\lambda\int_{\mathbf{R}^n}\Big|\nabla R(x)-(\frac{\lambda_1}{\lambda})^{\frac{n}{4}}\,\nabla R((\frac{\lambda_1}{\lambda})^{1/2}x)\Big|^2\,dx. \qquad (2.40)
\end{aligned}
$$

The first integral in (2.40) can be bounded above as follows:

$$
\begin{aligned}
&\lambda_1\int_{\mathbf{R}^n}\Big|R(x)-(\frac{\lambda_1}{\lambda})^{\frac{n}{4}}R((\frac{\lambda_1}{\lambda})^{1/2}x)\Big|^2\,dx\\
&\leqq 2\lambda_1\Big(\frac{\lambda_1}{\lambda}\Big)^{\frac{n}{2}}\int_{\mathbf{R}^n}|R(x)-R((\frac{\lambda_1}{\lambda})^{1/2}x)|^2\,dx+2\,\frac{\lambda_1\Big|\lambda_1^{\frac{n}{4}}-\lambda^{\frac{n}{4}}\Big|^2}{\lambda^{\frac{n}{2}}}\,\|R\|^2.
\end{aligned}
$$

Thus, the Fundamental Theorem of Calculus together with Minkowski's inequality yield

$$
\begin{aligned}
\int_{\mathbf{R}^n}\Big|R(x)-R((\frac{\lambda_1}{\lambda})^{1/2}x)\Big|^2\,dx &\leqq \int_{\mathbf{R}^n}\Big(\int_{(\frac{\lambda_1}{\lambda})^{1/2}}^1|\frac{d}{dt}R(tx)|\,dt\Big)^2\,dx\\
&\leqq\Big(\int_{(\frac{\lambda_1}{\lambda})^{1/2}}^1\Big(\int_{\mathbf{R}^n}|\frac{d}{dt}R(tx)|^2\,dx\Big)^{1/2}\,dt\Big)^2\\
&=\|x\cdot\nabla R\|^2\Big(\int_{(\frac{\lambda_1}{\lambda})^{1/2}}^1 t^{-\frac{n+2}{2}}\,dt\Big)^2=\frac{4|\lambda_1^{\frac{n}{4}}-\lambda^{\frac{n}{4}}|^2}{n^2\lambda_1^{\frac{n}{2}}}\|x\cdot\nabla R\|^2.
\end{aligned}
$$

Consequently, it transpires that

$$
\lambda_1\int_{\mathbf{R}^n}\Big|R(x)-(\frac{\lambda_1}{\lambda})^{\frac{n}{4}}R((\frac{\lambda_1}{\lambda})^{1/2}x)\Big|^2\,dx\leqq\frac{2\lambda_1\,|\lambda_1^{\frac{n}{4}}-\lambda^{\frac{n}{4}}|^2}{\lambda^{\frac{n}{2}}}\Big[\|R\|^2+\frac{4}{n^2}\|x\cdot\nabla R\|^2\Big]. \qquad (2.41)
$$

Similarly, the second integral on the right-hand side of (2.40) may be bounded above thusly:

$$
\begin{aligned}
&\lambda\int_{\mathbf{R}^n}\Big|\nabla R(x)-(\frac{\lambda_1}{\lambda})^{\frac{n}{4}}\,\nabla R((\frac{\lambda_1}{\lambda})^{1/2}x)\Big|^2\,dx\\
&\qquad\leqq\frac{2\lambda|\lambda_1^{\frac{n}{4}}-\lambda^{\frac{n}{4}}|^2}{\lambda^{\frac{n}{2}}}\sum_{i=1}^n\Big[\|\partial_{x_i}R\|^2+\frac{4}{n^2}\|x\cdot\nabla\partial_{x_i}R\|^2\Big]. \qquad (2.42)
\end{aligned}
$$

The inequalities (2.41) and (2.42) imply

$$
\begin{aligned}
[\rho_{\lambda_1}(G,G_{\lambda_1})]^2 &\leqq \frac{2\lambda_1|\lambda_1^{\frac{n}{4}}-\lambda^{\frac{n}{4}}|^2}{\lambda^{\frac{n}{2}}}\left[\|R\|^2+\frac{4}{n^2}\|x\cdot\nabla R\|^2\right] \\
&+\frac{2\lambda|\lambda_1^{\frac{n}{4}}-\lambda^{\frac{n}{4}}|^2}{\lambda^{\frac{n}{2}}}\sum_{i=1}^{n}\left[\|\partial_{x_i}R\|^2+\frac{4}{n^2}\|x\cdot\nabla\partial_{x_i}R\|^2\right] \\
&\leqq C_1(R)\frac{\lambda+\lambda_1}{\lambda^{\frac{n}{2}}}|\lambda_1^{\frac{n}{4}}-\lambda^{\frac{n}{4}}|^2. \qquad (2.43)
\end{aligned}
$$

It is now determined that there is a positive constant $C = C(\lambda, R)$ such that

$$|\lambda_1-\lambda| \leqq C\delta \qquad (2.44)$$

at least for small values of δ. In fact, from Young's inequality and (2.38),

$$
\begin{aligned}
|\lambda_1-\lambda| &= \frac{1}{\|\nabla R\|^2}\left|\|\nabla u_0\|^2-\|\nabla G\|^2\right| \leqq \frac{2\delta}{\|\nabla R\|^2}\|\nabla G\|^2 \\
&+\frac{1}{\|\nabla R\|^2}(1+\frac{1}{2\delta})\|\nabla u_0-\nabla G\|^2 \leqq \frac{1}{\|\nabla R\|^2}(2\lambda\|\nabla R\|^2+\frac{3}{2})\,\delta.
\end{aligned}
$$

The inequality (2.44) certainly implies that, $|\frac{\lambda}{\lambda_1}-1| \leqq 1$ and $|\lambda_1^{\frac{n}{4}}-\lambda^{\frac{n}{4}}| \leqq C_2\delta$, where $C_2 = C_2(\lambda, R) > 0$. From (2.43), it then follows that

$$\rho_{\lambda_1}(G,G_{\lambda_1}) \leqq C_3(R)\delta. \qquad (2.45)$$

From (2.44) we have that $\frac{\lambda_1}{\lambda} \leqq 1+\frac{C}{\lambda}\delta$. Therefore the assumption (2.38) implies

$$[\rho_{\lambda_1}(u_0,G)]^2 \leqq \max\left\{\frac{\lambda_1}{\lambda},1\right\}\left[\lambda\|u_0-G\|^2+\|\nabla u_0-\nabla G\|^2\right] \leqq C_4\delta^2, \qquad (2.46)$$

where $C_4 = C_4(\lambda, R) > 0$. Inequalities (2.45) and (2.46) lead to the conclusion $\rho_{\lambda_1}(u_0,G_{\lambda_1}) \leqq \widetilde{C}\delta$, and therefore that

$$I_{\lambda_1}(u(\cdot,t),G_1,\mu_1) \leqq \epsilon^2. \qquad (2.47)$$

Theorem 2.2. is now established. ∎

3 Behavior of the stability parameters $\alpha_0 = \alpha_0(t)$ and $\alpha_1 = \alpha_1(t)$

In the proof of Theorem 2.2, it was actually shown that there is a choice of $\alpha_0 = \alpha_0(t)$ and $\alpha_1 = \alpha_1(t)$ for which

$$\rho_\lambda(\phi(\cdot,t), G) = \left(\|\nabla\phi(\cdot+\alpha_0,t)e^{i\alpha_1} - \nabla G(\cdot)\|^2 + \lambda\|\phi(\cdot+\alpha_0,t)e^{i\alpha_1} - G(\cdot)\|^2\right)^{1/2} \leqq \epsilon \tag{3.1}$$

for all $t < t^*$, and that a choice of α_0 and α_1 for which (3.1) holds may be determined via the orthogonality conditions (see (2.20))

$$\operatorname{Im}\int_{\mathbb{R}^n} G^{p+1}(x)\left[e^{i\alpha_1(t)}\phi(x+\alpha_0(t),t)\right]dx = 0, \tag{3.2}$$

$$\operatorname{Re}\int_{\mathbb{R}^n} G^{p}(x)G_{x_i}(x)\left[e^{i\alpha_1(t)}\phi(x+\alpha_0(t),t)\right]dx = 0, \tag{3.3}$$

for $i = 1,..,n$ and $\phi(x,t) = \mu^{-\frac{n}{2}}(t)u(\mu(t)^{-1}x,t)$.

By an application of the implicit-function theorem as in [BS], it may be shown that as long as ϕ satisfies (3.1), there is a unique, continuously differentiable choice of the values $\alpha_0(t)$ and $\alpha_1(t)$ that achieve (3.2) and (3.3).

The principal result regarding the behavior of the parameters α_0 and α_1 is stated in the next theorem. Its proof will appear elsewhere.

Theorem 3.1 *Let $p = \frac{4}{n}$ and $G = G_\lambda$ be a ground-state solution of (1.4). For any $\epsilon > 0$, there exists $\delta = \delta(\epsilon) > 0$ such that if $\|u_0 - G\|_1 < \delta$, then there are C^1-mappings $\alpha_0 : (-t^*, t^*) \to \mathbb{R}^n$ and $\alpha_1 : (-t^*, t^*) \to \mathbb{R}$ such that*

(i) $\|\phi(\cdot+\alpha_0(t),t)e^{i\alpha_1(t)} - G(\cdot)\|_1 \leqq \epsilon$, for $t \in (-t^*, t^*)$, and

(ii) for $\alpha_0(t) = (\alpha_{0,1}(t), ..., \alpha_{0,n}(t))$ and $t \in [0, t^*)$,

$$\left|\alpha_1(t) + \lambda\int_0^t \mu^2(s)ds\right| \leqq C\epsilon\left[\int_0^t \mu^2(s)ds + \int_0^t \frac{|\mu'(s)|}{\mu(s)}ds\right],$$

$$\left|\alpha_{0,i}(t)\right| \leqq C\epsilon\mu(t)\left[\int_0^t \mu(s)ds + \int_0^t \frac{|\mu'(s)|}{\mu^2(s)}ds\right], \qquad i = 1, ..., n,$$

where C depends only on G.

References

[ABH] J.P. Albert, J.L. Bona and D. Henry, Sufficient conditions for stability of solitary-wave solutions of model equations for long waves, Physica D **24** (1987), 343-366.

[B] T. B. Benjamin, The Stability of solitary waves, Proc. Roy. Soc. London A **338** (1972), 153-183.

[BC] H. Berestycki and T. Cazenave, Instabilité des états stationnaires dans les équations de Schrödinger et de Klein-Gordon non linéaires, C. Rend. Acad. Sc. Paris **293** (1981), 489-492.

[BL1] H. Berestycki and P.-L. Lions, Existence of stationary states in nonlinear scalar field equations, in Bifurcation Phenomena in Mathematical Physics and Related Topics, (C. Bardos and D. Bessis, eds.), K. Reidel Publ. Co. Dordrecht, Boston, London (1976), 269-292.

[BL2] H. Berestycki and P.-L. Lions, Nonlinear scalar field equations I, Arch. Rat. Mech. Anal. **82** (1983), 313–346.

[BLP] H. Berestycki, P.-L. Lions and L. A. Peletier, An ODE approach to the existence of positive solutions for semilinear problems in $\mathbb{R}^n$, Indiana Univ. Math. J. **30** (1981), 141-157.

[Bo] J. L. Bona, On the stability theory of solitary waves, Proc. Roy. Soc. London A **344** (1975), 363–374.

[BS] J. L. Bona and A. Soyeur, On the stability of solitary-wave solutions of model equations for long waves, J. Nonlinear Science **4** (1994), 449–470.

[C] T. Cazenave, An introduction to nonlinear Schrödinger equations, Textos de Métodos Matemáticos, Universidade Federal do Rio de Janeiro **26**, third edition (1996).

[CGT] R. Y. Chiao, E. Garmire and C. H. Townes, Self-trapping of optical beams, Phys. Rev. Lett. **13** (1964), 479–482.

[CL] T. CAZENAVE AND P.-L. LIONS, Orbital stability of standing waves for some nonlinear Schrödinger equations, Comm. Math. Phys. **85** (1982), 549–561.

[CLe] E.A. CODDINGTON AND N. LEVINSON, Theory of ordinary differential equations, McGraw-Hill, New York 1955.

[G] R. T. GLASSEY, On the blowing up of solutions to the Cauchy problem for the nonlinear Schrödinger equation, J. Math. Phys. **18** (1977), 1794–1797.

[GSS] M. GRILLAKIS, J. SHATAH AND W. STRAUSS, Stability theory of solitary waves in the presence of symmetry I., J. Funct. Anal. **74** (1987), 160–197.

[GV] J. GINIBRE AND G. VELO, On a class of nonlinear Schrödinger equations I. The Cauchy problem, general case, J. Funct. Anal. **32** (1979), 1–32.

[K] M. K. KWONG, Uniqueness of positive solutions of $\Delta u - u + u^p = 0$ in $\mathbb{R}^n$, Arch. Rat. Mech. Anal. **105** (1989), 243–266.

[LBS] E. W. LAEDKE, L. BLAHA AND K. H. SPATSCHEK, Collapsing states of generalized Korteweg-de Vries equations, Physica D **40** (1989), 249–264.

[LBSK] E. W. LAEDKE, L. BLAHA, K. H. SPATSCHEK AND E. A. KUZNETSOV, On Stability of collapse in the critical case, J. Math. Phys. **33**(3) (1992), 967–973.

[LS] E. W. LAEDKE AND K. H. SPATSCHEK, Variational principles in soliton physics, Diff. Geometry, Calculus of Variations and their Application, Lecture Notes in Pure and Appl. Math. **100**, Dekker, New York (1985), 335–357.

[LSS] E. W. LAEDKE, K. H. SPATSCHEK AND L. STENFLO, Evolution theorem for a class of perturbed envelope soliton solutions, J. Math. Phys. **24**(12) (1983), 2764–2769.

[M] K. MCLEOD, Uniqueness of positive radial solutions of $\Delta u + f(u) = 0$ in $\mathbb{R}^n$, Trans. Amer. Math. Soc. **339** (1993), 495–505.

[N] H. NAWA, Limiting profiles of blow-up solutions of the nonlinear Schrödinger equation with critical power nonlinearity, Proc. Japan Acad. Ser. A **73**, (1997), 171–175.

[OT] T. Ogawa and Y. Tsutsumi, Blow-up of H^1-solutions for the one-dimensional nonlinear Schrödinger equation with critical power nonlinearity, Proc. Amer. Math. Soc. **111**(2) (1991), 487-496.

[S] W. Strauss, Existence of solitary waves in higher dimensions, Comm. Math. Phys. **55** (1977), 149–162.

[W1] M. I. Weinstein, Nonlinear Schrödinger equation and sharp interpolation estimates, Comm. Math. Phys. **87** (1983), 567–576.

[W2] M. I. Weinstein, Modulational stability of ground states of nonlinear Schrödinger equations, SIAM J. Math. Anal. **16** (1985), 472–491.

[W3] M. I. Weinstein, On the structure and formation of singularities in solutions to nonlinear dispersive evolution equations, Comm. PDE **11**(5) (1986), 545–565.

[W4] M. I. Weinstein, Liapunov stability of ground states of nonlinear dispersive evolution equations, Comm. Pure Appl. Math. **39** (1986), 51–68.

[W5] M. I. Weinstein, Solitary waves of nonlinear dispersive evolution equations with critical power nonlinearities, J. Diff. Eq. **69** (1987), 192–203.

[Z] V. E. Zakharov, Collapse of Langmuir waves, Sov. Phys JETP 35 (1972), 908–912.

J. Angulo, M. Scialom, IMECC-UNICAMP-Caixa-Postal 6065, 13081-970, Campinas, SP, Brazil.

J.L. Bona, Department of Mathematics and Texas Institute for Computational and Applied Mathematics, The University of Texas at Austin, Austin, TX 78712 USA.

F. Linares, IMPA, Estrada Dona Castorina 110, Rio de Janeiro, 22460-320, Brazil.

The Schrödinger Equation with Point Interaction in an Algebra of New Generalized Functions

Anatolij Antonevich

1 Introduction

Let us consider the Schrödinger operator in R^3 having the form

$$Lu \equiv -\Delta u + a\delta u, \tag{1.1}$$

where $-\Delta$ is the Laplace operator and δ is the Dirac function. A typical problem treated in Quantum Mechanics is to describe the motion of a non-relativistic particle moving in the field of external potential forces. The corresponding Hamiltonian is then given by $-\Delta + V$, where V is a measurable function. If we deal with the short-range forces near point 0, then V is supported in a small neighborhood of 0. In this situation it is natural to think about an idealization leading to zero-range potential supported at a point. Symbolically we obtain an operator of the form (1.1), where a is the coupling constant describing intensity of the forces. The models of this sort have been studied by R. de L. Kronig and W.G. Penney [17], H. Bethe and R. Peierls [8] and others.

The first mathematical treatment of such physical systems was given by F. Berezin and L. Faddeev in [7] who used Krein's theory of self-adjoint extensions of symmetric operators [1]. Let $\overset{\circ}{A}$ be the restriction of the Laplace operator to the domain $D(\overset{\circ}{A}) = \{u \in H^2 : u(0) = 0\}$. Then $\overset{\circ}{A}$ is a symmetric operator, the self-adjoint extensions of which can be considered as "perturbations" of $-\Delta$ by a zero-range (singular) potential supported at the point $0 \in R^3$. The explicit description of all self-adjoint extensions A^α of $\overset{\circ}{A}$ was obtained by Krein's method; in this case the operators A^α have a natural parametrization by one parameter $\alpha \in R$.

C. Friedman showed in [13] that the Schrödinger operator with point interaction can be approximated by operators of the form $-\Delta + V_\varepsilon(x)$, where $V_\varepsilon(x)$ are regular potentials. Similar results were obtained by E.Rosinger [19]. See [2] for a survey of these results.

But there are other approaches to investigate this problem as well. In particular the authors of [14,3] used rank-one perturbations of $-\Delta$ and the authors of [4,16] developed an approach based on the form-sum method.

The main obstacle to the solution of this problems is related to the multiplication of some distributions. In fact all these results include in an implicit form a definition of the product $(a\delta) \cdot E_\lambda$, where

$$E_\lambda = \frac{1}{4\pi}\frac{e^{-\sqrt{-\lambda}|x|}}{|x|}, \quad |x| = \sqrt{x_1^2 + x_2^2 + x_3^2}$$

(E_λ is the fundamental solution of the operator $-\Delta - \lambda$). It is impossible to define a product of arbitrary distributions but it is possible to introduce new objects instead of distributions which admit the well-defined multiplication and preserve the main properties of the distributions [15, 10, 11, 12, 6]. Colombeau's construction is more convenient. It is natural to consider the operator (1.1) in Colombeau's algebra of new generalized functions. This problem was posed by M. Oberguggenberger in his book [18]. In the present paper we obtain one possible solution of this problem.

2 New generalized functions generated by Sobolev spaces

Let $S(R^3)$ be the Schwartz space and F be the Fourier transform:

$$(Fu)(\xi) = \hat{u}(\xi) = \int e^{-i(x,\xi)}u(x)dx, \ \xi \in R^3.$$

The Sobolev norms $||u||_s$, $s \in R$, are defined on $S(R^3)$ by

$$||u||_s = \left(\int (1+|\xi|^2)^s |\hat{u}(\xi)|^2 d\xi\right)^{\frac{1}{2}}.$$

The Sobolev space $H^s(R^3) = H^s$ is the completion of $S(R^3)$ with respect to the norm $||u||_s$. We set

$$H^\infty = \bigcap_s H^s, \quad H^{-\infty} = \bigcup_s H^s.$$

Let

$$G_M = \{u = (u_\varepsilon), \varepsilon > 0, u_\varepsilon \in H^\infty : \forall s \ \exists q \ \exists c \text{ such that } ||u_\varepsilon||_s \leq c\frac{1}{\varepsilon^q}\},$$

$$N = \{u \in G_M : \forall p \, \forall s \, \exists c \text{ such that } ||u_\varepsilon||_s \le c\varepsilon^p\}.$$

The quotient space $G \equiv G(H^\infty) = G_M/N$ is an algebra with differentiation.
Let γ be an arbitrary function from H^∞ such that $\hat{\gamma}(\xi)$ is infinitely differentiable in some neighborhood of $0 \in R^3$ and $\hat{\gamma}(0) = 1$. We set $\gamma_\varepsilon(x) = \varepsilon^{-3}\gamma\left(\frac{x}{\varepsilon}\right)$. Then the mapping $\tilde{J}_\gamma u = u * \gamma_\varepsilon$, where $*$ is the convolution, generates a linear embedding J_γ of the space with differentiation $H^{-\infty}$ into the algebra G. If $\hat{\gamma}(\xi) = 1$ for $|\xi| < r$, where $r > 0$, then H_∞ is embedded as an algebra. The algebra $G(H^\infty(R^3))$ was introduced in [9] and [5].
Now let $L_2^s(R^3) \equiv L_2^s$ be the space $L_2(R^3, (1 + |\xi|^2)^s d\xi)$ with the weight $(1 + |\xi|^2)^s$ and

$$L_2^\infty = \bigcap_s L_2^s, \quad L_2^{-\infty} = \bigcup_s L_2^s.$$

We set

$$\hat{G}_M = \{f = (f_\varepsilon), \quad f_\varepsilon \in L_2^\infty : \forall s, \exists q \exists c \text{ such that} ||f_\varepsilon||_s \le c\frac{1}{\varepsilon^q}\},$$

$$\hat{N} = \{f \in \hat{G}_M : \forall s, \forall p \exists c \text{ such that} ||f_\varepsilon||_s \le c\varepsilon^p\}$$

(here $||f_\varepsilon||_s$ is the norm in L_2^s), and $\hat{G} = \hat{G}_M/\hat{N}$.
Theorem 1. *The space $\hat{G}$ is an algebra with respect to the convolution product. The Fourier transform F generates an isomorphism between the algebra G and the algebra $\hat{G}$.*
We denote by $\tilde{C}$ the algebra of generalized numbers, where $\tilde{C} = C_M/N_C$,

$$C_M = \{a = (a_\varepsilon), a_\varepsilon \in C : \exists q, c \text{ such that} |a_\varepsilon| \le c\frac{1}{\varepsilon^q}\},$$

$$N_C = \{a_\varepsilon \in C_M : \forall p \exists c \text{ such that} |a_\varepsilon| \le c\varepsilon^p\}.$$

Let $\tilde{C}^{-1}$ be the set of all invertible generalized numbers and $\tilde{C}_0 = \tilde{C} \setminus \tilde{C}^{-1}$.

3 Nonperturbed Schrödinger operator on G

Definition. *Let $A : G \to G$ be a linear operator. The set $\tilde{Sp}(A) = \{\lambda \in \tilde{C} : A - \lambda$ is irreversible$\}$ is called the (generalized) spectrum of A.*
Theorem 2. *The spectrum of operator $-\Delta$ acting on G is the set $\tilde{Sp}(-\Delta) = \tilde{R}_0^+ + \tilde{C}_0$.*

4 Resolvent of perturbed Schrödinger operator on G

Let us consider the operator (1.1) on G. For this purpose we replace the δ-function by an element from G associated with this function and having the form

$$\delta_\varphi(x) = \frac{1}{\varepsilon^3}\varphi\left(\frac{x}{\varepsilon}\right) \equiv \varphi_\varepsilon(x),$$

where $\varphi \in H^\infty, \hat{\varphi}\,(0) = 1$, and $\hat{\varphi}$ is an infinitely differentiable function in neighborhood of the point 0.

We shall consider the operators of the form

$$L_{\varphi,a_\varepsilon}u_\varepsilon = -\Delta u_\varepsilon + a_\varepsilon\varphi_\varepsilon(x)\int \varphi_\varepsilon(y)u_\varepsilon(y)dy. \tag{4.1}$$

On smooth functions in H^∞ the operators of the form (4.1) converge to the operator (1.1) in the sense of convergence in $H^{-\infty}$. For a fixed value ε an operator of the form (4.1) is a rank-one perturbation of the operator $-\Delta$. Therefore for fixed values of ε this operator can be investigated easily. The main problem is a clarification of the behavior of these operators as $\varepsilon \to 0$.

First of all, we construct formally the inverse operator of $L_{\varphi,a_\varepsilon} - \lambda$. Consider in H^∞ the equation

$$L_{\varphi,a_\varepsilon}u_\varepsilon - \lambda_\varepsilon u_\varepsilon = f_\varepsilon.$$

Applying the Fourier transform we obtain

$$(-\lambda_\varepsilon + |\xi|^2)\,\hat{u}_\varepsilon\,(\xi) + (2\pi)^{-3}a_\varepsilon\,\hat{\varphi}\,(\varepsilon\xi)\int \hat{\varphi}\,(\varepsilon\eta)\,\hat{u}_\varepsilon\,(-\eta)d\eta = \hat{f}_\varepsilon\,(\xi). \tag{4.2}$$

If there exists a solution u_ε of the equation (4.2), then

$$\hat{u}_\varepsilon\,(\xi) = \frac{\hat{f}_\varepsilon\,(\xi)}{-\lambda_\varepsilon + |\xi|^2} + C_\varepsilon\frac{\hat{\varphi}\,(\varepsilon\xi)}{-\lambda_\varepsilon + |\xi|^2}, \tag{4.3}$$

where C_ε is an constant. Substituting (4.3) into (4.2) we have the following equation for C_ε:

$$C_\varepsilon(1 + a_\varepsilon b_\varepsilon) = a_\varepsilon d_\varepsilon,$$

where

$$b_\varepsilon \equiv b_\varepsilon(\varphi, \lambda_\varepsilon) = (2\pi)^{-3}\int \frac{\hat{\varphi}\,(\varepsilon\eta)\,\hat{\varphi}\,(-\varepsilon\eta)}{-\lambda_\varepsilon + |\eta|^2}d\eta,$$

$$d_\varepsilon \equiv d_\varepsilon(f_\varepsilon, \varphi, \lambda_\varepsilon) = (2\pi)^{-3} \int \frac{\hat{\varphi}(\varepsilon\eta)\,\hat{f}_\varepsilon(-\eta)}{-\lambda_\varepsilon + |\eta|^2} d\eta.$$

If $1 + a_\varepsilon b_\varepsilon \neq 0$, we obtain

$$\hat{u}_\varepsilon(\xi) = \frac{\hat{f}_\varepsilon(\xi)}{-\lambda_\varepsilon + |\xi|^2} + \frac{a_\varepsilon}{1 + a_\varepsilon b_\varepsilon} d_\varepsilon \frac{\hat{\varphi}(\varepsilon\xi)}{-\lambda_\varepsilon + |\xi|^2}. \tag{4.4}$$

The next result clarifies the conditions under which the expression (4.4) generates an operator acting on G.

Theorem 3. *Let $a = (a_\varepsilon) \in \tilde{C}^{-1}$. The operator $L_{\varphi,a} - \lambda$ is invertible in G if and only if $\lambda \notin \tilde{R}_0^{+} + \tilde{C}_0$ and*

$$1 + a_\varepsilon b_\varepsilon \in \tilde{C}^{-1}. \tag{4.5}$$

Now we shall investigate the condition (4.5) in detail.

Lemma 1. *Let a family (λ_ε) be bounded, i.e. $|\lambda_\varepsilon| \leq C \in R^+$. Then*

$$b_\varepsilon = M_{-1}\frac{1}{\varepsilon} - \frac{\sqrt{-\lambda_\varepsilon}}{4\pi} + \varepsilon\lambda_\varepsilon M_1 + \ldots + \frac{1}{\varepsilon}(-\varepsilon\sqrt{-\lambda_\varepsilon})^k M_k + \ldots, \tag{4.6}$$

where

$$M_k \equiv M_k(\varphi) = \frac{1}{4\pi(k+1)!} \int \varphi_1(x)|x|^k dx, \quad k = -1, 1, 2, \ldots, \quad \varphi_1 = \varphi * \varphi,$$

and the series (4.6) converges.

To obtain more concrete results we shall assume, that $a_\varepsilon = a_0 + a_1\varepsilon + a_2\varepsilon^2 + \ldots$ and $\mu_\varepsilon = \sqrt{-\lambda_\varepsilon} = \mu_0 + \mu_1\varepsilon + \mu_2\varepsilon^2 + \ldots$. Then

$$1 + a_\varepsilon b_\varepsilon = a_0 M_{-1}\frac{1}{\varepsilon} + (1 + a_1 M_{-1} - \frac{a_0\mu_0}{4\pi}) +$$

$$+[a_0(-\frac{\mu_1}{4\pi} + \mu_0^2 M_1) - \frac{a_1\mu_0}{4\pi} + a_2 M_{-1}]\varepsilon + \ldots . \tag{4.7}$$

Therefore for the invertibility of the element $1 + a_\varepsilon b_\varepsilon$, it is necessary and sufficient that at least one coefficient of the asymptotic series (4.7) is nonzero. Here one can arrive at a number of qualitatively different cases, which are described in the following lemma.

Lemma 2.

1) If $a_0 M_{-1} \neq 0$, then the element $1 + a_\varepsilon b_\varepsilon$ is invertible in $\tilde{C}$ for any λ and $h_\varepsilon = \frac{a_\varepsilon}{1+a_\varepsilon b_\varepsilon} \to 0$ as $\varepsilon \to 0$.

2) If $M_{-1} = 0, a_0 \neq 0$, then there exists a unique value λ^*, for which $1 + a_\varepsilon b_\varepsilon = 0$. For λ^* we have

$$\mu_0^* = \frac{4\pi}{a_0} + \cdots .$$

For the rest of the values of λ the element $1 + a_\varepsilon b_\varepsilon$ is invertible and

$$\lim_{\varepsilon \to 0} h_\varepsilon = \begin{cases} \frac{4\pi}{4\pi\alpha - \mu_0}, & if\ \mu_0 \neq \mu_0^*, \\ \infty, & if\ \ \mu_0 = \mu_0^*, \end{cases}$$

where $\alpha = \frac{4\pi}{a_0}$.

3) If $a_0 = 0$ and $1 + a_1 M_{-1} \neq 0$, then the element $1 + a_\varepsilon b_\varepsilon$ is invertible and $h_\varepsilon \to 0$ as $\varepsilon \to 0$ for all values of λ.

4) If $a_0 = 0, a_1 \neq 0$ and $1 + a_1 M_{-1} = 0$, then there exists a unique value λ^{**} such that $1 + a_\varepsilon b_\varepsilon = 0$. For this value we have

$$\mu_0^{**} = -4\pi a_2 (M_{-1})^2.$$

For the rest of the values of λ the element $1 + a_\varepsilon b_\varepsilon$ is invertible and

$$\lim_{\varepsilon \to 0} h_\varepsilon = \begin{cases} \frac{4\pi}{4\pi\alpha - \mu_0}, & if\ \mu_0 \neq \mu_0^{**}, \\ \infty, & if\ \mu_0 = \mu_0^{**}, \end{cases}$$

where $\alpha = -a_2 (M_{-1})^2$.

We remark that if $\varphi(x) \in C$ then it is possible that $M_{-1} = 0$, but if $\varphi(x) \in R$ then $M_{-1} > 0$.

One more case when $1 + a_\varepsilon b_\varepsilon = 0$ is described in Lemma 3. Let

$$N_k(t) \equiv N_k(\varphi, t) = \frac{1}{4\pi(k+1)!} \int e^{-t|x|} (|x|)^k \varphi_1(x) dx, k = -1, 0, 1, 2....$$

Let $\varphi(x) \in R$. If $t > 0$ then $0 < N_k(t) < M_k$ and $N_k(0) = M_k$.

Lemma 3. *Let $\varphi(x) \in R, a_\varepsilon = a_0 + a_1 \varepsilon + a_2 \varepsilon^2 + ...,$ and $a_1 \in R,\ a_1 M_1 + 1 \neq 0$.*

1) If $a_0 \neq 0$ or $a_0 = 0,\ a_1 > -(M_1)^{-1}$ then $1 + a_\varepsilon b_\varepsilon \neq 0$ for all $\lambda \in \widetilde{C}$.

2) If $a_0 = 0$ and $a_1 < -(M_1)^{-1}$ then there exists a unique λ such that $1 + a_\varepsilon b_\varepsilon(\lambda) = 0$. For this λ^{***} we have

$$\lambda^{***} = -(\mu^{***})^2,\ \mu^{***} = \mu_{-1}^{***} \frac{1}{\varepsilon} + \mu_0^{***} + ..., \tag{4.8}$$

where μ_{-1}^{***} is such that $N_{-1}(\mu_{-1}^{***}) a_1 + 1 = 0$.

Theorem 4. *Let the coefficient a admit the representation* $a = a_0 + a_1\varepsilon + a_2\varepsilon^2 + \ldots$, *and* $\varphi(x) \in R$. *Then*

$$\tilde{Sp}\,(L_{\varphi,a}) = \tilde{R}_0^{+} + \tilde{C}_0\,.$$

if λ^* *and* λ^{***} *do not exist, and*

$$\tilde{Sp}\,(L_{\varphi,a}) = (\tilde{R}_0^{+} \bigcup \lambda^*) + \tilde{C}_0\,.$$

or

$$\tilde{Sp}\,(L_{\varphi,a}) = (\tilde{R}_0^{+} \bigcup \lambda^{***}) + \tilde{C}_0\,.$$

if λ^* *or* λ^{***} *does exist.*

In this case, λ^* *or* λ^{***} *is an eigenvalue of the operator* $L_{\varphi,a}$.

5 Operator associated with the resolvent

We shall say that an operator A is associated with an operator $B : D(B) \to H^l$ with a domain $D(B) \subset H^s$ in the sense of (γ, s, l, l_1)-convergence if

$$\lim_{\varepsilon\to 0} A_\varepsilon(u * \gamma_\varepsilon) \stackrel{def}{=\!=} v$$

exists in H^{l_1} and belongs to H^l if and only if $u \in D(B)$ and $Bu = v$.

For example, for any γ the operator (4.1) is associated with the operator $-\Delta + a_0\,\delta : H^2 \to H^{-2}$ in the sense of $(\gamma, 2, -2, -2)$-convergence.

In the applications the operators for which there exist associated operators are of prime interest. Let us consider this question for the resolvent of the operator $L_{\varphi,a}$.

Theorem 5. *Let* $L_{\varphi,a}$ *be an operator of the form (4.1) and* $\tilde{R}_\lambda \equiv (L_{\varphi,a} - \lambda)^{-1}$ *its resolvent in* G. *Let* $\lambda \in C, \lambda \notin R_0^+$ *and* $\lambda \neq \lambda_0^*$ *or* $\lambda \neq \lambda_0^{**}$, *if the latter numbers exist. Then for any* γ *there exists an operator* R_λ *in* L_2 *which is associated with* $\tilde{R}_\lambda$ *in the sense of* $(\gamma, 0, 0, 0)$*-convergence and has the following form:*

a) $R_\lambda = (-\Delta - \lambda_0)^{-1}$ *when if* $h_\varepsilon \to 0$.

b) $R_{\lambda,\alpha} f(x) = (-\Delta - \lambda)^{-1} f(x) + \frac{4\pi}{4\pi\alpha - \sqrt{-\lambda}} E_\lambda(x) \int f(y) E_\lambda(y) dy$ *when* $\lim h_\varepsilon \neq 0$.

In case a) we have

$$Im(R_\lambda) = H^2 \subset L_2;$$

In case b) we have

$$Im(R_{\lambda,\alpha}) = \{u = u_0 + \frac{4\pi}{4\pi\alpha - \sqrt{-\lambda}} u_0(0) E_\lambda : u_0 \in H^2\}$$

and the operator $R_{\lambda,\alpha}$ is the resolvent of A^α.

Let us compare the statements of Theorem 4 with the previous results.

Case a) corresponds to the situation where the perturbation of the operator does not influence the resolvent in the limit.

Case b) corresponds to the situation where in the limit the resolvent of a nontrivial self-adjoint extension of the operator $\mathring{A}$ arises.

In [13] it was noted that if $a\varphi(x) \geq 0$, then the resolvent of a perturbed operator converges to $(-\Delta - \lambda)^{-1}$, i.e., the case a) holds. Further, it was noted that a nontrivial extension, i.e., resonances (case b)), can arise for $\varphi(x) \leq 0$ and infinity small values of a of the form $a = a_1\varepsilon + a_2\varepsilon^2 + \cdots$ (the case 2) of Theorem 4). In case 2 of Lemma 2 we have a new situation in principle: if $M_{-1} = 0$, then resonances can arise for all finite values of a of the form $a = a_0 + a_1\varepsilon + a_2\varepsilon^2$, where $a_0 \neq 0$. But this is possible only if $\varphi(x) \in C$.

6 Operators associated with the perturbed Schrödinger operators

Now we consider operators on H_0 associated with the operators of the form (4.1). A type of association relation has an important function in this case. We shall use the $(\gamma, 0, 0, -2)$ and $(\gamma, 0, 0, 0)$-convergence.

Denote

$$b_\varepsilon = b_\varepsilon(\varphi, \gamma, \lambda) = \int \frac{\varphi(\varepsilon\eta)\gamma(-\varepsilon\eta)}{\lambda + |\eta|^2} d\eta.$$

Similarly to Lemma 1 we have

$$b_\varepsilon = M_{-1}\frac{1}{\varepsilon} - \frac{\mu}{4\pi} + \varepsilon\mu^2 M_1(\psi) + ...$$

where $\psi = \varphi * \gamma$.

Theorem 6. *Let $a_\varepsilon = a_0 + a_1\varepsilon + a_2\varepsilon^2 + ...$ and $M_{-1}(\varphi) \neq 0$.*

The Operator $L_{\varphi,a}$ is associated in the sense of $(\gamma, 0, 0, -2)$-convergence with the following operators:

1) If $a_0 \neq 0$, then $L_{\varphi,a} \to \mathring{A}$.

2) If $a_0 = 0,\ 1 + a_1 M_{-1} \neq 0$, then $L_{\varphi,a} \to -\Delta$.

3) If $a_0 = 0,\ 1 + a_1 M_{-1} = 0$, then $L_{\varphi,a} \to (\mathring{A})^*$.

Proof. It easy to check that $A_\varepsilon \to B$ iff $A_\varepsilon - \lambda \to B - \lambda$ and $Ker(B-\lambda) = 0$. Then we will consider the operator $L_{\varphi,a} - \lambda$, where $\lambda \notin R_0^+$. We have

$$w_\varepsilon = (L_{\varphi,a} - \lambda)(u * \gamma_\varepsilon) = (-\Delta - \lambda)(u * \gamma_\varepsilon) + a_\varepsilon \varphi_\varepsilon(x) \int \varphi_\varepsilon(y)(u * \gamma_\varepsilon)(y)dy. \quad (6.1)$$

If $u \in H^0$, then

$$(-\Delta - \lambda)(u * \gamma_\varepsilon) \to (-\Delta - \lambda)u \text{ as } \varepsilon \to 0 \text{ in } H^{-2}.$$

Clearly the limit of (6.1) exists iff the limit of second term in (6.1) exists. In this case, we have $w_\varepsilon \to v + c\delta$, where $v = (-\Delta - \lambda)u, c \in C$. By the definition of $(\gamma, 0, 0, -2)$-convergence we have $v + c\delta = v_0 \in H^0$ and

$$u = u_0 - cE_\lambda, \quad (6.2)$$

where $u_0 = (-\Delta - \lambda)^{-1} v_0 \in H^2 \subset H^0$. Substituting (6.2) in (6.1) we obtain

$$w_\varepsilon = (-\Delta - \lambda)(u_0 * \gamma_\varepsilon) - c(\gamma_\varepsilon + a_\varepsilon b_\varepsilon \varphi_\varepsilon(x)) + a_\varepsilon \varphi_\varepsilon \int \varphi_\varepsilon(y)(u_0 * \gamma_\varepsilon)(y)dy. \quad (6.3)$$

Let $a_0 \neq 0$. Then $a_\varepsilon b_\varepsilon \to \infty$ and we conclude that $c = 0$ and $w_\varepsilon \to v_0 + a_0 u_0(0)\delta$. Since $v_0 \in H^0$ and $\delta \notin H^0$, it follows that $u_0(0) = 0$ and $L_{\varphi,a} \to \overset{\circ}{A}$.
Let $a_0 = 0$. We have

$$w_\varepsilon \to v_0 - c(1 + a_1 M_{-1}(\psi))\delta.$$

Therefore $c(1 + a_1 M_{-1}(\psi)) = 0$. If $1 + a_1 M_{-1}(\psi) \neq 0$ we obtain $c = 0$ and $L_{\varphi,a} \to -\Delta$. If $1 + a_1 M_{-1}(\psi) = 0$, then c may be an arbitrary constant and $L_{\varphi,a} \to B$, where

$$D(B) = u_0 + cE_\lambda : u_0 \in H^2, c \in C, Bu = -\Delta u_0.$$

Note that in this case we have $B = (\overset{\circ}{A})^*$.
Denote

$$S_\varepsilon(u) = \int \varphi_\varepsilon(y)(u * \gamma_\varepsilon)(y)dy.$$

Theorem 7. *Let the conditions of Theorem 6 hold. The Operator $L_{\varphi,a}$ is associated in the sense of $(\gamma, 0, 0, 0)$-convergence with the following operators:*
1) If $a_0 \neq 0$, then $L_{\varphi,a} \to B$, where

$$D(B) = \{u \in H^2 : S_\varepsilon(u) = o(\varepsilon^{\frac{3}{2}})\}, \quad Bu = -\Delta u.$$

2) Let $a_0 = 0, 1 + a_1 M_{-1}(\psi) \neq 0$.

If $a_1 \neq 0$, *then* $L_{\varphi,a} \to \overset{\circ}{A}$.

If $a_1 = 0$, *then* $L_{\varphi,a} \to -\Delta$.

3) Let $a_0 = 0, 1 + a_1 M_{-1}(\psi) = 0$.

If $\gamma \neq \varphi$, *then* $L_{\varphi,a} \to \overset{\circ}{A}$.

If $\gamma = \varphi$, *then* $L_{\varphi,a} \to A^\alpha$, *where*

$$D(A^\alpha) = \{u = u_0 + \frac{4\pi}{4\pi\alpha - \mu} E_\lambda u_0(0) : u_0 \in H^2\}, \quad A^\alpha u = -\Delta u_0.$$

Proof. It is clear that $\lim L_{\varphi,a}$ in the sense of $(\gamma, 0, 0, 0)$-convergence may be only some restriction of the $\lim L_{\varphi,a}$ in the sense of $(\gamma, 0, 0, -2)$-convergence. Let $a_0 = 0, a_1 \neq 0, 1 + a_1 M_{-1} \neq 0$. Then $c = 0, u = u_0$ and

$$w_\varepsilon = (-\Delta - \lambda)(u * \gamma_\varepsilon) + a_\varepsilon \varphi_\varepsilon S_\varepsilon(u).$$

Since $||\varphi_\varepsilon||_0 = c_1 \varepsilon^{-\frac{3}{2}}$ and $S_\varepsilon(u) = u(0) + O(\varepsilon)$, we see that

$$||a_\varepsilon S_\varepsilon(u) \varphi_\varepsilon||_0 = c_1 a_1 u(0) \varepsilon^{-\frac{1}{2}} + O(\varepsilon^{\frac{1}{2}}).$$

If $a_1 \neq 0$, then it follows that $u(0) = 0$ and $L_{\varphi,a} \to \overset{\circ}{A}$.

If $a_1 = 0$, then $L_{\varphi,a} \to -\Delta$.

Let $a_0 = 0, 1 + a_1 M_{-1}(\psi) = 0$.

If $\gamma \neq \varphi$ then $||\gamma_\varepsilon + a_\varepsilon b_\varepsilon \varphi_\varepsilon||_0 \sim c_2 \varepsilon^{-\frac{3}{2}}$ and as before we obtain $c = 0, u(0) = 0$ and $L_{\varphi,a} \to \overset{\circ}{A}$.

Let $\gamma = \varphi$. Then

$$w_\varepsilon = v_0 * \gamma_\varepsilon + \varphi_\varepsilon \{[-c(-\frac{1}{4\pi} a_1 \mu_0 + a_2 M_{-1}) + a_1 u_0(0)]\varepsilon + O(\varepsilon^2)\}. \tag{6.4}$$

Hence the limit of (6.4) in H^0 exists iff

$$c = \frac{a_1}{a_2 M_{-1} - \frac{a_1 \mu}{4\pi}} u_0(0) = \frac{4\pi}{4\pi\alpha - \mu} u_0(0).$$

Now let $a_0 \neq 0$. Then

$$w_\varepsilon = v_0 * \gamma_\varepsilon + \varphi_\varepsilon[(a_0 + a_1\varepsilon) S_\varepsilon(u_0) + O(\varepsilon^2)]$$

since it is necessary not only to fulfill the condition $u_0(0) = 0$ but also $S_\varepsilon(u_0) = o(\varepsilon^{\frac{3}{2}})$.

Acknowledgment

The author gratefully acknowledges the Erwin Schrödinger International Institute for Mathematical Physics for invitation and the financial support during his stay at the Institute. The author would like to thank V. Koshmanenko, M. Oberguggenberger and E. Rosinger for interesting discussions. This investigation was partially supported by the Fundamental Research Fund of the Republic of Belarus.

References

[1] N. AKHIEZER, I.M.GLAZMAN, Theory of Linear Operators in Hilbert Space. Vol. II. Pitman Advanced Publ. Boston, 1981.

[2] S. ALBEVERIO, F. GESZTESY, HOEGH-KROHN, H. HOLDEN, Solvable models in quantum mechanics, Springer-Verlag, New York, 1988.

[3] S. ALBEVERIO, V. KOSHMANENKO, Some remarks on the Gesztesy-Simon version of rank-one perturbation. Preprint Nr. 727/5/96 Universität Bielefeld.

[4] S. ALBEVERIO, V. KOSHMANENKO, Form-sum approximations of singular perturbation of self-adjoint operators. Preprint Nr. 771/4/97 Universität Bielefeld.

[5] A. ANTONEVICH, N. FURSENKO, A model elliptic equation with a δ-shaped coefficient. Differentsial'nye Uravneniya, 30, No. 8, 1317-1324.

[6] A. ANTONEVICH, YA. RADYNO, On a general method of construction algebras of generalized functions. Soviet Math. Dokl. 1991, v. 43, No.3, 680-684.

[7] F. BEREZIN AND L. FADDEEV, A remark on Schrödinger equation with a singular potential, Soviet. Math. Dokl. 2(1961), 372-375.

[8] H. BETHE, R. PEIERLS, Quantum theory of the diplon. Proc. Roy. Soc. (London) 148A (1935), 146-156.

[9] H.A. BIAGIONI, M. OBERGUGGENBERGER, Generalized solutions to the Korteweg-de Vries and the regularized long-wave equations. SIAM J. Math. Anal. Vol. 23, No. 4, July 1992, 923-940.

[10] CHR. CHRISTOV, B. DAMIANOV, Asymptotic functions- a new class of generalized functions. Bulgar. J.Phys. p.1. 1978, No. 6. 543-556, p. II, 1979, No.1. 3-23, p. III, 1979, No.3. 245-256, p. IV, 1979 No.4, 377-397.

[11] J.L. COLOMBEAU, New generalized functions and multiplication of distributions. North-Holland Math. Stud. 84, North-Holland, Amsterdam, 1984.

[12] YU. V. EGOROV, On the theory of generalized functions. Uspekhi Math. Nauk. v. 45. No. 5, 1990, 3-40.

[13] C. FRIEDMAN, Perturbations of the Schrödinger equation by potentials with small support. J. Funct. Anal. 10(1972), 346-360.

[14] F. GESZTESY, B. SIMON, Rank-one perturbations at infinite coupling. J. Funct. Anal. 128, 1995, 245-252.

[15] V.K. IVANOV, The multiplication of distributions and the regularization of divergent integrals. Izv. VUZ. Math. 1971, No. 1. 41-49.

[16] V.D. KOSHMANENKO, Singular bilinear forms in perturbations theory of self-adjoint operators. Naukova dumka, Kiev, 1993.

[17] R. DE L. KRONIG, W.G. PENNEY, Quantum mechanics of electrons in crystal lattices. Proc. Roy. Soc. (London) 130A (1931), 499-513.

[18] M. OBERGUGGENBERGER, Multiplication of distributions and applications to partial differential equations. Pitman Research Notes Math. v. 259. Longman, Harlow, 1992.

[19] E.E. ROSINGER, Distributions and nonlinear partial differential equations. Lect. N. Math., 684, 1978.

A. ANTONEVICH
Belarussian State University, Minsk, Belarus, E-mail: anton@mmf.bsu.unibel.by

Polynomial A Priori Estimates for Some Evolution PDE and Generalized Solutions

HEBE A. BIAGIONI AND TODOR GRAMCHEV

1 Introduction

The main purpose of this paper is to derive a priori estimates for the solutions to some evolution equations with elliptic dissipative terms and polynomial nonlinearities with the subsequent applications for solving the initial value problem (IVP) in suitable algebras of generalized functions. Such equations contain as particular cases the semilinear heat equation, the Navier-Stokes equation for an incompressible fluid, Burgers' equation, the Kuramoto-Sivashinsky equation. In order to simplify the presentation we will concentrate on two model cases: a semilinear heat equation on the line and Kuramoto-Sivashinsky type equations in $\mathbb{R}^n$, $n = 1, 2, 3$. The proof of such estimates exploits the well-known smoothing property of the linearized operator, depending on the order of the strong parabolicity of its real part, and the Leibnitz rule for higher order derivations. Broadly speaking, our motivation comes from the mathematical study of IVP for semilinear evolution PDE with singular initial data. The difficulties in extending the concept of weak distributional solutions for nonlinear equations are related to the impossibility of multiplying two arbitrary distributions. In case of the semilinear heat equations, it is an established tradition to study weak solutions and their internal regularity in x for positive time when the initial data are measures or belong to negative Sobolev or Besov spaces, under suitable restrictions on the growth of the nonlinear terms. In the pioneering paper [14] it was proved that the semilinear heat equation $u_t - \Delta u + u|u|^{s-1} = 0$, $s > 1$ with positive Dirac measures as initial data admits a weak solution if and only if $s < (n+2)/n$. For further developments concerning the existence and/or the uniqueness of weak (local and/or global) solutions to semilinear heat equations, the Navier-Stokes equation and Burgers' type equations, with initial data measures or

in negative Sobolev spaces, we refer to [2], [4], [18], [19], [23], [20], [22], [7], [27], [28], [6] and the references therein. In particular, [23],[27], [28], [7], [6] allow initial data more singular than measures, e.g., fractional derivatives of finite Radon measures. For similar results on higher order equations we refer to the recent papers [8], [9]. One common feature of most of the aforementioned works is the reduction of the IVP to integral equations in weighted Kato-Fujita type Hölder spaces with the use of $L^p - L^q$ estimates on the heat semigroup. Finally, we mention that, by means of different methods and techniques, solutions to degenerate parabolic equations with measures as initial or right-hand side data are of considerable interest as well, e.g., see [13], [1] and the references therein.
As it concerns the study of semilinear evolution PDE with strongly singular initial data in algebras of generalized functions of Colombeau type, we stress that such equations must admit global in time solutions without any smallness requirement on the initial data. Indeed, this follows from the fact that one of the principal ideas in the study of nonlinear evolution equations in this framework is to replace the singular initial data u^0 with smooth approximations u^0_ε, $\varepsilon \searrow 0$, such that for every $\alpha \in \mathbb{Z}^n_+$, $\|\partial^\alpha u^0_\varepsilon\|_{L^p} = O(\varepsilon^{-N_\alpha})$ for some $N_\alpha \in \mathbb{R}$ (see [15], [25], [5], [29] for details). The first results for semilinear heat equations in the setting of algebras of generalized functions are for monotone semilinear parabolic equations in bounded domains cf. [16], [24], recovering the classical solutions when the initial data are L^p functions, while, for Burgers' equation and conservation laws with nonlinear terms growing at most linearly at infinity generalized solutions have been studied in [11] and [26], respectively. Generalized solutions for the one-dimensional Kuramoto-Sivashinsky equation have been constructed in [10].
The plan of this paper is as follows: in Section 2 we show polynomial L^2-based estimates for a semilinear heat equation on the line; in Section 3 we do the same for n-dimensional Kuramoto-Sivashinsky type equations. We mention that we can show such a priori estimates for larger classes of evolution PDE, like Burgers' equation, when the "gain" of smoothness by the dissipative term is less than 1. The proofs become more involved, since we need to employ Gagliardo-Nirenberg type estimates and rather sophisticated L^p based estimates using the Leibnitz rule for fractional derivatives cf. [21]. Finally, in Section 4 we apply these a priori estimates to obtain unique generalized solutions in the sense of Colombeau. Actually, we define what seems to be a new algebra of generalized functions, namely, its elements having

representatives which are L^{∞}_{loc}-based on the first variable and L^2 on the second. We will denote by $|\cdot|_r$ the L^r-norm and by $\|\cdot\|_r$ the norm in the Sobolev space H^r, with $\|\cdot\| := \|\cdot\|_0$; $H^r_p(\mathbb{R}^n)$ denote the Sobolev spaces based on L^p, $H^r(\mathbb{R}^n) := H^r_2(\mathbb{R}^n)$; the sup-norm in $C([0,T] : H^r(\mathbb{R}^n))$ will be denoted by $\|\cdot\|_{r,T}$; $H^{\infty}(\mathbb{R}^n) = \cap_{s\in\mathbb{R}} H^s(\mathbb{R}^n)$.

2 Semilinear heat equation on the line

This section deals with the model equation

$$u_t - u_{xx} + u^3 = f, \qquad t > 0,\ x \in \mathbb{R} \tag{1}$$

$$u(0) = u^0. \tag{2}$$

This equation is well studied in the case $f = 0$. Throughout this section u^0, f, and therefore the solutions u, will be assumed real valued. The next polynomial L^2 estimates are quite natural; nevertheless we are not aware of their proof in the literature.

Theorem 1 *For every real valued $u^0 \in H^{\infty}(\mathbb{R})$ and every real valued right-hand side $f \in C^{\infty}([0,+\infty[: H^{\infty}(\mathbb{R}))$ the IVP (1), (2) admits a unique global real valued solution $u \in C^{\infty}([0,+\infty[: H^{\infty}(\mathbb{R}))$. Furthermore, there exists a positive constant C such that the following estimates hold:*

$$\|u(t)\|_1^2 \le e^t\Big(\|u^0\|_1^2 + \int_0^t \|f(\tau)\|_1^2\, d\tau\Big); \tag{3}$$

$$\|u(t)\|_r \le C\Big(\|u^0\|_r + \int_0^t \|f(\tau)\|_r\, d\tau + t^{1/4}\|u\|_{r-1,t}^3\Big), \tag{4}$$

$r \ge 2$, $t > 0$, and if $u_j = H[u_j^0, f_j]$ is the unique solution to (1), (2) then

$$\|u_1(t) - u_2(t)\|^2 \le e^t\Big(\|u_1^0 - u_2^0\|^2 + \int_0^t \|f_1(\tau) - f_2(\tau)\|^2 d\tau\Big); \tag{5}$$

$$\|u_1(t) - u_2(t)\|_r \le C\Big(\|u_1^0 - u_2^0\|_r + \int_0^t \|f_1(\tau) - f_2(\tau)\|_r\, d\tau + \Omega_r[u_1,u_2](t)\Big) \tag{6}$$

where $\Omega_r[u_1,u_2](t) = t^{1/4}\|u_1 - u_2\|_{r-1,t}(\|u_1\|_{r,t} + \|u_2\|_{r,t})^2$, $r \in \mathbb{N}$, $t > 0$.

Proof. We obtain, after multiplying equation (1) by $u(t)$, applying the Schwarz inequality, integrating from 0 to t and the Gronwall inequality, that

$$\|u(t)\|^2 \leq e^t(\|u^0\|^2 + \int_0^t \|f(\tau)\|^2 \, d\tau), \qquad t > 0. \tag{7}$$

Letting $v(t) = u_x(t)$, we obtain by differentiating equation (1), $v_t - v_{xx} + 3u^2 v = f_x$. Multiplying this equation by $v(t)$, integrating from 0 to t we show, in view of fact that $u^2 \geq 0$, the same estimate for u_x as in (7) with $(u^0)'$ and f_x.

Next, we rewrite the IVP as the integral equation

$$u(t) \;=\; L[u^0, f](t) - K[u](t) \tag{8}$$

where

$$L[u^0, f](t) \;=\; E(t) * u^0 + \int_0^t E(t-\tau) * f(\tau) d\tau \tag{9}$$

$$K[u](t) \;=\; \int_0^t E(t-\tau) * u^3(\tau) d\tau, \tag{10}$$

and $E(t)$ stands for the fundamental solution of the heat equation. We have by using the Young inequality and the L^p estimates of $E(t)$,

$$\begin{aligned} \|D^r K[u](t)\| &\leq \int_0^t \|\partial_x E(t-\tau)\| \, |D^{r-1}(u^3(\tau))|_1 d\tau \\ &\leq C \sum_{j=[\frac{r}{2}]}^{r-1} \int_0^t \frac{\|D^j u(\tau)\|}{(t-\tau)^{3/4}} \sum_{\mu+\nu=r-1-j} |D^\mu u(\tau)|_\infty |D^\nu u(\tau)|_\infty d\tau. \end{aligned} \tag{11}$$

Since $r \geq 2$, we have $|D^\mu h|_\infty \leq \|h\|_{\mu+1} \leq \|h\|_{r-1}$ if $0 \leq \mu \leq r-1-j$, $j \geq [r/2] \geq 1$, and the estimate (11) leads to the inductive polynomial estimate (4).

Let $u_j = H[u_j^0, f_j]$, $j = 1, 2$, $w = u_1 - u_2$, $w^0 = u_1^0 - u_2^0$, $g = f_1 - f_2$. Clearly w satisfies $w_t - w_{xx} + w(u_1^2 + u_1 u_2 + u_2^2) = g$, $w(0) = w^0$ and multiplication by $w(t)$, integration from 0 to t and Gronwall inequality give (5).

The subsequent estimates are obtained by using the integral representation

$$w(t) = L[w^0, g](t) - \int_0^t E(t-\tau) * w(\tau)(u_1^2(\tau) + u_1(\tau)u_2(\tau) + u_2^2(\tau)) \, d\tau. \tag{12}$$

We get, once more taking advantage of the smoothing effect of the dissipation by using the Young inequality,

$$\|w(t)\|_r \;\leq\; \|L[w^0, g](t)\|_r + Ct^{1/4} \sum_{j=0}^{r-1} \|w\|_{j,t}(\|u_1\|_{r,t} + \|u_2\|_{r,t})^2 \tag{13}$$

by differentiating (12). The proof is complete.

3 Kuramoto-Sivashinsky type equations in $\mathbb{R}^n$

This section deals with the equation

$$u_t + P(D)u + \chi(u^2) = f, \qquad t > 0,\ x \in \mathbb{R}^n \tag{14}$$

$$u(0) = u^0 \tag{15}$$

where $P(D) = \Delta^2 + P_3(D)$, $P_3(\xi)$ is a real polynomial of degree ≤ 3, χ is a constant real-vector field in $\mathbb{R}^n$.

Theorem 2 *Let $n \leq 3$. Under the hypotheses above, for every $u^0 \in H^\infty(\mathbb{R}^n)$ and every right-hand side $f \in C^\infty([0, +\infty[: H^\infty(\mathbb{R}^n))$ we claim that the IVP (14), (15) admits a unique global solution $u \in C^\infty([0, +\infty[: H^\infty(\mathbb{R}^n))$. Furthermore, there exist two constants $C > 0$ and $c \in \mathbb{R}$ such that the following estimates hold:*

$$\|u(t)\|^2 \leq e^{ct}\left(\|u^0\|^2 + \int_0^t \|f(\tau)\|^2\, d\tau\right); \tag{16}$$

$$\|u(t)\|_r \leq Ce^{ct}\left(\|u^0\|_r + \int_0^t \|f(\tau)\|_r\, d\tau + t^{\frac{1}{2}-\frac{n}{8}}\|u\|^2_{r-1,t}\right), \tag{17}$$

$r \in \mathbb{N}$, $t > 0$ and, if $u_j = K[u_j^0, f_j]$ is the unique solution to (14), (15), $j = 1, 2$,

$$\|u_1(t) - u_2(t)\|^2 \leq Ce^{c(t+\int_0^t |\nabla u_1(\tau)|_\infty d\tau)}(\|u_1^0 - u_2^0\|^2 + \int_0^t \|(f_1 - f_2)(\tau)\|^2 d\tau) \tag{18}$$

$$\|u_1(t) - u_2(t)\|_r \leq Ce^{ct}\Big(\|u_1^0 - u_2^0\|_r + \int_0^t \|(f_1 - f_2)(\tau)\|_r\, d\tau + t^{\frac{1}{2}-\frac{n}{8}}\|u_1 - u_2\|_{r-1,t}(\|u_1\|_{r-1,t} + \|u_2\|_{r-1,t})\Big) \tag{19}$$

$r \in \mathbb{N}$, $t > 0$.

Proof. First we note that, as a consequence of Theorem 4.1 in [9], or by means of the Parseval identity, for every $b > -\mathcal{P}$, $\mathcal{P} := \inf_{\xi \in \mathbb{R}^n}(|\xi|^4 + P_3(\xi))$ and for every $\alpha \in \mathbb{Z}_+^n$ there exists $C > 0$ such that the fundamental solution $E(t) = E_P(t)$ of $\partial_t + P(D)$ satisfies

$$\|\partial_x^\alpha E(t)\| \leq Ce^{bt}t^{-\frac{|\alpha|}{4}-\frac{n}{8}}, \qquad t > 0. \tag{20}$$

We also observe that, in view of the identity $< g, \chi(g^2) >= 0$, $g \in H^s(\mathbb{R}^n)$, $s > n/2$, we obtain, after multiplying the equation by $u(t)$ and applying the Schwarz inequality,

$$\begin{aligned} \frac{1}{2}\frac{d}{dt}(\|u(t)\|^2) &= - < P(D)u(t), u(t) > + < u(t), f(t) > \\ &\leq (\frac{1}{2} - \mathcal{P})\|u(t)\|^2 + \frac{1}{2}\|f(t)\|^2, \end{aligned} \tag{21}$$

for every $t > 0$, which implies, after integration from 0 to t and the Gronwall inequality, that for $c > 1 - 2\mathcal{P}$

$$\|u(t)\|^2 \leq e^{ct}(\|u^0\|^2 + \int_0^t \|f(\tau)\|^2 \, d\tau), \qquad t > 0. \tag{22}$$

Next, we rewrite the IVP as the integral equation

$$u(t) = L[u^0, f](t) - K[u](t) \tag{23}$$

where

$$L[u^0, f](t) = E(t) * u^0 + \int_0^t E(t-\tau) * f(\tau) d\tau \tag{24}$$

$$K[u](t) = \int_0^t \chi E(t-\tau) * u^2(\tau) d\tau. \tag{25}$$

Let D^r stand for r times differentiation ∂_x^α for $|\alpha| = r$. We have from (20)

$$\begin{aligned} \|D^r K[u](t)\| &\leq \int_0^t \|D\chi E(t-\tau)\| \, |D^{r-1}(u^2(\tau))|_1 d\tau \\ &\leq Ce^{bt} \sum_{j=0}^{r-1} \int_0^t \frac{\|D^j u(\tau)\| \|D^{r-1-j} u(\tau)\|}{(t-\tau)^{1/2+n/8}} d\tau \end{aligned} \tag{26}$$

which, in view of $n \leq 3$, leads to the following inductive estimates for every $r \in \mathbb{N}$, $t > 0$

$$\|u(t)\|_r \leq \|L[u^0, f](t)\|_r + Ce^{bt} t^{\frac{1}{2}-\frac{n}{8}} \sum_{j=0}^{r-1} \|u\|_{j,t} \|u\|_{r-1-j,t}. \tag{27}$$

Let $u_j = K[u_j^0, f_j]$, $j = 1, 2$, $w = u_1 - u_2$, $w^0 = u_1^0 - u_2^0$, $g = f_1 - f_2$. Clearly w satisfies $w_t + P(D)w + \chi(w(u_1 + u_2)) = g$, $w(0) = w^0$. Multiplying the previous equation by $w(t)$, integrating from 0 to t and using the conservation law, we get

$$\begin{aligned} \|w(t)\|^2 &\leq Ce^{ct}(\|w^0\|^2 + \int_0^t \|g(\tau)\|^2 \, d\tau) \\ &+ \int_0^t < w^2(t), \chi u_1(t) >). \end{aligned} \tag{28}$$

Now, as in [10], we get by Gronwall

$$\|w(t)\|^2 \leq Ce^{c(t+\int_0^t |\nabla u_1(\tau)|_\infty d\tau)}(\|w^0\|^2 + \int_0^t \|g(\tau)\|^2 \, d\tau). \tag{29}$$

The integral representation

$$w(t) = L[w^0, g](t) - \int_0^t \chi E(t-\tau) * w(\tau)(u_1(\tau) + u_2(\tau)) \, d\tau \tag{30}$$

gives, taking advantage once more of the smoothing effect of the dissipation by using the Young inequality,

$$\begin{aligned} \|w(t)\|_r &\leq \|L[w^0, g](t)\|_r \\ &+ Ce^{bt} t^{\frac{1}{2}-\frac{n}{8}} \sum_{j=0}^{r-1} \|w\|_{j,t}(\|u_1\|_{r-1-j,t} + \|u_2\|_{r-1-j,t}). \end{aligned} \tag{31}$$

The proof is complete.

4 Generalized solutions

This section will show an immediate application of the a priori estimates in the previous two sections. We will exhibit the difference between the monotone case (where no log-like conditions will be required) and the Kuramoto-Sivashinksy equation, where a log-like condition will be imposed. We set:

$$\begin{aligned} \mathcal{E}^{M,\infty,2}_{loc,gl}[\mathbb{R}^+ \times \mathbb{R}^n] &= \{u = (u_\epsilon)_{\epsilon>0} : \forall \epsilon > 0 \; u_\epsilon \in C^\infty(\mathbb{R}^+ : H^\infty(\mathbb{R}^n)) \text{ and} \\ &\quad \forall k, r \in \mathbb{Z}_+ \; \forall K \subset \mathbb{R}^+ \text{ compact } \exists N \in \mathbb{N} : \\ &\quad \sup_{t \in K} \|\partial_t^k u_\epsilon(t, \cdot)\|_r = \mathcal{O}(\epsilon^{-N}), \; \epsilon \to 0\}; \\ \mathcal{N}^{\infty,2}_{loc,gl}(\mathbb{R}^+ \times \mathbb{R}^n) &= \{u \in \mathcal{E}^{M,\infty,2}_{loc,gl}[\mathbb{R}^+ \times \mathbb{R}^n] : \forall k, r \in \mathbb{Z}_+ \; \forall q \in \mathbb{N} \\ &\quad \sup_{t \in K} \|\partial_t^k u_\epsilon(t, \cdot)\|_r = \mathcal{O}(\epsilon^q), \epsilon \to 0\}. \end{aligned}$$

The set $\mathcal{E}^{M,\infty,2}_{loc,gl}[\mathbb{R}^+ \times \mathbb{R}^n]$ is an algebra with partial derivatives and $\mathcal{N}^{\infty,2}_{loc,gl}(\mathbb{R}^+ \times \mathbb{R}^n)$ is an ideal of $\mathcal{E}^{M,\infty,2}_{loc,gl}[\mathbb{R}^+ \times \mathbb{R}^n]$ invariant by partial derivatives. The quotient $\mathcal{G}^{\infty,2}_{loc,gl}(\mathbb{R}^+ \times \mathbb{R}^n) := \mathcal{E}^{M,\infty,2}_{loc,gl}[\mathbb{R}^+ \times \mathbb{R}^n] / \mathcal{N}^{\infty,2}_{loc,gl}(\mathbb{R}^+ \times \mathbb{R}^n)$ is an algebra with partial derivatives and there is an injective map $\iota : C^\infty(\mathbb{R}^+ : H^{-\infty}(\mathbb{R}^n)) \to \mathcal{E}^{M,\infty,2}_{loc,gl}[\mathbb{R}^+ \times \mathbb{R}^n]$, $w \mapsto (w(t, \cdot) * \rho_\epsilon)_\epsilon$, which turns $C^\infty(\mathbb{R}^+ : H^\infty(\mathbb{R}^n))$ into a subalgebra of $\mathcal{G}^{\infty,2}_{loc,gl}(\mathbb{R}^+ \times \mathbb{R}^n)$.

Here $\rho_\epsilon = \epsilon^{-n}\rho(\epsilon^{-1}x)$, with $\rho \in \mathcal{S}(\mathbb{R}^n)$ such that $\int_{\mathbb{R}^n} \rho(x)dx = 1$, $\displaystyle\int_{\mathbb{R}^n} x^\alpha \rho(x)dx = 0$ $\forall \alpha \in \mathbb{Z}^n_+$, $|\alpha| > 0$.

We state our results for the model equations in the previous sections (here $\mathcal{G}_2(\mathbb{R})$ is the algebra defined with the global L^2-estimates, see also [12]).

Theorem 3 *Let $u^0 \in \mathcal{G}_2(\mathbb{R})$, $f \in \mathcal{G}^{\infty,2}_{loc,gl}(\mathbb{R}^+ \times \mathbb{R})$. Then there exists a unique solution $u \in \mathcal{G}^{\infty,2}_{loc,gl}(\mathbb{R}^+ \times \mathbb{R})$ to the (IVP) (1), (2). If the initial data $u^0 \in H^r(\mathbb{R})$, $f \in C(\mathbb{R}^+ : H^r(\mathbb{R}))$, $r \in \mathbb{Z}_+$ (resp. $u^0 \in H^\infty(\mathbb{R})$, $f \in C^\infty(\mathbb{R}^+ : H^\infty(\mathbb{R}))$) then the corresponding generalized solution is associated (resp. coincides) with the classical solution.*

Theorem 4 *Let $1 \leq n \leq 3$, $u^0 \in \mathcal{G}_2(\mathbb{R}^n)$, $f \in \mathcal{G}^{\infty,2}_{loc,gl}(\mathbb{R}^+ \times \mathbb{R}^n)$. Then there exists a solution $u \in \mathcal{G}^{\infty,2}_{loc,gl}(\mathbb{R}^+ \times \mathbb{R}^n)$ to the IVP (14), (15). Furthermore, there is at most one solution $u \in \mathcal{G}^{\infty,2}_{loc,gl}(\mathbb{R}^+ \times \mathbb{R}^n)$ with a representative $(u_\epsilon)_\epsilon \in \mathcal{E}^{M,\infty,2}_{loc,gl}[\mathbb{R}^+ \times \mathbb{R}^n]$ satisfying for all $T > 0$*

$$\int_0^T |\partial_x u_\epsilon(\tau,\cdot)|_\infty d\tau = \mathcal{O}(\log\frac{1}{\epsilon}),\ \epsilon \to 0. \tag{32}$$

In particular, if u^0, f have representatives $(u^0_\epsilon)_\epsilon$, $(f_\epsilon)_\epsilon$ satisfying

$$||u^0_\epsilon||_{[\frac{n}{2}]+2-j} + \sup_{0<t<T}\int_0^t ||f_\epsilon(\tau)||_{[\frac{n}{2}]+2-j}\, d\tau = \mathcal{O}((\log\frac{1}{\epsilon})^{\frac{1}{2^j}}) \tag{33}$$

for all $T > 0$, $j = 0, 1, \ldots, [\frac{n}{2}] + 2$, then there exists a unique solution to the IVP (14), (15).

The two theorems follow from the a priori estimates (5), (6), (18) and (19).

Acknowledgments. The research on this subject was partially carried out while both authors were participating in the project Nonlinear Theory of Generalized Functions at the E. Schrödinger International Institute for Mathematical Physics, Vienna, October-November 1997, organized by M. Oberguggenberger. The authors would like to thank M. Oberguggenberger for the invitation to participate and for the useful discussions on various aspects of the theory of generalized functions. The research of the first author is partially supported by FAPESP and CNPq, Brazil while the research of the second author is partially supported by FAPESP, Brazil; a Coordinated Research Project of the University of Cagliari; 40% of MURST, Italy and NFSR of Bulgaria.

References

[1] ANDREUCCI, D.: Degenerate parabolic equations with initial data measures, *Trans. Amer. Math. Soc.*, 349 (1997), 3911-3923.

[2] BARAS P. AND PIERRE, M.: Problèmes paraboliques non linéaires avec données mesures, *Appl. Anal.*, 18 (1984), 111-149.

[3] BEKHIRANOV, D.: The initial value problem for the generalized Burgers' equation, *Differential & Integral Equations* 9, (1996), 1253-1265.

[4] BENFATTO, G; ESPOSITO, R. AND PULVIRENTI, M: Planar Navier-Stokes flow for singular initial data, *Nonl. Anal. TMA*, 9 (1985), 533-545.

[5] BIAGIONI, H. A.: A nonlinear theory of generalized functions, *Lecture Notes in Math.* 1421, Springer, Berlin, 1990.

[6] BIAGIONI, H.A., CADEDDU, L. AND GRAMCHEV, T.: Parabolic equations with conservative nonlinear term and singular initial data, *Proc. 2nd World Congress of Nonlinear Analysts, Athens, Greece, July 10-17, 1996, Nonl. Anal. TMA*, 30 (1997) 2489-2496.

[7] BIAGIONI, H.A. AND GRAMCHEV, T.: On the 2D Navier-Stokes equation with singular data, *Proc. Fourth Workshop on Partial Differential Equations, Rio de Janeiro, July 17-21, 1995, Matemática Contemporânea*, 10 (1996), 1-20.

[8] BIAGIONI, H.A. AND GRAMCHEV, T.: Evolution PDE with elliptic dissipative terms: critical index for singular initial data, self-similar solutions and analytic regularity, *C.R. Acad. Sci. Paris*, to appear.

[9] BIAGIONI, H.A. AND GRAMCHEV, T.: Multidimensional Kuramoto-Sivashinsky type equations: singular initial data and analytic regularity, *Proc. Fifth Workshop on Partial Differential Equations, Rio de Janeiro, July 14-18, 1997, Matemática Contemporânea*, to appear.

[10] BIAGIONI, H.A. AND IORIO JR., R.J.: Generalized solutions to the Kuramoto-Sivashinsky equation, *Integral Transforms and Special Functions*, vol. 6, no. 1-4 (1997), 1-8.

[11] BIAGIONI, H.A. AND OBERGUGGENBERGER, M.: Generalized solutions to Burgers' equation, *J. Differential Equations*, 97 (1992), 263-287.

[12] BIAGIONI, H.A. AND OBERGUGGENBERGER M.: Generalized solutions to the Korteweg-de Vries and the regularized long wave equations, *SIAM J. Math. Anal.* 23 (1992), 923-940.

[13] BOCCARDO, L., DALL'AGLIO, A., GALLOUËT, T. AND ORSINA, L.: Nonlinear parabolic equations with measure data, *J. Funct. Anal.*, 147 (1997), 237-258.

[14] BREZIS, H. AND FRIEDMAN, H.: Nonlinear parabolic equations involving measures as initial conditions, *J. Math. Pures Appl.*, 62 (1983), 73-97.

[15] COLOMBEAU, J.-F.: Multiplication of distributions; a tool in mathematics, numerical engineering and theoretical physics, generalized functions, *Lecture Notes in Math.*, 1532 Springer, Berlin 1992.

[16] COLOMBEAU, J.-F. AND LANGLAIS, M.: Generalized solutions of nonlinear parabolic equations with distributions as initial conditions, *J. Math. Anal. Appl.*, 145 (1990), 186-196.

[17] DIX, D.: Nonuniqueness and uniqueness in the initial value problem for Burgers' equation, *SIAM J. Math. Anal.*, 27 (1996), 709-724.

[18] GIGA, Y.; MIYAKAWA, T. AND OSADA, H.: Two-dimensional Navier-Stokes flow with measures as initial vorticity, *Arch. Rational Mech. Anal.* 104 (1988), 223-250.

[19] GIGA, Y; MIYAKAWA, T.: Navier-Stokes flow in $\mathbb{R}^3$ with measures as initial vorticity and Morrey spaces, *Comm. in Partial Differential Equations*, 14 (1989), 577-618.

[20] KATO, T.: The Navier-Stokes equation for an incompressible fluid in $\mathbb{R}^2$ with a measure as the initial vorticity, *Differential & Integral Equations* 7 (1994), 949-966.

[21] KATO, T. AND PONCE, G.: Commutator estimates and the Euler and Navier-Stokes equations, *Comm. on Pure and Appl. Math.* 41 (1988), 891-907.

[22] KATO, T. AND PONCE, G.: The Navier-Stokes equation with weak initial data, *International Mathematics Research Notes* 10 (1994), 1-10.

[23] KOZONO, H. AND YAMAZAKI, M.: Semilinear heat equations and the Navier-Stokes equation with distributions in new function spaces as initial data, *Comm. in Partial Differential Equations*, 19 (1994), 959-1014.

[24] LANGLAIS, M.: Generalized functions solutions of monotone and semilinear parabolic equations, *Mh. Math.*, 110 (1990), 117-136.

[25] OBERGUGGENBERGER, M.: Multiplication of distributions and applications to partial differential equations, Pitman Researchs in Mathematics Series 259, New York, 1992.

[26] OBERGUGGENBERGER, M. AND WANG Y.G.: Generalized solutions to conservation laws, *Z. Anal. Anwendungen*, 13 (1994), No. 1, 7-18.

[27] RIBAUD, F.: Analyse de Littlewood-Paley pour la résolution d'équations paraboliques semi-linéaires, *Thèse de Docteur en Sciences, Orsay* 1996.

[28] RIBAUD, F.: Semilinear parabolic equations with distributions as initial data, *Discrete Contin. Dynam. Systems*, 3 (1997) No. 3, 305-316.

[29] ROSINGER, E. E.: Generalized solutions of nonlinear partial differential equations, *North Holland Math. Studies* 146, Amsterdam (1987).

H. A. BIAGIONI

University of Campinas, Department of Mathematics, 13081-970, Campinas, SP, Brazil, E-mail: hebe@ime.unicamp.br

T. GRAMCHEV

University of Cagliari, Department of Mathematics, 09124 Cagliari, Italy, E-mail: todor@vaxcal.unica.it

Shift Differentials of Maps in BV Spaces

ALBERTO BRESSAN AND MARTA LEWICKA

1 Introduction

The aim of this note is to provide a brief outline of the theory of shift-differentials, introduced in [2], and show how their construction can be extended to the case of vector valued functions.

In the following, we consider the space BV of scalar integrable functions having bounded variation, endowed with the $\mathbf{L}^1$ norm. We recall that, given a map $\Phi : X \mapsto Y$ between normed linear spaces, its differential at a point x_0 is the continuous linear map $\Lambda : X \mapsto Y$ such that

$$\lim_{h \to 0} \frac{\|\Phi(x_0 + h) - \Phi(x_0) - \Lambda(h)\|_Y}{\|h\|_X} = 0. \tag{1.1}$$

This concept of differential (see for example [6]) is one of the cornerstones of mathematical analysis, providing a basic tool in the study of regular maps. For maps which do not admit a first-order linear approximation, various concepts of weak or generalized differentials can be found in the literature [4], [7], [9], [11]. The present paper intends to provide some further contribution in this direction.

The primary motivation for the introduction of shift-differentials comes from the theory of hyperbolic conservation laws [8] [10]. As a simple example, consider Burgers' equation

$$u_t + [u^2/2]_x = 0 \tag{1.2}$$

with the family of initial conditions

$$u^\theta(0, x) = \theta x \cdot \chi_{[0,1]}(x). \tag{1.3}$$

By χ_I we denote here the characteristic function of the interval I. Assuming $\theta > 0$, the corresponding solution of (1.2)-(1.3) is

$$u^\theta(t,x) = \frac{\theta x}{1+\theta t} \cdot \chi_{[0,\ \sqrt{1+\theta t}]}(x). \tag{1.4}$$

Observe that the map $\theta \mapsto u^\theta(0,\cdot)$ describes a smooth curve in $\mathbf{L}^1$, namely a segment. However, for $t > 0$, the map $\theta \mapsto u^\theta(t,\cdot)$ is Lipschitz continuous but nowhere differentiable because the location $x^\theta(t) = \sqrt{1+\theta t}$ of the shock varies with θ. Therefore, the limit

$$\lim_{\Delta\theta \to 0} \frac{u^{\theta+\Delta\theta} - u^\theta}{\Delta\theta}$$

is not well defined as an element of the space $\mathbf{L}^1$.

More generally, call $S : \mathbf{L}^1 \times [0,\infty[\mapsto \mathbf{L}^1$ the semigroup [5] [8] generated by the nonlinear conservation law

$$u_t + f(u)_x = 0. \tag{1.5}$$

In other words, let $t \mapsto u(t,\cdot) = S_t\bar{u}$ be the unique entropic solution of (1.5) with initial condition $u(0,x) = \bar{u}(x)$. Then for each $t > 0$ the flow map $\bar{u} \mapsto S_t\bar{u}$ is a contraction in the space $\mathbf{L}^1$, in general nondifferentiable in the usual sense.

To cope with this situation, a new differential structure on the space BV of integrable functions with bounded variation can be introduced. Let a function $u \in BV$ be given. In order to define a "tangent space" at u, we follow a procedure which is now standard in differential geometry. On the family of continuous maps $\gamma : [0,\theta^*] \mapsto \mathbf{L}^1$, with $\gamma(0) = u$ and $\theta^* > 0$ (possibly depending on γ), consider the equivalence relation

$$\gamma \sim \gamma' \qquad \text{iff} \qquad \lim_{\theta\to 0} \frac{1}{\theta} \|\gamma(\theta) - \gamma'(\theta)\|_{\mathbf{L}^1} = 0. \tag{1.6}$$

Every equivalence class can be regarded as a "first-order tangent vector" at the point u. Call $\widehat{T}_u$ the set of all these equivalence classes. Observe that, if Φ is any Lipschitz continuous map from $\mathbf{L}^1$ into itself, then $\gamma \sim \gamma'$ implies $\Phi \circ \gamma \sim \Phi \circ \gamma'$. Therefore Φ induces a well defined map $D\Phi(u) : \widehat{T}_u \mapsto \widehat{T}_{\Phi(u)}$. In principle, this map $D\Phi$ could be regarded as an abstract differential of Φ at the point u. However, this does not seem a fruitful point of view. Indeed, the set $\widehat{T}_u$ of all equivalence classes is extremely large and cannot be adequately described. Therefore, one usually works with a particular subset $T_u \subset \widehat{T}_u$ of tangent vectors which admit a suitable representative.

The standard choice at this stage is to consider a family T_u of tangent vectors which can be put in a one-to-one correspondence with $\mathbf{L}^1(\mathbb{R})$. More precisely, T_u is defined as the family of all equivalence classes of the maps

$$\theta \mapsto \gamma_v(\theta) \doteq u + \theta v, \qquad v \in \mathbf{L}^1(\mathbb{R}). \tag{1.7}$$

As we saw earlier, this choice is not adequate for describing a first-order variation of the flow map S_t for (1.2). For example, taking

$$\bar{u}(x) \doteq x \cdot \chi_{[0,1]},$$

the path $\vartheta \mapsto \gamma(\vartheta) \doteq u^{1+\vartheta}(0, \cdot)$ in (1.3) determines an equivalence class in $T_{\bar{u}}$. On the other hand, the equivalence class of the path $\vartheta \mapsto S_t \circ \gamma(\vartheta) = u^{1+\vartheta}(t, \cdot)$ defined at (1.4) determines some tangent vector which is not in the space $T_{S_t\bar{u}} \doteq \mathbf{L}^1$.

In this paper, given $u \in BV$, we study a different space T_u of tangent vectors, which can be put into a one-to-one correspondence with $\mathbf{L}^1(\mathrm{D}u)$. Here $\mathrm{D}u$ denotes the (signed) Radon measure corresponding to the distributional derivative of u. The basic idea is the following. In the special case where v is continuously differentiable with compact support, to v we associate the equivalence class of the map $\theta \mapsto u^\theta$, where u^θ is implicitly defined as

$$u^\theta(x + \theta v(x)) = u(x) \tag{1.8}$$

for all $\theta \geq 0$ sufficiently small. We then show that this correspondence can be uniquely extended to the whole space $\mathbf{L}^1(\mathrm{D}u)$. Observe that in (1.7) the graph of $u^\theta = \gamma_v(\theta)$ is obtained by lifting the graph of u vertically by θv. On the other hand, in (1.8), the graph of u is shifted horizontally by θv. This motivates the term "shift-differential" used in the sequel.

In Section 2 of this paper we review the definition and the basic properties of shift-differentials, for maps within the space of scalar BV functions. All results are taken from [2], to which we refer for details of the proofs.
In Section 3 we extend the notion of shift-differential to the case of vector-valued BV functions. If $u : \mathbb{R} \mapsto \mathbb{R}^n$ is a BV function, a shift tangent vector is defined in terms of:

(i) A decomposition of u into n scalar components,

(ii) A n-tuple of functions $(v_1, \ldots, v_n) \in \mathbf{L}^1(\mathrm{D}u)$, determining the rate at which each component of u is shifted.

This approach is strongly motivated by the study of hyperbolic systems of conservation laws. The main result in [2] shows that the flow generated by a single conservation law is differentiable "almost everywhere". We conjecture that a similar result holds for $n \times n$ strictly hyperbolic systems of conservation laws, with the above definition of shift-differential. For piecewise Lipschitz solutions to general $n \times n$ systems of conservation laws, a detailed analysis of first order variations can be found in [3]. Shift differentials of an approximate flow, defined within a class of piecewise constant functions, played a key role in [1] to construct the semigroup generated by a 2×2 system of conservation laws.

2 Shift tangent vectors

Throughout the following, our basic function space is the normed space

$$X \doteq (\mathbf{L}^1(\mathbb{R}) \cap \mathbf{BV}(\mathbb{R});\ \|\cdot\|_{\mathbf{L}^1(\mathbb{R})}). \tag{2.1}$$

Here $\mathbf{L}^1(\mathbb{R})$ refers to the standard Lebesgue measure. The elements of X are thus equivalence classes of functions. For sake of definiteness, we shall always deal with left continuous representatives. By $\mathbf{Lip}(\mathbb{R})$ we denote the space of Lipschitz continuous functions. The Lipschitz constant and the $\mathbf{L}^\infty$ norm of v are denoted by $\mathrm{Lip}(v)$ and $\|v\|_{\mathbf{L}^\infty}$, respectively. For $u \in \mathbf{BV}(\mathbb{R})$, by $\mathbf{L}^1(\mathrm{D}u)$ we denote the space of functions integrable w.r.t. the Radon measure $\mathrm{D}u$.

Given $u \in X$, for every $v \in \mathbf{L}^1(\mathrm{D}u)$ we shall construct a first-order variation $\theta \mapsto u^\theta$ of u, consistent with the definition (1.8) in the smooth case. The idea is to approximate v by a family of Lipschitz continuous functions v^θ and define u^θ implicitly by

$$u^\theta(x + \theta v^\theta(x)) = u(x). \tag{2.2}$$

Observe that the condition $\mathrm{Lip}(\theta v^\theta) < 1$ is essential in order that the function u^θ in (2.2) be well defined. Some key properties of the approximating functions v^θ are singled out by the next definition.

Definition 1. Let $u \in X$ and $v \in \mathbf{L}^1(\mathrm{D}u)$. Consider a family of functions $v^\theta \in \mathbf{L}^1(\mathrm{D}u) \cap \mathbf{Lip}(\mathbb{R})$, with $\theta \in (0, \theta^*]$. We say that v^θ *hat-converges* to v, and write $v^\theta \stackrel{\wedge}{\longrightarrow} v$, provided that

$$\lim_{\theta \to 0} \int |v^\theta(x) - v(x)| \, d\mu_u(x) = 0, \qquad \limsup_{\theta \to 0} \mathrm{Lip}(\theta v^\theta) < 1, \qquad \lim_{\theta \to 0} \|\theta v^\theta\|_{\mathbf{L}^\infty} = 0. \tag{2.3}$$

Here and in the sequel, $\mu_u \doteq |\mathrm{D}u|$ denotes the Radon measure of total variation of u. By definition, $\mathbf{L}^1(\mathrm{D}u) = \mathbf{L}^1(\mu_u)$. Observe that the last two conditions in (2.3) essentially depend on the parametrization of the family $\{v^\theta\}$. Indeed, from any sequence of bounded Lipschitz functions $v_n \to v$, it is always possible to construct a hat-convergent family (satisfying even (2.4) below) by a suitable parametrization.

Lemma 1. *For any $u \in X$ and $v \in \mathbf{L}^1(\mathrm{D}u)$ there exists a family of functions $v^\theta \in \mathbf{C}_c^\infty(\mathbb{R})$, $\theta \in (0, \theta^*]$ satisfying*

$$\mathrm{Lip}(\theta v^\theta) \leq \sqrt{\theta}, \qquad \|\theta v^\theta\|_{\mathbf{L}^\infty} \leq \sqrt{\theta}, \qquad \lim_{\theta \to 0} v^\theta = v \quad \textit{in } \mathbf{L}^1(\mathrm{D}u). \tag{2.4}$$

In particular, $v^\theta \stackrel{\wedge}{\longrightarrow} v$.

If $\mathrm{Lip}(\theta v^\theta) \leq \alpha < 1$, then the function $y_\theta(x) \doteq x + \theta v^\theta(x)$ is Lipschitz continuous with $\mathrm{Lip}(y_\theta) \leq 1 + \alpha$. Its inverse $x_\theta(y)$ is also Lipschitz continuous, with constant $\mathrm{Lip}(x_\theta) \leq 1/(1-\alpha)$. Hence, the function u^θ in (2.2) is well defined. In particular, if $v^\theta \stackrel{\wedge}{\longrightarrow} v$, then for all $\theta > 0$ sufficiently small the definition (2.2) is meaningful. We now introduce a convenient notation for the functions u^θ, obtained by shifting horizontally the graph of u.

Definition 2. If $u \in X$ and $\mathrm{Lip}(\theta v^\theta) \leq \alpha < 1$, we denote by $v^\theta \star u$ the function implicitly defined by

$$(v^\theta \star u)(x + \theta v^\theta(x)) = u(x). \tag{2.5}$$

With the above notations, the basic definition of a shift tangent vector can now be introduced.

Definition 3. Fix $u \in X$ and consider a path $\theta \mapsto u^\theta \in \mathbf{L}^1(\mathbb{R})$, defined on some interval $\theta \in (0, \theta^*]$. We say that the path u^θ *generates the shift tangent vector* $v \in \mathbf{L}^1(\mathrm{D}u)$ if, for some functions $v^\theta \overset{\wedge}{\longrightarrow} v$, one has

$$\lim_{\theta \to 0} \frac{1}{\theta} \| u^\theta - v^\theta \star u \|_{\mathbf{L}^1(\mathbb{R})} = 0. \tag{2.6}$$

((2.6) implies that $u^\theta \to u$ as $\theta \to 0$.)

Roughly speaking, this means that in first approximation the functions u^θ are obtained by shifting the graph of u horizontally by the amount θv. For the above definition to be meaningful, there must be a one-to-one correspondence between shift tangent vectors and functions in $\mathbf{L}^1(\mathrm{D}u)$. In other words, all families v^θ hat-converging to the same function v should determine paths belonging to the same equivalence class. This is indeed the content of the following theorem.

Theorem 1. *Let $u \in X$, $v, \tilde{v} \in \mathbf{L}^1(\mathrm{D}u)$ and assume $v^\theta \overset{\wedge}{\longrightarrow} v$, $\tilde{v}^\theta \overset{\wedge}{\longrightarrow} \tilde{v}$. Then*

$$\lim_{\theta \to 0} \frac{1}{\theta} \int |(v^\theta \star u)(y) - (\tilde{v}^\theta \star u)(y)| \, dy = \int |v(x) - \tilde{v}(x)| \, d\mu_u(x). \tag{2.7}$$

Theorem 1 yields the uniqueness of shift tangent vectors. Indeed, assume that the path $\theta \mapsto u^\theta$ generates the shift tangent vectors $v, \tilde{v} \in \mathbf{L}^1(\mathrm{D}u)$. Choosing functions $v^\theta \overset{\wedge}{\longrightarrow} v$, $\tilde{v}^\theta \overset{\wedge}{\longrightarrow} \tilde{v}$, from Theorem 1 it now follows

$$\begin{aligned}
\|v - \tilde{v}\|_{\mathbf{L}^1(\mathrm{D}u)} &= \lim_{\theta \to 0} \frac{1}{\theta} \|v^\theta \star u - \tilde{v}^\theta \star u\|_{\mathbf{L}^1(\mathbb{R})} \\
&\leq \lim_{\theta \to 0} \frac{1}{\theta} \|v^\theta \star u - u^\theta\|_{\mathbf{L}^1(\mathbb{R})} + \lim_{\theta \to 0} \frac{1}{\theta} \|u^\theta - \tilde{v}^\theta \star u\|_{\mathbf{L}^1(\mathbb{R})} \\
&= 0.
\end{aligned}$$

Therefore, v and $\tilde{v}$ must coincide almost everywhere w.r.t. the measure $\mathrm{D}u$.

On the other hand, let $v^\theta \overset{\wedge}{\longrightarrow} v$, $\tilde{v}^\theta \overset{\wedge}{\longrightarrow} v$. Then by (2.7) the paths $\gamma(\theta) \doteq v^\theta \star u$, $\tilde{\gamma}(\theta) \doteq \tilde{v}^\theta \star u$ are equivalent. This shows that the construction of a tangent vector does not depend on the choice of the family v^θ in (2.3).

Example 1. Let $u(x) = x \cdot \chi_{[0,1]}(x)$ and define the parametrized curve

$$u^\theta(x) = (1 + \theta)^2 x \cdot \chi_{[0,1+\theta]}(x).$$

Then the curve $\theta \mapsto u^\theta$ generates the shift tangent vector $v \in \mathbf{L}^1(\mathrm{D}u)$, with

$$v(x) = \begin{cases} -2x & \text{if} \quad x \in [0,1), \\ 1 & \text{if} \quad x = 1. \end{cases}$$

Indeed, one can check that (2.6) holds, choosing for example,

$$v^\theta(x) = \max\{-2x,\ 1 + \frac{x-1}{2\theta}\} \qquad\qquad (0 \le x \le 1)$$

and setting $v^\theta(x) = 0$ for $x \le 0$, $v^\theta(x) = 1$ for $x \ge 1$ (we consider $0 < \theta \le \frac{1}{2}$).

It is interesting to relate the standard notion of differentiability to the existence of a shift tangent vector. Recall that a continuous path $\theta \mapsto u^\theta$ with $u^0 = u \in X$ is shift differentiable if there exists $v \in \mathbf{L}^1(\mathrm{D}u)$ such that (2.6) holds for some $v^\theta \overset{\wedge}{\longrightarrow} v$. On the other hand, the path is differentiable in the usual sense if there exists $w \in \mathbf{L}^1(\mathbb{R})$ such that

$$\lim_{\theta \to 0} \left\| \frac{u^\theta - u}{\theta} - w \right\|_{\mathbf{L}^1(\mathbb{R})} = 0. \tag{2.8}$$

Theorem 2. *Consider a map $\theta \mapsto u^\theta \in X$, with $u^0 = u$, generating the shift tangent vector $v \in \mathbf{L}^1(\mathrm{D}u)$. Then this map is differentiable at $\theta = 0$ in the usual sense if and only if $v \cdot \mathrm{D}u$ is absolutely continuous w.r.t. Lebesgue measure. In this case we have $v \cdot \mathrm{D}u = -w \cdot dx$ with w defined by (2.8).*
Vice versa, let (2.8) hold. Then the curve $\theta \mapsto u^\theta$ generates a shift tangent vector $v \in \mathbf{L}^1(\mathrm{D}u)$ if and only if the measure $w \cdot dx$ is absolutely continuous w.r.t. the measure $\mathrm{D}u$, and the equality $w \cdot dx = -v \cdot \mathrm{D}u$ holds.

Starting from the definition of shift tangent vector, one can introduce the notion of shift-differentiability for a map with values in the space X as in (2.1). In the following, we consider a locally Lipschitz continuous operator Φ mapping the function space X into itself.

Definition 4. We say that Φ is *shift-differentiable at the point $u \in X$ along the direction $v \in \mathbf{L}^1(\mathrm{D}u)$* if there exists $w \in \mathbf{L}^1(\mathrm{D}\Phi(u))$ such that

$$\lim_{\theta \to 0} \frac{1}{\theta} \| w^\theta \star \Phi(u) - \Phi(v^\theta \star u) \|_{\mathbf{L}^1(\mathbb{R})} = 0 \tag{2.9}$$

for some $v^\theta \overset{\wedge}{\longrightarrow} v$, $w^\theta \overset{\wedge}{\longrightarrow} w$. In this case, we call w the shift-derivative of Φ at the point u along v and write $w = \overset{\leftrightarrow}{\nabla}_v \Phi(u)$.

Remark 1. The Lipschitz continuity of Φ and the discussion following Theorem 1 imply that, if Φ is shift-differentiable at u along the direction v, then (2.9) holds for all $\tilde{v}^\theta, \tilde{w}^\theta$ hat-converging to v and $\overset{\leftrightarrow}{\nabla}_v\Phi(u)$ respectively. Moreover, if a shift-derivative exists, it is necessarily unique.

Definition 5. We say that Φ is *shift-differentiable* at u if there exists a continuous linear map $\Lambda: \mathbf{L}^1(\mathrm{D}u) \to \mathbf{L}^1(\mathrm{D}\Phi(u))$ such that the following holds. For all $v \in \mathbf{L}^1(\mathrm{D}u)$ there exist $v^\theta \xrightarrow{\wedge} v$ and $w^\theta \xrightarrow{\wedge} \Lambda v$ such that

$$\lim_{\theta\to 0}\frac{1}{\theta}\|w^\theta \star \Phi(u) - \Phi(v^\theta \star u)\|_{\mathbf{L}^1(\mathbb{R})} = 0. \tag{2.10}$$

In this case we say that Λ is the *shift-differential* of Φ at u and write $\Lambda = \overset{\leftrightarrow}{\nabla}\Phi(u)$.

In other words Φ is shift-differentiable at u if it is shift-differentiable along each direction $v \in \mathbf{L}^1(\mathrm{D}u)$ and the map $v \mapsto \overset{\leftrightarrow}{\nabla}_v\Phi(u)$ is linear and continuous.

Remark 2. By Theorem 1, the shift-differential of Φ is a linear continuous map Λ whose norm is bounded by the local Lipschitz constant of Φ.

3 The vector valued case

In this section we consider the notion of shift tangent vector at a point $u \in X$, where X is now the space of all vector valued, integrable functions $u : \mathbb{R} \mapsto \mathbb{R}^n$ of bounded variation, endowed with the $\mathbf{L}^1$ topology. As mentioned in the Introduction, to define such a tangent vector we first need to split the vector u into n scalar components, then we have to specify the rate at which each component shifts.
The first task is accomplished by assigning to each $x \in \mathbb{R}$ a (positively oriented) basis of unit vectors in $\mathbb{R}^n$, say $r_1(x), \ldots, r_n(x)$. Denoting by $l_1(x), \ldots, l_n(x)$ the dual basis, for every $x \in \mathbb{R}$ we thus have

$$r_1(x) \wedge \cdots \wedge r_n(x) > 0, \tag{3.1}$$

$$|r_i(x)| = 1, \qquad \langle l_i(x),\ r_i(x)\rangle = \delta_{ij} \qquad \text{for all } i, j. \tag{3.2}$$

We then assign n scalar functions $v_1, \ldots, v_n$ determining the shifts. In this way, we thus determine a matrix valued function $A : \mathbb{R} \mapsto \mathbb{R}^{n\times n}$ having the r_i, l_i as right and left eigenvectors, and the v_i as eigenvalues. In other words:

$$A(x)w = \sum_{i=1}^{n} v_i(x)\langle l_i(x), w\rangle r_i(x), \qquad x \in \mathbb{R},\ w \in \mathbb{R}^n. \tag{3.3}$$

In the following we shall assume that, for every $i = 1, \ldots, n$,

(A1) the functions $r_i,\ \ l_i : \mathbb{R} \mapsto \mathbb{R}^n$ are Borel measurable and uniformly bounded,

(A2) $v_i \in \mathbf{L}^1(\mu_u)$.

We recall that $\mu_u \doteq |Du|$ denotes the Radon measure of total variation of u. In analogy with Definition 1, we introduce

Definition 5. We say that the family of matrix valued functions $A^\theta : \mathbb{R} \mapsto \mathbb{R}^{n\times n}$, parametrized by $\theta \in (0, \theta^*]$, *hat-converges* to A (and write $A^\theta \overset{\wedge}{\longrightarrow} A$), provided that:

(i) for each $\theta \in (0, \theta^*]$, $\ A^\theta$ has the form:

$$A^\theta(x)w = \sum_{i=1}^{n} v_i^\theta(x)\langle l_i^\theta(x), w\rangle r_i^\theta(x),$$

where $\{r_i^\theta(x)\}_{i=1}^n$ and $\{l_i^\theta(x)\}_{i=1}^n$ are the corresponding bases of right and left eigenvectors of $A^\theta(x)$, normalized as in (3.2),

(ii) For each given θ, the functions $r_i^\theta,\ l_i^\theta : \mathbb{R} \mapsto \mathbb{R}^n$ and $v_i^\theta : \mathbb{R} \mapsto \mathbb{R}$ are Lipschitz continuous and bounded. Moreover, they remain constant outside a compact interval K^θ,

(iii) for each $i = 1 \ldots n$ one has

$$\lim_{\theta\to 0} \int_{\mathbb{R}} \Big\{ |r_i^\theta(x) - r_i(x)| + |l_i^\theta(x) - l_i(x)| \ + |v_i^\theta(x) - v_i(x)| \Big\}\, d\mu_u(x) = 0, \tag{3.4}$$

$$\lim_{\theta\to 0} \theta^{1/4} \left(\mathrm{Lip}(r_i^\theta) + \mathrm{Lip}(l_i^\theta) + \mathrm{Lip}(v_i^\theta) + \|v_i^\theta\|_{\mathbf{L}^\infty} \right) = 0, \tag{3.5}$$

$$\lim_{\theta\to 0} \max_j \{\|v_j^\theta\|_{\mathbf{L}^\infty}\} \int_{\mathbb{R}} \Big(|r_i^\theta(x) - r_i(x)| + |l_i^\theta(x) - l_i(x)| \Big) d\mu_u(x) = 0, \tag{3.6}$$

$$\|r_i^\theta\|_{\mathbf{L}^\infty} \equiv 1, \qquad\qquad \sup_{i,\theta} \|l_i^\theta\|_{\mathbf{L}^\infty} < \infty. \tag{3.7}$$

The next lemma ensures that some hat-convergent family always exists.

Lemma 2. *Let the matrix valued function A in (3.3) satisfy* **(A1)-(A2)**. *Then there exists $\theta^* > 0$ and a family of matrix valued functions A^θ, with $\theta \in (0, \theta^*]$, such that $A^\theta \overset{\wedge}{\longrightarrow} A$.*

Proof. Denote by R the matrix valued function, having $r_1, \ldots, r_n$ as column vectors. By (3.2), its inverse R^{-1} is the matrix having $l_1, \ldots, l_n$ as row vectors. By (3.1) and the assumption **(A1)**, there exists some $Q > 0$ such that $\det R(x) \in [1/Q,\ Q]$ for all x. Fix $\theta \in (0,1)$. By Lusin's theorem, there exists a compact set B_θ such that

$$\mu_u(\mathbb{R} \setminus B_\theta) < \theta \tag{3.8}$$

and such that the restriction of R to B_θ is continuous. The complementary set $\mathbb{R} \setminus B_\theta$ consists of two open half-lines $(-\infty, b_0)$ and (a_0, ∞) and a countable number of open intervals (a_k, b_k), $k \geq 1$. For each such k, let $\widetilde{R}^k : [a_k, b_k] \mapsto \mathbb{R}^{n \times n}$ be a continuous matrix valued function with the following properties:

$$\begin{aligned}
&\widetilde{R}^k(a_k) = R(a_k), \qquad \widetilde{R}^k(b_k) = R(b_k),\\
&\text{each column vector } \ \tilde{r}_i^k(x) \ \text{ has unit length },\\
&\|\widetilde{R}^k(x) - R(a_k)\| \leq C\|R(b_k) - R(a_k)\|,\\
&\det \widetilde{R}^k(x) \in [1/Q_1,\ Q_1],
\end{aligned} \tag{3.9}$$

for some constants $Q_1 > Q$ and $C > 0$ depending only on Q, and for all $x \in [a_k, b_k]$. We now define

$$\widetilde{R}^\theta(x) \doteq \begin{cases} R(x) & \text{if} \quad x \in B_\theta,\\ R(b_0) & \text{if} \quad x \in (-\infty,\ b_0),\\ R(a_0) & \text{if} \quad x \in (a_0,\ \infty),\\ \widetilde{R}^k & \text{if} \quad x \in (a_k, b_k). \end{cases}$$

Let $\widetilde{L}^\theta \doteq \left(\widetilde{R}^\theta\right)^{-1}$. Then $\widetilde{R}^\theta$ and $\widetilde{L}^\theta$ are continuous, constant outside the compact interval $[b_0, a_0]$, bounded by a number which does not depend on θ, and such that each column vector $\tilde{r}_i^\theta(x)$ has unit length. Repeating this construction for every $\theta > 0$, by (3.8) and (3.9) we thus obtain functions $\tilde{r}_i^\theta$, $\tilde{l}_i^\theta$, $i = 1, \ldots, n$, normalized as in (3.2), such that

$$\lim_{\theta \to 0} \left(\|\tilde{r}_i^\theta - r_i\|_{\mathbf{L}^1(|Du|)} + \|\tilde{l}_i^\theta - l_i\|_{\mathbf{L}^1(|Du|)} \right) = 0. \tag{3.10}$$

By a mollification and renormalization procedure, we can also assume that all the functions $x \mapsto \tilde{r}_i^\theta(x)$ and $x \mapsto \tilde{l}^\theta(x)$ are smooth.

Next, we approximate each v_i in $\mathbf{L}^1(\mu_u)$ by a sequence $\{\tilde{v}_i^k\}_{k\geq 1}$ of smooth functions with compact support.

Define a sequence $\{\theta_k\}_{k\geq 1}$, strictly decreasing to 0 and such that, for each k,

$$\max_j \{\mathrm{Lip}(\tilde{r}_j^{1/k})\} + \max_j \{\mathrm{Lip}(\tilde{l}_j^{1/k})\} \leq \theta_k^{-1/8}. \tag{3.11}$$

Moreover, define

$$r_i^\theta(x) \doteq \tilde{r}_i^{1/k}, \qquad l_i^\theta(x) \doteq \tilde{l}_i^{1/k}, \qquad \text{for } x \in (\theta_{k-1}, \theta_k].$$

After having constructed the eigenvector functions r_i^θ, l_i^θ, we now determine a second sequence $\{\theta_k'\}_{k\geq 1}$, strictly decreasing to 0, with the property that for each k,

$$\max_j \{\mathrm{Lip}(\tilde{v}_j^k)\} + \max_j \{\|\tilde{v}_j^k\|_{\mathbf{L}^\infty}\} \leq \Big(\theta^{1/4} + \max_j \{\|\tilde{r}_j^\theta - r_j\|_{\mathbf{L}^1(\mathrm{D}u)}\}$$
$$+ \max_j \{\|\tilde{l}_j^\theta - l_j\|_{\mathbf{L}^1(\mathrm{D}u)}\}\Big)^{-1/2} \quad \text{for all } \theta \in (0, \theta_k']. \tag{3.12}$$

One can now define:

$$v_i^\theta(x) = \tilde{v}_i^k \ \text{ for } x \in (\theta_{k-1}',\ \theta_k'].$$

This completes the construction of the approximating family A^θ. All requirements of Definition 5 are then satisfied. Indeed, (3.4) follows from (3.10) and the convergence $\tilde{v}_i^k \to v_i$ in $\mathbf{L}^1(\mu_u)$. The bounds (3.11) and (3.12) together imply (3.5), while (3.6) is also a consequence of (3.12). By construction, $|r_i^\theta(x)| = 1$ for all i, θ, x. This property, together with the uniform boundedness of the matrices $\widetilde{L}^\theta$, yields (3.7).

Let now A be a matrix valued function, described in terms of its eigenvectors and eigenvalues as in (3.3). Let the conditions **(A1)-(A2)** be satisfied and assume $A^\theta \xrightarrow{\wedge} A$. We wish to construct a path $\theta \mapsto u^\theta \in \mathbf{L}^1(\mathbb{R}, \mathbb{R}^n)$, originating from $u^0 = u$ and corresponding to the shift tangent vector determined by A. Roughly speaking, u^θ should thus be obtained from u by shifting each component $u_i \doteq \langle l_i, u\rangle r_i$ by the amount θv_i. Some notation must first be introduced. For a fixed $\theta > 0$ consider the points $P_k^\theta \doteq \frac{k}{2}\theta^{3/4}$, $k \in \mathbb{Z}$. Moreover, for each integer k let I_k^θ and J_k^θ be the open intervals centered at P_k^θ and with lengths $\frac{1}{2}\theta^{3/4}$ and $\theta^{3/4}$, respectively.

Definition 6. Let $A^\theta \xrightarrow{\wedge} A$, $\theta \in (0, \theta^*]$. Then for each $\theta \in [0, \theta^*]$ we define the function $u^\theta \doteq A^\theta \star u$ by the formula:

$$\begin{aligned} u^\theta(x) &= \sum_{i=1}^{n} \left\langle l_i^\theta(P_k^\theta),\ u(x - \theta v_i^\theta(P_k^\theta)) \right\rangle r_i^\theta(P_k^\theta), \qquad \text{for } x \in I_k^\theta \text{ and } \theta > 0, \\ u^0(x) &= u(x), \qquad \text{for } x \in \mathbb{R}. \end{aligned} \tag{3.13}$$

Note that, restricted to each interval I_k^θ, one has $u^\theta = w_k^\theta(\theta, \cdot)$, where w_k^θ is the solution to the linear hyperbolic system with constant coefficients:

$$w_t + A^\theta(P_k^\theta) w_x = 0, \quad w(0, \cdot) = u.$$

The values of u^θ on I_k^θ depend only on the values of u on J_k^θ. From the assumptions (ii) and (iii) in Definition 5 it follows:

Lemma 3. *The path $\theta \mapsto u^\theta \doteq A^\theta \star u$ is continuous at $\theta = 0$, namely,*

$$\lim_{\theta \to 0} \int_{\mathbb{R}} |u^\theta - u|\, dx = 0.$$

The following is a natural generalization of Definition 3.

Definition 7. Let $A : \mathbb{R} \mapsto \mathbb{R}^{n \times n}$ satisfy **(A1)-(A2)**. We say that a path $\theta \mapsto w^\theta \in \mathbf{L}^1(\mathbb{R}, \mathbb{R}^n)$, defined on some interval $[0, \theta^*]$, with $w^0 = u$, generates the shift tangent vector determined by the matrix valued function A, if for some hat-converging family $A^\theta \stackrel{\wedge}{\longrightarrow} A$ one has

$$\lim_{\theta \to 0} \frac{1}{\theta} \int_{\mathbb{R}} |w^\theta - A^\theta \star u|\, dx = 0. \tag{3.14}$$

To justify the above definition, one needs to check that the equivalence class of the path $A^\theta \star u$ does not depend on the choice of the approximating family A^θ. This is the content of the next theorem.

Theorem 3. *Let A be a matrix valued function, satisfying (A1) and (A2). Let $A^\theta \stackrel{\wedge}{\longrightarrow} A$ and $\widetilde{A}^\theta \stackrel{\wedge}{\longrightarrow} A$. Then the paths $u^\theta = A^\theta \star u$ and $\tilde{u}^\theta = \widetilde{A}^\theta \star u$ satisfy*

$$\lim_{\theta \to 0} \frac{1}{\theta} \int_{\mathbb{R}} |u^\theta - \tilde{u}^\theta|\, dx = 0. \tag{3.15}$$

Toward a proof of Theorem 3, two technical lemmas will be used.

Lemma 4. *Let $a, b, y, z \in \mathbb{R}$ and $a < b$. Then*

$$\int_a^b |u(x-y)-u(x-z)|\,dx \le |y-z|\cdot \text{Tot.Var.}\Big\{u;\ [a-\max\{|y|,|z|\},\ b+\max\{|y|,|z|\}]\Big\}.$$

Lemma 5. *Let $\{r_i\}_{i=1}^n$ and $\{\tilde{r}_i\}_{i=1}^n$ be two bases of $\mathbb{R}^n$ and $\{l_i\}_{i=1}^n$, $\{\tilde{l}_i\}_{i=1}^n$ their corresponding dual bases, normalized as in (3.2) and bounded by a number M. Let $\lambda_1, \ldots, \lambda_n$ be real numbers. Fix an open interval I and let J be a second open interval, with the same center as I but twice as long. Then for every $t \ge 0$ such that $|I| > 2t\max_i|\lambda_i|$ one has*

$$\begin{aligned}\int_I \left| \sum_{i=1}^n \langle l_i,\ u(x-t\lambda_i)\rangle r_i - \sum_{i=1}^n \langle \tilde{l}_i,\ u(x-t\lambda_i)\rangle \tilde{r}_i \right| dx \\ \le 2Mt\cdot\left(\max_i|\lambda_i|\right)\cdot\left(\sum_{i=1}^n |l_i-\tilde{l}_i| + \sum_{i=1}^n |r_i-\tilde{r}_i|\right)\cdot \text{Tot.Var.}\{u;\ J\}.\end{aligned}$$

Proof of Theorem 3. Note first that $\int_{\mathbb{R}} |u^\theta - \tilde{u}^\theta|dx \le \mathrm{I}_1 + \mathrm{I}_2$, where

$$\begin{aligned}\mathrm{I}_1 \doteq \sum_k \int_{I_k^\theta} \Bigg| &\sum_{i=1}^n \langle l_i^\theta(P_k^\theta),\ u(x-\theta v_i^\theta(P_k^\theta))\rangle r_i^\theta(P_k^\theta) \\ &- \sum_{i=1}^n \langle l_i^\theta(P_k^\theta),\ u(x-\theta\tilde{v}_i^\theta(P_k^\theta))\rangle r_i^\theta(P_k^\theta)\Bigg|\, dx,\end{aligned}$$

$$\begin{aligned}\mathrm{I}_2 \doteq \sum_k \int_{I_k^\theta} \Bigg| &\sum_{i=1}^n \langle l_i^\theta(P_k^\theta),\ u(x-\theta \tilde{v}_i^\theta(P_k^\theta))\rangle r_i^\theta(P_k^\theta) \\ &- \sum_{i=1}^n \langle \tilde{l}_i^\theta(P_k^\theta), u(x-\theta\tilde{v}_i^\theta(P_k^\theta))\rangle \tilde{r}_i^\theta(P_k^\theta)\Bigg|\, dx.\end{aligned}$$

Let M be a common bound for all functions l_i^θ, r_i^θ. Using Lemma 4 we estimate the first integral:

$$\begin{aligned}\mathrm{I}_1 &\le M\sum_{i=1}^n\sum_k \int_{I_k^\theta} |u(x-\theta v_i^\theta(P_k^\theta)) - u(x-\theta\tilde{v}_i^\theta(P_k^\theta))|\,dx \\ &\le M\theta\sum_{i=1}^n\sum_k |v_i^\theta(P_k^\theta) - \tilde{v}_i^\theta(P_k^\theta)|\cdot \text{Tot.Var.}\{u;\ J_k^\theta\}\end{aligned}$$

$$
\begin{aligned}
&\leq M\theta \sum_{i=1}^{n} \left\{ \sum_{k} \int_{J_k^\theta} |J_k^\theta| \Big(\mathrm{Lip}(v_i^\theta) + \mathrm{Lip}(\tilde{v}_i^\theta) \Big)\, d\mu_u(x) \right. \\
&\qquad\qquad \left. + \sum_{k} \int_{J_k^\theta} |v_i^\theta(x) - \tilde{v}_i^\theta(x)|\, d\mu_u(x) \right\} \\
&\leq 2M\theta \sum_{i=1}^{n} \Big\{ |J_k^\theta| \Big(\mathrm{Lip}(v_i^\theta) + \mathrm{Lip}(\tilde{v}_i^\theta) \Big) \cdot \mathrm{Tot.Var.}\{u; \mathbb{R}\} \\
&\qquad\qquad + \int_{\mathbb{R}} |v_i^\theta(x) - v_i(x)|\, d\mu_u(x) + \int_{\mathbb{R}} |\tilde{v}_i^\theta(x) - v_i(x)|\, d\mu_u(x) \Big\}
\end{aligned}
$$

Concerning the second integral, using Lemma 5 we obtain

$$
\begin{aligned}
\mathrm{I}_2 &\leq 2M\theta \cdot \max_i \{ \|v_i^\theta\|_{\mathbf{L}^\infty} \} \\
&\quad \cdot \sum_{k} \left(\sum_{i=1}^{n} |r_i^\theta(P_k^\theta) - \tilde{r}_i^\theta(P_k^\theta)| + \sum_{i=1}^{n} |l_i^\theta(P_k^\theta) - \tilde{l}_i^\theta(P_k^\theta)| \right) \cdot \mathrm{Tot.Var.}\{u;\ J_k^\theta\} \\
&\leq 2M\theta \cdot \max_i \{ \|v_i^\theta\|_{\mathbf{L}^\infty} \} \\
&\quad \cdot \sum_{i=1}^{n} \sum_{k} \left\{ \int_{J_k^\theta} |J_k^\theta| \Big(\mathrm{Lip}(r_i^\theta) + \mathrm{Lip}(\tilde{r}_i^\theta) + \mathrm{Lip}(l_i^\theta) + \mathrm{Lip}(\tilde{l}_i^\theta) \Big) d\mu_u(x) \right. \\
&\qquad\qquad \left. + \int_{J_k^\theta} \Big(|r_i^\theta(x) - \tilde{r}_i^\theta(x)| + |l_i^\theta(x) - \tilde{l}_i^\theta(x)| \Big)\, d\mu_u(x) \right\} \\
&\leq 4M\theta \cdot \max_i \{ \|v_i^\theta\|_{\mathbf{L}^\infty} \} \\
&\quad \cdot \sum_{i=1}^{n} \Big\{ |J_k^\theta| \Big(\mathrm{Lip}(r_i^\theta) + \mathrm{Lip}(\tilde{r}_i^\theta) + \mathrm{Lip}(l_i^\theta) + \mathrm{Lip}(\tilde{l}_i^\theta) \Big) \cdot \mathrm{Tot.Var.}\{u;\ \mathbb{R}\} \\
&\qquad\qquad + \int_{\mathbb{R}} |r_i^\theta(x) - r_i(x)|\, d\mu_u(x) + \int_{\mathbb{R}} |l_i^\theta(x) - l_i(x)|\, d\mu_u(x) \\
&\qquad\qquad + \int_{\mathbb{R}} |\tilde{r}_i^\theta(x) - r_i(x)|\, d\mu_u(x) + \int_{\mathbb{R}} |\tilde{l}_i^\theta(x) - l_i(x)|\, d\mu_u(x) \Big\}
\end{aligned}
$$

The above estimates together yield

$$
\lim_{\theta \to 0} \frac{1}{\theta} (\mathrm{I}_1 + \mathrm{I}_2) = 0,
$$

proving Theorem 3.

References

[1] BRESSAN, A. & COLOMBO, R.M.: The semigroup generated by 2×2 conservation laws, Arch. Rational Mech. Anal. 133 (1995), 1-75.

[2] BRESSAN, A. & GUERRA, G.: Shift-differentiability of the flow generated by a conservation law, Discr. Cont. Dynam. Syst. 3 (1997), 35-58.

[3] BRESSAN, A. & MARSON, A.: A variational calculus for shock solutions of systems of conservation laws, Comm. Part. Diff. Eq. 20 (1995), 1491-1553.

[4] CLARKE, F.H.: Optimization and Nonsmooth Analysis, Wiley Interscience, 1983.

[5] CRANDALL, M.: The semigroup approach to first order quasilinear equations in several space variables, Israel J. Math. 26 (1977), 1097-1119.

[6] DIEUDONNÉ, J.: Foundations of Modern Analysis, 4th ed, Academic Press, New York 1963.

[7] EVANS, L.C. & GARIEPY, R.F.: Measure Theory and Fine Properties of Functions, CRC Press, Boca Raton, 1992.

[8] KRUŽKOV, S.: First order quasilinear equations with several space variables, Math. USSR Sbornik 10 (1970), 217-243.

[9] RUDIN, W.: Real and Complex Analysis, 2nd ed., McGraw Hill, New York, 1974.

[10] SMOLLER, J.: Shock waves and reaction diffusion equations, Springer-Verlag, New York, 1982.

[11] ZIEMER, W.P.: Weakly Differentiable Functions: Sobolev Spaces and Functions of Bounded Variation, Springer-Verlag, New York, 1989.

A. BRESSAN

S.I.S.S.A., Via Beirut 4, Trieste 34014 Italy, E-mail: bressan@sissa.it

M. LEWICKA

S.I.S.S.A., Via Beirut 4, Trieste 34014 Italy, E-mail: lewicka@sissa.it

Calculation of the Singularity Dynamics for Quadratic Nonlinear Hyperbolic Equations. Example: the Hopf Equation

VLADIMIR G. DANILOV AND GEORGII A. OMEL'YANOV

1 Introduction

One of the most interesting applications of the theory of generalized functions is the problem of propagation and interaction of singularities. Nonsmooth solutions of nonlinear equations are very important from the physical viewpoint, since they describe such actual phenomena as shock waves, typhoons, tsunami waves, and so on. There are some different approaches to these problems [1–15]. In this paper we discuss the approach related to Maslov's ideas [1, 2]. At present, this approach is most attractive from the viewpoint of physical application, since in this approach the dynamics of a singularity can be described by using only the parameters of the solution calculated on the front of the singularity. However, Hugoniot-type equations describing these parameters form an infinite system. This system is not triangular in the sense that for each N the first N equations contain $M > N$ unknown functions. Thus, if we want to use this approach for numerical modeling, we must choose and justify a method for truncating this system so to obtain a well-posed system of equations. In other words, we arrive at a problem similar to the well-known problem of truncating of Bogolyubov–Born–Green–Kirkwood–Yvon chains. Any general solution of such problems is still unknown.

The simplest method for truncating of infinite chains is to neglect the additional functions. For the problem about the motion of a shock wave of small amplitude A_0, this procedure of "neglecting" extra functions looks natural. Since in this case the truncated chain is the model equation for the leading term of the asymptotics w.r.t. small $|A_0| \ll 1$. This formal asymptotics of the chain of the Hugoniot conditions was first studied by Maslov [1]. Comparatively recently, Ravindran and

Prasad [16] used the Maslov ideas for numerical modeling of the shock wave of large amplitude ($|A_0| \sim O(1)$). The surprising thing is that the comparison of the obtained trajectory with the true trajectory shows that for not very small times these trajectories coincide sufficiently well. Then the truncated system was used in [17] for calculating the motion of a typhoon eye. The comparison of the computed trajectory with the true trajectory (obtained by natural measurements of the actual typhoon FORREST) also showed that they coincide well during some hours. These examples make it possible to hope that the passage to a truncated system can be justified rigorously for all types of singularities.

In this paper we present such justification by using, as an example, the simplest problem about the motion of a shock wave for the Hopf equation. Our approach is based on an analysis of the algebraic structure of the truncated system by using the Feynman formula for the solution representation via the T-product. Here we describe the main ideas used to derive the Hugoniot-type conditions and to justify the truncation procedure (for the complete proof, see [20]).

We suppose that our approach can be also used for justifying the truncation procedure for chains, describing the dynamics of singularities of other types for partial differential equations with quadratic nonlinearity.

2 Shock wave propagation and Hugoniot-type conditions

Now we shall consider the process of shock wave propagation. This means that we shall construct a generalized solution of the Hopf equation

$$u_t + \frac{1}{2}(u^2)_x = 0, \qquad u\Big|_{t=0} = H^0(x) + \theta(x - \varphi^0)A^0(x). \tag{1}$$

Here $x \in \mathbf{R}^1$, $H^0(x), A^0(x) \in C^\infty$, $\varphi^0 = \text{const}$, and $\theta(\tau)$ is the Heaviside function, $\theta(\tau) = 0$ for $\tau < 0$, and $\theta(\tau) = 1$ for $\tau > 0$. We suppose that

$$H^0_x > 0 \quad \text{for } x \le \varphi^0, \quad H^0_x + A^0_x > 0 \quad \text{for } x \ge \varphi^0, \quad \text{and} \quad A^0(\varphi^0) < 0. \tag{2}$$

Our first goal is to prove the existence of the self-similar solution for (1)

$$u(x,t) = H(x,t) + \theta(x - \varphi(t))A(x,t). \tag{3}$$

Substituting (3) into (1), we obtain (in the D′-sense)

$$H_t + \frac{1}{2}(H^2)_x + (A_t + (HA + \frac{1}{2}A^2)_x)\theta(x-\varphi) - A(\varphi_t - H - \frac{1}{2}A)\delta(x-\varphi) = 0, \quad (4)$$

where we use the relation $\theta^2(\tau) = \theta(\tau)$. It is clear that (4) can be satisfied iff the coefficients of the functions $\delta(x-\varphi)$ and $\theta(x-\varphi)$ are zero. This leads to the well-known Hugoniot–Rankine condition

$$\varphi_t = (H + A/2)\Big|_{x=\varphi} \quad (5)$$

and to the two Hopf equations

$$H_t + \frac{1}{2}(H^2)_x = 0, \qquad x < \varphi(t), \quad t > 0, \quad (6)$$

$$g_t + \frac{1}{2}(g^2)_x = 0, \qquad x > \varphi(t), \quad t > 0, \quad (7)$$

where $g = H + A$. Using the method of characteristics, one can see that the solutions of (6), (7) have been determined by the initial data

$$H\Big|_{t=0} = H^0(x), \quad x < \varphi^0, \quad g\Big|_{t=0} = g^0(x) = H^0(x) + A^0(x), \quad x > \varphi^0. \quad (8)$$

However, in contrast with the classical situation, we must now solve these Cauchy problems together with the ordinary differential equation (5). In some cases, this can be done easily. However, for general initial data and more complicated problems, such as the Hopf equation, solving the partial differential equations is very complicated and technical.

Now we will consider the Maslov scheme [1] for calculating the shock wave propagation, using only ordinary differential equations. To this end, let us introduce necessary conditions for the existence of the solution (3).

It is well-known that to solve a linear problem of determining the trajectory of a singularity (for example, of the shock wave type), it suffices to construct an approximating solution that satisfies the equation up to a sufficiently smooth function [18, 19]. It follows from the possibility to decompose the solution (3) into the smooth $H(x,t)$ and nonsmooth part $\theta(x-\varphi)A(x,t)$. But in our case, these parts interact since equation (1) is nonlinear. So, instead of the usual condition, we introduce a stronger one. Namely, we assume that our approximating solution u_M satisfies the following relation

$$\frac{\partial u_M}{\partial t} + \frac{1}{2}\frac{\partial}{\partial x}(u_M^2) = O(x-\varphi)^M, \qquad x \to \varphi, \quad (9)$$

where the estimate is uniform on each compact subset in R^1 and M is a sufficiently large positive number.

To construct u_M, let us decompose the functions $A(x,t)$ and $H(x,t)$ at the point $x = \varphi$ by using the Taylor formula. Then we obtain the following representation:

$$u = \sum_{i=0}^{M+1} H_i(t)(x-\varphi)^i + \sum_{i=0}^{M+1} A_i(t)\theta(t-\varphi)(x-\varphi)^i + r_{M+2}, \tag{10}$$

where $r_{M+2} \in C^{M+1}$ is the remainder of order $(x-\varphi)^{M+2}$. Let us write $u_M = u - r_{M+2}$. Then from (6) and (7) we obtain the following chain of equations for the coefficients H_i and $g_i = H_i + A_i$:

$$\begin{aligned} \frac{dg_i}{dt} + \frac{i+1}{2}\Big\{\sum_{j=0}^{i} g_j g_{i+1-j} + g_{i+1}A_0\Big\} &= 0, \\ \frac{dH_i}{dt} + \frac{i+1}{2}\sum_{j=0}^{i}(H_j H_{i+1-j} - H_{i+1}A_0) &= 0, \end{aligned} \tag{11}$$

where $i = 0, 1, \ldots, M$.

In turn, we rewrite the equation for φ as follows: $\varphi_t = (H_0 + g_0)/2$. The system (11) is called the chain of Hugoniot-type conditions. It is easy to see that for each N this system is underdetermined.

3 Estimations for discrepancies

Let us fix a number $N < M$ and neglect the extra functions H_{N+1} and g_{N+1}. The solution of such truncated system we denote by ψ, $\widehat{H}_k$ and $\widehat{g}_k$, $k = 0, \ldots, N$. Our main goal is to derive estimates for the discrepancy σ between the trajectories $\varphi(t)$ and $\psi(t)$. This can be done using the equation

$$\sigma_t = (\omega_0(t) + w_0(t))/2, \qquad \sigma\big|_{t=0} = 0, \tag{12}$$

that follows from (5). Here $\omega_0 = H_0 - \widehat{H}_0$, $w_0 = g_0 - \widehat{g}_0$. In its turn, to calculate ω_0 and w_0, we need to consider the systems of equations for vectors of all differences $\Omega = (\omega_0, \ldots, \omega_N)$ and $W = (w_0, \ldots, w_N)$, where $\omega_k = H_k - \widehat{H}_k$, $w_k = g_k - \widehat{g}_k$. Note that the structure of this system is such that the estimate for the differences ω_0 and w_0 significantly differs from the estimate for the total vectors Ω and W (see below).

Using (11) and the truncated chain, we readily obtain the system for Ω and W:

$$\frac{d}{dt}\begin{pmatrix}\Omega\\ W\end{pmatrix}+\left\{\begin{pmatrix}B & 0\\ 0 & C\end{pmatrix}+\begin{pmatrix}B_1 & 0\\ 0 & -B_1\end{pmatrix}\right.$$
$$\left.+\begin{pmatrix}0 & \mathcal{A}_1\\ \mathcal{A}_2 & 0\end{pmatrix}\right\}\begin{pmatrix}\Omega\\ W\end{pmatrix}=\begin{pmatrix}F_1\\ F_2\end{pmatrix}. \tag{13}$$

Here F_i are $(N+1)$-component vectors with only one nonzero component

$$F_1=\frac{N+1}{2}A_0H_{N+1}\begin{pmatrix}0\\ \vdots\\ 0\\ 1\end{pmatrix},\qquad F_2=-\frac{N+1}{2}A_0g_{N+1}\begin{pmatrix}0\\ \vdots\\ 0\\ 1\end{pmatrix}, \tag{14}$$

B is the $(N+1)\times(N+1)$ triangular matrix of the form (note that we begin numeration of matrix elements and vectors with 0):

$$B=\frac{1}{2}\begin{pmatrix}H_1 & 0 & \dots & 0 & 0\\ 2H_2 & 2(H_1+\widehat{H}_1) & \dots & 0 & 0\\ 3H_3 & 3(H_2+\widehat{H}_2) & \dots & 0 & 0\\ \dots & \dots & \dots & \dots & \dots\\ NH_N & N(H_{N-1}+\widehat{H}_{N-1}) & \dots & N(H_1+\widehat{H}_1) & 0\\ 0 & (N+1)(H_N+\widehat{H}_N) & \dots & (N+1)(H_2+\widehat{H}_2) & (N+1)(H_1+\widehat{H}_1)\end{pmatrix}. \tag{15}$$

We also use the notation

$$\mathcal{A}_1=-\frac{1}{2}\begin{pmatrix}H_1 & 0 & \dots & 0\\ 2H_2 & 0 & \dots & 0\\ \dots & \dots & \dots & \dots\\ NH_N & 0 & \dots & 0\\ 0 & 0 & \dots & 0\end{pmatrix},\qquad B_1^0=\begin{pmatrix}0 & 1 & 0 & \dots & 0 & 0\\ 0 & 0 & 2 & \dots & 0 & 0\\ \dots & \dots & \dots & \dots & \dots & \dots\\ 0 & 0 & 0 & \dots & 0 & N\\ 0 & 0 & 0 & \dots & 0 & 0\end{pmatrix}, \tag{16}$$

such that $B_1=-\frac{1}{2}\widehat{A}_0B_1^0$, and C and $\mathcal{A}_2$ have the same form as B and $\mathcal{A}_1$ (with g_k and $\widehat{g}_k$ instead of H_k and $\widehat{H}_k$). Let

$$\mathcal{A}=\begin{pmatrix}0 & \mathcal{A}_1\\ \mathcal{A}_2 & 0\end{pmatrix},\quad \mathcal{B}=\begin{pmatrix}B & 0\\ 0 & C\end{pmatrix},\quad \mathcal{B}_1=\begin{pmatrix}B_1 & 0\\ 0 & -B_1\end{pmatrix},\quad \mathcal{F}=\begin{pmatrix}F_1\\ F_2\end{pmatrix}.$$

Denote $L_k=\{\ell_k=(\ell_k^{(1)},\ell_k^{(2)})\in R^{2N+2},\ell_k^{(i)}=(\ell_{k,0}^{(i)},\dots,\ell_{k,N}^{(i)}),\ell_{k,j}^{(i)}=0$ for all $j=0,1,\dots,k-1$ and $i=1,2\}$.

It is easy to establish

Proposition 1
1) *For each instant of time* $\tau' \in [0,t]$, *and for* $k \le N$

$$\exp\{-\mathcal{A}(\tau')\}|_{L_k} = 1, \qquad \exp\{-\mathcal{B}(\tau')\}L_k \subset L_k,$$

2) *for all* τ'

$$\| \exp\{-\mathcal{B}(\tau')\|_{\mathbf{R}^{2N+2}} \le \exp\big(-\tfrac{1}{2}\min\{g_1(\tau'), H_1(\tau')\}\big).$$

3) *For all sequences of numbers* $\{\tau_1, \tau_2, \dots, \tau_k\}$ *and* $k \ge N+1$

$$\mathcal{B}_1(\tau_1)\mathcal{B}_1(\tau_2)\dots\mathcal{B}_1(\tau_k) = 0.$$

Let us recall some well-known results. If we have a system

$$\dot{X} = D(t)X + f(t), \quad X\Big|_{t=0} = X_0,$$

where D is an $r \times r$ matrix with elements from $C[0,T]$, then the solution X can be presented in the form of the chronological product (T-product) [21]

$$X = e^{\int_0^t D(\tau)\,d\vec{\tau}} X_0 + \int_0^t e^{\int_\tau^t D(\tau')\,d\vec{\tau}'} f(\tau)\,d\tau. \tag{17}$$

The exponents in the last formula are defined as the limits

$$e^{\int_\tau^t D(\tau')\,d\vec{\tau}'} q(\tau) = \Big\{ \lim_{\Delta t \to 0} e^{D(t)\Delta t} \cdot e^{D(t-\Delta t)\Delta t} \cdot \dots \cdot e^{D(\tau+\Delta t)\Delta t} \Big\} q(\tau). \tag{18}$$

In our case we have $r = 2N+2$, $f = \mathcal{F}$, and $D = -(\mathcal{B} + \mathcal{B}_1 + \mathcal{A})$, and we can set $X_0 = 0$ [20].
Applying the Campbell–Hausdorff formula (see [23]) and theorem due to O. Kreiss and S. Strang [22], we obtain the following statement.

Proposition 2 *Let the matrices* $\mathcal{B}(t)$, $\mathcal{B}_1(t)$, *and* $\mathcal{A}(t)$ *be bounded. Then*

$$\begin{pmatrix} \Omega \\ W \end{pmatrix} = \int_0^t G(\tau,t)\mathcal{F}(\tau)\,d\tau, \qquad G(\tau,t) = e^{-\int_\tau^t (\overset{3}{\mathcal{B}}(\tau') + \overset{2}{\mathcal{B}}_1(\tau') + \overset{1}{\mathcal{A}}(\tau'))\,d\vec{\tau}'}, \tag{19}$$

where the numbers over the matrices mean the order of action.

By expanding $\exp\{-\int_\tau^t (\overset{3}{\mathcal{B}} + \overset{2}{\mathcal{B}}_1 + \overset{1}{\mathcal{A}})\,d\vec{\tau}'\}$ into the Taylor series with respect to $-\int_\tau^t \overset{2}{\mathcal{B}}_1\,d\vec{t}'$, we obtain the Feynman representation [24] (see also [25, 26])

$$G = \sum_{j=0}^{\infty} \frac{(-1)^j}{j!} \widehat{S}_j, \tag{20}$$

where

$$\begin{aligned}
\widehat{S}_j &= \int_\tau^t \cdots \int_\tau^t e^{-\int_\tau^t (\overset{3}{\mathcal{B}}(\tau')+\overset{1}{\mathcal{A}}(\tau'))\,d\tau''} \overset{2}{\mathcal{B}_1}(\tau_1) \cdot \ldots \cdot \overset{2}{\mathcal{B}_1}(\tau_j)\, d\tau_1 \cdot \ldots \cdot d\tau_j \\
&= \sum_{\mathcal{P}_j} \int \cdots \int_{\mathcal{P}_j(\sigma_j)} e^{-\int_{\tau_{p(j)}}^t (\overset{3}{\mathcal{B}}+\overset{1}{\mathcal{A}})\,d\tau''} \overset{2}{\mathcal{B}_1}(\tau_{p(j)}) \cdot \ldots \\
&\qquad \cdot e^{-\int_{\tau_{p(k)}}^{\tau_{p(k+1)}} (\overset{3}{\mathcal{B}}+\overset{1}{\mathcal{A}})\,d\tau''} \overset{2}{\mathcal{B}_1}(\tau_{p(k)}) \cdot \ldots \\
&\qquad \cdot \overset{2}{\mathcal{B}_1}(\tau_{p(1)}) e^{-\int_\tau^{\tau_{p(1)}} (\overset{3}{\mathcal{B}}+\overset{1}{\mathcal{A}})\,d\tau''} d\tau_{p(1)} \cdot \ldots \cdot d\tau_{p(j)},
\end{aligned} \tag{21}$$

and σ_j is the j-dimensional simplex $\sigma_j = \{\tau \le \tau_1 \le \ldots \le \tau_j \le t\}$ in $R_{\tau_1} \times \ldots \times R_{\tau_j}$, $\mathcal{P}_j(1, \ldots, j) = (p(1), \ldots, p(j))$ determines the ordered substitution, and $\mathcal{P}_j(\sigma_j) = \{\tau \le \tau_{p(1)} \le \cdots \le \tau_{p(j)} \le t\}$.

Now we note that to estimate the difference between the trajectories $\varphi(t)$ and $\psi(t)$ it is enough to know only the components ω_0 and w_0 of representations (19), (20). It is easy to prove that the first and the $(N+2)$-th components of the vectors $\widehat{S}_j\mathcal{F}$ are zero for all $j \le N-1$. To consider such components for $j \ge N$, let us obtain a more convenient representation for the vectors $\widehat{S}_j\mathcal{F}$. First, let us consider the vector $\widehat{S}_N\mathcal{F}$. Certainly, it suffices to consider only one term in the sum (21), say, with $\mathcal{P}(1, \ldots, N) = (1, \ldots, N)$. As long as we do not apply N matrices $\mathcal{B}_1$, all matrices $\exp\{-\int_{\tau_j}^{\tau_{j+1}} A\, d\tau''\}$ are equal to the unit matrix. So, the term under consideration can be presented in the form

$$\int_{\sigma_N} e^{-\int_{\tau_N}^t (\overset{2}{\mathcal{B}}+\overset{1}{\mathcal{A}})\,d\tau''} V^I(\tau_N, \ldots, \tau_1)\, d\tau_1 \ldots d\tau_N, \tag{22}$$

where

$$V^I(\tau_N, \ldots, \tau_1) = \left\{\mathcal{B}_1(\tau_N) e^{-\int_{\tau_{N-1}}^{\tau_N} \overset{2}{\mathcal{B}}(\tau')\,d\tau''} \mathcal{B}_1(\tau_{N-1}) \cdot \ldots \cdot \mathcal{B}_1(\tau_1) e^{-\int_\tau^{\tau_1} \overset{2}{\mathcal{B}}\,d\tau''}\right\} \mathcal{F}.$$

Similarly, we rewrite the next term in (20) (for $j = N+1$):

$$\begin{aligned}
\widehat{S}_{N+1}\mathcal{F} = \sum_{\mathcal{P}_{N+1}} \int_{\mathcal{P}_{N+1}(\sigma_{N+1})} & e^{-\int_{\tau_{p(N+1)}}^t (\overset{3}{\mathcal{B}}+\overset{1}{\mathcal{A}})\,d\tau''} \overset{2}{\mathcal{B}_1}(\tau_{p(N+1)}) e^{-\int_{\tau_{p(N)}}^{\tau_{p(N+1)}} (\overset{3}{\mathcal{B}}+\overset{1}{\mathcal{A}})\,d\tau''} \\
\cdot V^{\mathcal{P}_{N+1}}(\tau_{p(N)}, \ldots, \tau_{p(1)})\, & d\tau_{p(1)} \cdot \ldots \cdot d\tau_{p(N+1)},
\end{aligned} \tag{23}$$

where the vector $V^{\mathcal{P}_{N+1}}(\tau_{p(N)}, \ldots, \tau_{p(1)})$ is defined as follows:

$$\begin{aligned}
& V^{\mathcal{P}_{N+1}}(\tau_{p(N)}, \ldots, \tau_{p(1)}) \\
&= \left\{\mathcal{B}_1(\tau_{p(N)}) e^{-\int_{\tau_{p(N-1)}}^{\tau_{p(N)}} \mathcal{B}\,d\tau''} \mathcal{B}_1(\tau_{p(N-1)}) \cdot \ldots \cdot \mathcal{B}_1(\tau_{p(1)}) e^{-\int_\tau^{\tau_{p(1)}} \mathcal{B}\,d\tau''}\right\} \mathcal{F}.
\end{aligned} \tag{24}$$

Similar formulas can be obtained for all terms in (20).

Proposition 3 *The following formula holds:*

$$
\begin{aligned}
&V^{\mathcal{P}_j}(\tau_{p(N)},\dots,\tau_{p(1)})\\
&=\prod_{i=1}^{N}\Big(\frac{-\widehat{A}_0(\tau_{p(i)})}{2}\Big)K\widetilde{\mathcal{B}}_N e^{-\int_{\tau_{p(N-1)}}^{\tau_{p(N)}}\mathcal{B}\,d\tau'}\widetilde{\mathcal{B}}_{N-1}\cdot\ldots\cdot\widetilde{\mathcal{B}}_1 e^{-\int_{\tau}^{\tau_{p(1)}}\mathcal{B}\,d\tau'}\mathcal{F}, \qquad (25)
\end{aligned}
$$

where K is the matrix with the coefficients $K_{i,j}$ such that $K_{i,j}=0$ for $i\neq j$ and $K_{i,i}=N!N^i/i!$ and $K_{N+1+i,N+1+i}=(-1)^N K_{i,i}$, $i=0,1,\dots,N$, $\widetilde{\mathcal{B}}_i$ are block-matrices with constant elements b^i_{lk} such that $0\le b^i_{lk}\le 1$, $l,k=0,1,\dots,N$, $i=1,\dots,N$.

The estimate for the component $V_\ell^{\mathcal{P}_j}$ of the vector $V^{\mathcal{P}_j}$, $j\geq N$, readily follows from representation (25):

$$
|V_\ell^{\mathcal{P}_j}|\le\Big(\frac{a_0}{2}\Big)^N\frac{N!N^\ell}{\ell!}e^{-\int_{\tau}^{\tau_{p(N)}}\min\{H_1,g_1\}\,d\tau'/2}\|\mathcal{F}\|,\qquad \ell=0,1,\dots,N, \qquad (26)
$$

where $a_0=\max_{0\le\tau\le t}|\widehat{A}_0|$. We have the same estimate for the components $V_{N+1+\ell}^{\mathcal{P}_j}$, $\ell=0,1,\dots,N$.

Now we can obtain the final estimate for the discrepancies ω_0 and w_0 [20].

Theorem 1 *Let $N\geq 1$. Then the following inequality holds:*

$$
\delta_0\le\frac{(N+1)!}{N}\Big(\frac{a_0}{2}\Big)^N f_1\zeta_N\Big\{e^{\lambda t}\sum_{k=0}^{N}\frac{(-1)^k t^{N-k}}{(N-k)!\lambda^k}-\frac{(-1)^N}{\lambda^N}\Big\}, \qquad (27)
$$

where $\delta_0=\max_{0\le t'\le t}\{|\omega(t')|+|w_0(t')|\}$,

$$
\begin{aligned}
\lambda &= Na_0/2, \qquad \zeta_N=\max\{\max_{0\le t'\le t}|H_{N+1}(t')|,\max_{0\le t'\le t}|g_{N+1}(t')|\}, \qquad (28)\\
f_1 &= \exp\Big\{\int_0^t[\sqrt{g_1H_1}-\min\{g_1,H_1\}]\,d\tau/2\Big\}.
\end{aligned}
$$

To simplify the analysis, we coarsen the estimate (27), rewriting it as follows:

$$
\begin{aligned}
\delta_0 &\le f_1\zeta_N\Big(\frac{a_0}{2}t\Big)^{N+1}(1+O(t)) \quad\text{for}\quad t\ll 1,\\
&\qquad\qquad\qquad\qquad\qquad\qquad\qquad\qquad (29)\\
\delta_0 &\le f_1\zeta_N\frac{N+1}{N}\Big(\frac{a_0}{2}te^{a_0t/2}\Big)^N(1+O(1/t)) \quad\text{for}\quad t\gg 1.
\end{aligned}
$$

Now we can estimate the difference σ between the trajectories $\varphi(t)$ and $\psi(t)$. By integrating (12) and using (27), we obtain the main result of this paper.

Theorem 2 *Let $N \geq 1$. Then the following estimate holds for the difference σ between the true trajectory $x = \varphi(t)$ of the shock wave and the trajectory $x = \psi(t)$ calculated for the truncated system:*

$$|\sigma| \leq (-1)^N f_1 \zeta_N \frac{(N+1)!}{\lambda^{N+2}} \left(\frac{a_0}{2}\right)^{N+1} \left\{ e^{\lambda t} \sum_{k=0}^{N} (-1)^k \frac{N-k+1}{k!} (\lambda t)^k - \lambda t - N - 1 \right\},$$

where the variables a_0, f_1, ζ_N, and λ are defined in (26), (28).

Obviously, it follows from the last estimate that

$$|\sigma| \leq 2 \frac{f_1 \zeta_N}{a_0(N+2)} \left(\frac{a_0}{2} t\right)^{N+2} (1 + O(t)), \quad t \ll 1, \tag{30}$$

$$|\sigma| \leq 2 \frac{N+1}{N^2} \frac{f_1 \zeta_N}{a_0} \left(\frac{a_0}{2} t e^{a_0 t/2}\right)^{N} (1 + O(1/t)), \quad t \gg 1. \tag{31}$$

These formulas allow us to explain the notion of "small" t more precisely: namely, the value σ is sufficiently small only for $t < t^* = 2\kappa/a_0$, where κ is the solution of equation $\kappa \exp(\kappa) = 1$.

Finally, we see that the truncated system provides a sufficiently adequate description of the motion of a shock wave only for sufficiently small time less than or equal to t^*. We also see that the critical value of time t^* is independent of the order of approximation N, while at $t \sim t^*$ the error starts to grow rapidly with growth of N.

Acknowledgment The authors are cordially grateful to Professor J.-A. Marti for his hospitality at the University of Antilles and Guyana, where this work was done, and for fruitful discussions. One of the authors (G.O.) cordially thanks also Professor M. Oberguggenberger and the International Erwin Schrödinger Institute for Mathematical Physics for the opportunity to participate in the project "Nonlinear Theory of Generalized Functions".

This research was partially supported by the Russian Foundation for Basic Research under grant N 96-01-01492.

References

[1] MASLOV, V.P.: Propagation of shock waves in an isoentropic nonviscous gas, Itogi Nauki i Tekhniki, 8, VINITI, Moscow, 1977 (In Russian.)

[2] MASLOV, V.P.: Three algebras corresponding to nonsmooth solutions of systems with quasilinear hyperbolic equations, Russ. Math. Surv., **35**:2 (1980), 252–253.

[3] COLOMBEAU, J.F.: Elementary Introduction to New Generalized Functions, North Holland, Amsterdam 1985.

[4] EGOROV, YU.V.: A contribution to the theory of generalized functions, Russian Math. Surveys, **45**:5 (1990), 1–49.

[5] MARTI, J.-A.: Fundamental structures and asymptotic microlocalization in sheaves of generalized functions. To appear in: Integral Transforms and Special Functions.

[6] PILIPOVIĆ, S.: Colombeau's generalized functions and the pseudo-differential operators, Lecture Notes of Tokyo University, Tokyo (1994).

[7] BIAGIONI, H.A.: A Nonlinear Theory of Generalized Functions, Springer-Verlag, Berlin–Heidelberg–New York, 1990.

[8] OBERGUGGENBERGER, M.: Multiplication of Distributions and Applications to Partial Differential Equations, Longman, New York 1992.

[9] DANILOV, V.G, MASLOV, V.P., AND SHELKOVICH, V.M.: Algebras of singularities of singular solutions to quasilinear strongly hyperbolic first-order equations, To appear in Theoret. and Math. Phys.

[10] ARAGONA, J., AND VILLAREAL, F.: Colombeau's theory and shock waves in a problem of hydrodynamics, Jour. dÁnal. Math., **61** (1993), 113–144.

[11] BIAGIONI, H.A., AND OBERGUGGENBERGER, M.: Generalized solutions to Burgers' equation, J. Diff. Eq., **97**:2 (1992), 263–287.

[12] BIAGIONI, H.A., AND OBERGUGGENBERGER, M.: Generalized solutions to the Korteweg-de Vries and the regularized long-wave equations, SIAM J. Math. Anal., **23**:4 (1992), 923–940.

[13] CAURET, J.J., COLOMBEAU, J.F., AND LE ROUX, A.Y.: Discontinuous generalized solutions of nonlinear nonconservative hyperbolic equations, J. Math. Anal. Appl., **139** (1989), 552–573.

[14] DAPIĆ, N., AND PILIPOVIĆ, S.: Approximated shock wave solutions to generalized Hopf equation, to appear.

[15] NEDELJKOV, M.: Infinitely narrow soliton solutions to scalar conservation laws in Colombeau sense, Proceedings of the International Conference on Generalized Functions - Linear and Nonlinear Problems, Novi Sad, Yugoslavia, 1996, to appear.

[16] RAVINDRAN, R., PRASAD, P., A new theory of shock dynamics, Appl. Math. Lett., **3**:2 (1990), 107–109.

[17] BULATOV, V.V., VLADIMIROV, YU.V., DANILOV, V.G., AND DOBROKHOTOV, S.YU.: Propagation of a pointwise algebraic singularity for two-dimensional nonlinear equations of hydrodynamics, Mat. Zametki, **55**:3 (1994), 11–20.

[18] COURANT, R.: Partial Differential Equations, Interscience, New York and London, 1962.

[19] MASLOV, V.P.: Perturbation Theory and Asymptotic Methods, Dunod, Paris, 1972.

[20] DANILOV, V.P., AND OMEL'YANOV, G.A.: Truncation of a chain of Hugoniot-type conditions for shock waves and its justification for the Hopf equation, Vienna, ESI-Preprint **502** (1997).

[21] DOLLARD, J.D.: Product integration. In: " Encyclopedia of Mathematics and Its Applications," **10**, Addison–Wesley, 1979.

[22] RICHTMYER, R.D., AND MORTON, K.W.: Difference Methods for Initial Value Problems, Wiley, 1967.

[23] KARASEV, M.V., AND M. V. MOSOLOVA, M.V.: Infinite products and T-products, Teoret. Mat. Fiz., **28**:2 (1976); English transl. in Theoret. and Math. Phys.

[24] FEYNMAN, R.P.: An operator calculus having applications in quantum electrodynamics, Phys. Rev., **84** (1951), 108–128.

[25] TROTTER, E.F.: Approximation of semi-groups of operators, Pacific J. Math., **8** (1958), 887–919.

[26] NELSON, E.: Notes on noncommutative integration, J. Funct. Anal., **15** (1974), 103–116.

V.G.DANILOV AND G.A.OMEL'YANOV
Moscow State Institute of Electronics and Mathematics B.Trekhsvyatitel'skii per. 3/12, Moscow 109028, Russia, E-mail: omel@amath.msk.ru

Vanishing Viscosity Boundary Layers for Nonlinear Hyperbolic Systems

Olivier Guès

1 Introduction

In this note, we present some recent results in the theory of boundary layers in viscous perturbations of nonlinear hyperbolic systems near a boundary. The first section concerns the semilinear case and includes the case where the boundary is characteristic for the hyperbolic operator. The second section is devoted to the case of quasilinear hyperbolic systems, when the boundary is not characteristic.

Let us fix some general notations. For every $u \in \mathbb{R}^N$ let $\mathcal{H}(u)$ be the following first-order operator in $\mathbb{R}_t \times \mathbb{R}_x^n$

$$\mathcal{H}(u) = \partial_t + \sum_{j=1}^{n} A_j(u)\partial_j \tag{1}$$

where $A_j(.)$, $j = 0, ..., n$, are $N \times N$ matrices depending smoothly on $u \in \mathbb{R}^N$. We denote by $x = (x_1, ..., x_n)$ the space coordinates in $\mathbb{R}^n$ and by $t \in \mathbb{R}$ the time coordinate. We suppose that $\mathcal{H}(u)$ is a symmetrizable hyperbolic operator in the sense of Friedrichs ([7]):

Assumption 1.1 *There exists a C^∞, $N \times N$, symmetric matrix $S(.)$ such that $S(u)$ is positive definite and $SA_j(u)$ is symmetric for every $u \in \mathbb{R}^N$ and $j = 1, ..., N$.*

Let us also fix some smooth function $F(t, x, u) \in C^\infty(\mathbb{R}^{n+1} \times \mathbb{R}^N : \mathbb{R}^N)$ which will be a nonlinear source term in our problem. We suppose that $F(t, x, 0) \in H^\infty(\mathbb{R}^{n+1})$, and that $F(t, x, 0)|_{t<0} \equiv 0$. For any $\varepsilon > 0$ and $T > 0$, we call $P(\varepsilon, T)$ the following nonlinear mixed problem, where for any $T > -1$ we call $\Omega_T = \{(t, x) \in \mathbb{R}^{1+n} \mid -1 < t < T,\ x_n > 0\}$ and $\Gamma_T = \{(t, x) \in \mathbb{R}^{1+n} \mid -1 < t < T,\ x_n = 0\}$:

$$P(\varepsilon, T) \begin{cases} -\varepsilon \Delta_x u^\varepsilon + \mathcal{H}(u^\varepsilon)u^\varepsilon = F(t, x, u) & \text{in} \quad \Omega_T \\ u^\varepsilon{}_{|\Gamma_T} = 0, \quad u^\varepsilon{}_{|\Omega_0} = 0. \end{cases} \tag{2}$$

From the classical theory of nonlinear parabolic systems ([22], [20]) we know that there exist some $T_\varepsilon > 0$ and a unique $u^\varepsilon \in H^\infty(\Omega_{T_\varepsilon})$ solution of $P(\varepsilon, T_\varepsilon)$. We are interested in the two following natural questions:

1 — Does there exist $T > 0$ independent of ε such that $T_\varepsilon > T$ for any $\varepsilon \in (0, 1]$?

2 — How to describe precisely the behavior of u_ε as $\varepsilon \to 0$ at least for small time (of order O(1))?

2 The semilinear case

In this paragraph we will consider the simpler case of a *semilinear* hyperbolic problem perturbed by viscosity, which is the case where the matrices $A_j(u)$ in the problem $P(\varepsilon, T)$ do not depend on u. To this end we introduce the linear hyperbolic operator $L := \mathcal{H}(0) = \partial_t + A_1(0)\partial_1 + \ldots + A_n(0)\partial_n$ and consider the corresponding mixed problem:

$$S(\varepsilon, T) \begin{cases} -\varepsilon\Delta_x u^\varepsilon + Lu^\varepsilon = F(t, x, u) & \text{in} \quad \Omega_T \\ {u^\varepsilon}_{|\Gamma_T} = 0, \quad {u^\varepsilon}_{|\Omega_0} = 0. \end{cases} \tag{3}$$

Let us mention that more general semilinear hyperbolic systems, with nonconstant coefficients and more general viscosity perturbations can be treated as well ([16], [17]), but for the simplicity of the exposition we will concentrate on this simpler case. The first difficulty in the analysis is to find the boundary conditions for the limit hyperbolic problem obtained at the limit when $\varepsilon = 0$. Let us describe these boundary conditions. Since by assumption 1.1 the matrix A_n in the hyperbolic operator L has real eigenvalues, we call $\mathcal{N}$ the invariant subspace of $\mathbb{R}^N$ associated to nonpositive eigenvalues:

$$\mathcal{N} = \sum_{\lambda \leq 0} \ker(A_n(0) - \lambda I). \tag{4}$$

This invariant linear space has the following important property:

Lemma 2.1 *([2], [3], [27]) $\mathcal{N}$ is a maximal negative subspace*[1] *for the quadratic form $q(\xi) = < S(0)A(0)\xi, \xi >$.*

[1]A linear subspace V in $\mathbb{R}^N$ is *maximal negative* for q if: $q(\xi) \leq 0, \forall \xi \in \mathbb{R}^N$, and V has maximal dimension for this property.

Actually, let us introduce the matrix R such that $S(0) = {}^tRR$. If $\xi \in \mathcal{N}$, one has $q(\xi) =< RA(0)R^{-1}.R\xi, R\xi >\leq 0$, since $R\xi$ is in the nonpositive invariant space of $RA(0)R^{-1}$.

A consequence of this lemma is the following result, which is the first step in the description of the behavior of u^ε.

Theorem 2.2 *There exist T_0 and a unique $u^0 \in H^\infty(\Omega_{T_0})$ solution of the following semilinear hyperbolic mixed problem:*

$$\begin{cases} Lu^0 = F(t,x,u^0) \text{ in } \Omega_{T_0} \\ u^0 \in \mathcal{N} \text{ on } \Gamma_{T_0}, \quad u^0_{|\Omega_0} = 0. \end{cases} \tag{5}$$

Note that the boundary Γ_T can be characteristic (which means that $\det A_n = 0$) for the hyperbolic operator L. Theorem 2.2 is a consequence of a general theorem by O. Guès ([19]).

We shall call $\mathbf{P}_-$ the orthogonal projector of $\mathbb{R}^N$ onto $\sum_{\lambda<0} \ker(A_n - \lambda I)$ which is the negative linear invariant subspace of A_n, and we call $\mathbf{P}_0$ the orthogonal projector of $\mathbb{R}^N$ onto $\ker A_n$. Let us call $\mathbf{V}(t,y,.) \in C^\infty([0,+\infty[, \mathbb{R}^N)$ the unique solution of the following linear two points boundary value problem, where (t,y) are parameters.

$$\begin{cases} \mathbf{V}' = A_n \mathbf{V}, \\ \mathbf{V}(0) = -\mathbf{P}_- u^0(t,y,0). \end{cases} \tag{6}$$

Since the initial value $\mathbf{V}(0)$ belongs to the negative invariant subspace of $A_n(0)$, the solution $\mathbf{V}$ is a sum of exponential factors which decay rapidly to 0 at infinity.

Let us now introduce the following semilinear system which is of parabolic-hyperbolic type, where the unknown is the function $\mathbf{W}(t,y,\theta)$, depending on (t,y) and on a fast variable $\theta \in [0,+\infty[$. This function $\mathbf{W}(t,y,\theta)$ is valued in $\mathbb{R}^N$, but is required to be in fact valued in $\ker A_n$, and this is expressed by the first equation of the system. We set $G(t,y,w) := F(t,(y,0),u^0(t,y,0)+w) - F(t,(y,0),u^0(t,y,0))$.

$$\begin{cases} (\mathbf{P}_0 - I)\mathbf{W} = 0\,, \\ -\partial_\theta \mathbf{W} + \mathbf{P}_0 L \mathbf{P}_0 \mathbf{W} = \mathbf{P}_0 G(t,y,\mathbf{W}) \text{ in } \Gamma_T \times [0,+\infty[, \\ \mathbf{W}_{|\theta=0} = -\mathbf{P}_0 u^0_{|\Gamma_T}, \quad \mathbf{W}_{|t<0} = 0. \end{cases} \tag{7}$$

Problem (7) is well-posed:

Theorem 2.3 *([16], [17]) There exist T_1 ($T_1 \leq T_0$) and a unique $\mathbf{W} \in H^\infty(\Gamma_{T_1} \times [0,+\infty[)$ solution of problem (7).*

Note that $\mathbf{W}$ goes to 0 as $\theta \to \infty$ since it belongs to the Sobolev space H^∞, but one can also prove that it decays rapidly to 0 as $\theta \to \infty$ ([17]). We can now state the main result which describes the behaviour of u^ε on Ω_{T_1} as $\varepsilon \to 0$.

Theorem 2.4 *([16], [17]) There exists $\varepsilon_0 > 0$ such that, for every $\varepsilon \in]0, \varepsilon_0]$, problem $S(\varepsilon, T_1)$ has a unique solution $u^\varepsilon \in H^\infty(\Omega_{T_1})$. Furthermore u^ε admits the following asymptotic expansion*

$$u^\varepsilon(t,x) = u^0(t,x) + \mathbf{V}(t,y,x_n/\varepsilon) + \mathbf{W}(t,y,x_n/\sqrt{\varepsilon}\,) + \sqrt{\varepsilon}\, R_\varepsilon, \tag{8}$$

where $R_\varepsilon = \mathcal{O}(1)$ in $L^\infty(\Omega_{T_1}) \cap H^{1/2}(\Omega_{T_1})$ as $\varepsilon \to 0$.

Let us remark that an obvious corollary given by Theorem 2.4 is that the life span of the smooth solution u^ε of $S(\varepsilon, T_1)$ does not go to zero as $\varepsilon \to 0$. However we do not know how to prove this simple fact without performing a deeper analysis involving the boundary layers. A simple change of variables shows that the terms $\mathbf{V}(t,y,x_n/\varepsilon)$ and $\mathbf{W}(t,y,x_n/\sqrt{\varepsilon}\,)$ go to 0 in $L^2(\Omega_{T_1})$ as $\varepsilon \to 0$, which means that u^ε goes to u^0 in $L^2(\Omega_{T_1})$. In fact the convergence is a little better, as stated in the next corollary ([16], [17]).

Corollary 2.5 *$u^\varepsilon \to u^0$ in $H^s(\Omega_{T_1})$ for all $s < 1/2$, when $\varepsilon \to 0$. Moreover u^ε remains in a bounded set of $L^\infty(\Omega_{T_1})$ and in $H^{1/2}(\Omega_{T_1})$ as $\varepsilon \to 0$.*

In the *linear case*, part of these results was already known: the boundary conditions in the limit hyperbolic problem was obtained by C. Bardos, D. Brezis, H. Brezis ([2]), and the convergence u^ε to u^0 in $H^s(\Omega_T)$ for all $s < 1/2$ was obtained by C. Bardos and J. Rauch ([3], 1982). Several examples and related results were also given in the book by J.-L. Lions ([24], 1973). Of course there are very numerous other classical works on closed problems of singular perturbations ([31], [32], [23], [8]...)

3 The quasilinear case

In this section we go back to the quasilinear problem $P(\varepsilon, T)$. We make, however, an important additional hypothesis.

Assumption 3.1 $\det A_n(0) \neq 0$.

This is a very important assumption which states that for every state u belonging to some small enough neighborhood of 0, the boundary $\{x_n = 0\}$ is noncharacteristic for the hyperbolic operator $\mathcal{H}(u)$. Observe that we make no further assumptions about the eigenvalues of the matrix $A_n(.)$ (in particular, $\mathcal{H}(u)$ is not supposed to be strictly hyperbolic).

Under the Assumption 3.1, one can prove the convergence of u^ε to the solution of some hyperbolic limit problem, and we begin by describing this limit. Let us consider the following two-point boundary value problem where the unknown is $V \in C^\infty([0, +\infty[, \mathbb{R}^N)$ and $v \in \mathbb{R}^N$ is a parameter:

$$\begin{cases} V'' = A_n(V+v)V', \\ V(0) = -v, \quad \lim_{+\infty} V = 0, \\ V \in C^\infty([0,+\infty[, \mathbb{R}^N). \end{cases} \tag{9}$$

Let us then introduce the set

$$\Sigma = \{v \in \mathbb{R}^N \text{ such that (9) has a solution } V(v,.)\}.$$

Lemma 3.2 *There is a neighborhood ω of 0 in $\mathbb{R}^N$ such that:*
1) $\omega \cap \Sigma$ is a smooth manifold passing through 0,
2) for all $u \in \omega \cap \Sigma$ the tangent space $\mathcal{T}_u\Sigma$ to Σ at u is a maximal negative subspace for the quadratic form $q(\xi) = < SA_n(u)\xi, \xi >$, and $\mathcal{T}_0\Sigma = \mathcal{N}$.

This lemma is a nonlinear version of Lemma 2.1. The manifold Σ was introduced by M. Gisclon and D. Serre ([9], [10], [30]) in the case of a system of conservation laws. The general case can be found in [14]. Let us just give the argument of point 2) which is very simple. The point is that when $u = 0$, something stronger than 2) holds true: the quadratic form $q(.,.)$ is positive *definite* on $\mathcal{T}_0\Sigma = \mathcal{N}$. Then by continuity, the quadratic form remains positive (and definite) on $\mathcal{T}_u\Sigma$ when u stays in a small enough neighborhood of 0. Let us also mention that the function $V(v,z)$ defined in this way by (9), is a smooth function of (v,z) in $(\omega \cap \Sigma) \times [0, +\infty[$. A consequence of Lemma 3.2 is the following result which is an example of a well-posed quasilinear *noncharacteristic* hyperbolic mixed problem ([28], [26], [25], [19]).

Theorem 3.3 *There exist $T_0 > 0$ and a unique $u \in H^\infty(\Omega_{T_0})$ solution of the following quasilinear hyperbolic mixed problem*

$$\begin{cases} \mathcal{H}(u^0)u^0 = F(t,x,u^0) \quad in \quad \Omega_{T_0} \\ u^0 \in \Sigma \quad on \quad \Gamma_{T_0}, \quad {u^0}_{|\Omega_0} = 0. \end{cases} \tag{10}$$

Now, for every $(t, x) = (t, y, 0)$ belonging to the boundary Γ_{T_0}, the dynamical system (9) has a unique solution corresponding to the choice $v = u^0(t, x)$ because of the definition of Σ. We denote this solution by $V(u^0(t, 0, y), .) = \mathbf{V}(t, y, .)$.

Theorem 3.4 *There exists $T_1 > 0$ and $\varepsilon_0 > 0$ such that for every $\varepsilon \in]0, \varepsilon_0]$, problem $P(\varepsilon, T_1)$ has a unique solution $u^\varepsilon \in H^\infty(\Omega_{T_1})$. This solution admits the following asymptotic expansion*

$$u^\varepsilon(t, x) = u^0(t, x) + \mathbf{V}(t, y, x_n/\varepsilon) + \varepsilon R_\varepsilon, \tag{11}$$

where $R_\varepsilon = \mathcal{O}(1)$ in $L^\infty(\Omega_{T_1}) \cap H^{1/2}(\Omega_{T_1})$ as $\varepsilon \to 0$.

This result was first obtained in one space dimension by M. Gisclon and D. Serre ([9], [30]), and extended to several dimensions by E. Grenier and O. Guès ([14]).

Corollary 3.5 *u^ε converges to u^0 in $H^s(\Omega_{T_1})$ for $s < 1/2$, and remains bounded in $L^\infty(\Omega_{T_1})$ and $H^{1/2}(\Omega_{T_1})$.*

Boundary layers instabilities There is an important remark about Theorem 3.4. When $t \leq 0$ the boundary layer profile $\mathbf{V}$ is 0. When $t > 0$ the size of the boundary layer profile $\mathbf{V}$ begins to increase. When the size of $\mathbf{V}$ is small enough, (smaller than some critical value) some energy estimates are available which enables one to control the difference $u^\varepsilon - (u^0 + \mathbf{V}(t, y, x_n/\varepsilon))$, and prove Theorem 3.4. But when the size of $\mathbf{V}$ is too big, these energy estimates break down. This is the reason why $T_1 > 0$ must be taken small enough in Theorem 3.4: to insure that the size of $\mathbf{V}$ is small enough. On the other hand, when the size of $\mathbf{V}$ is too big, the solution u^ε can become strongly unstable, the convergence behavior of corollary (3.5) being no more true. In the paper [14] one can find an explicit example where, as soon as the size of the boundary layer profile becomes bigger than the critical value, a very strong instability arises in the solution u^ε itself, which blows up as $\varepsilon \to 0$ instead of converging to u^0.

The characteristic case The characteristic case is the case where the matrix $A_n(u)$ has a nontrivial kernel. The very particular case of a *totally characteristic boundary* has been studied by E. Grenier ([11], [12], [13]). A very difficult problem is the case of the vanishing viscosity for the Navier-Stokes equations, for which the formal limit is the Euler system with the natural slip condition at the boundary, which is a characteristic quasilinear hyperbolic mixed problem. Convergence results have been obtained in the analytic case by K. Asano, R. E. Caflisch and M. Sanmartino ([1], [4], [5]), but strong instabilities are expected for Sobolev regularity (H. Schichting [29]). Other results have been obtained for rotating fluids leading to anisotropic vanishing viscosity by T. Colin, P. Fabrie, E. Grenier, N. Masmoudi, ([6], [15]).

References

[1] K. ASANO: Zero viscosity limit of the incompressible Navier-Stokes equation, I and II, unpublished.

[2] C. BARDOS, D. BREZIS, H. BREZIS: Perturbations singulières et prolongements maximaux d'opérateurs positifs, Arch. Rational Mech. Anal., 53 (1973), 69-100.

[3] C. BARDOS, J. RAUCH: Maximal positive boundary value problems as limits of singular perturbation problems, Trans. Amer. Math. Soc., 270, (1982), 377-408.

[4] R. E. CAFLISCH, M. SANMARTINO: "Zero viscosity limit for analytic solutions of the Navier-Stokes equation on a half-space I. Existence for Euler and Prandtl equations," Comm. Math. Phys. (submitted).

[5] R. E. CAFLISCH, M. SANMARTINO: Zero viscosity limit for analytic solutions of the Navier-Stokes equation on a half-space II. Construction of the Navier-Stokes solution, Comm. Math. Phys. (submitted).

[6] T. COLIN, P. FABRIE: Rotating fluid at high Rossby number driven by a surface stress: existence and convergence, preprint université de Bordeaux 1, to appear.

[7] K. O. FRIEDRICHS: Symmetric positive linear differential equations, Comm. Pure Appl. Math. 7, (1954), 345-392.

[8] K. O. FRIEDRICHS: Well-posed problems of mathematical physics, in mimeog. lectures notes, New-York Univ.

[9] M. GISCLON: Etude des conditions aux limites pour un système strictement hyperbolique, via l'approximation parabolique, Journal de Mathématiques Pures et Appliquées, 75, (1996), 485-508.

[10] M. GISCLON, D. SERRE: Etude des conditions aux limites pour un système strictement hyperbolique, via l'approximation parabolique, C. R. Acad. Sci. Paris, t. 319, Série I, 1994, 377-382.

[11] E. GRENIER: Couches limites de systèmes paraboliques nonlinéaires caractéristiques, C. R. Acad. Sci. Paris, t 323 (1996), 1013-1017.

[12] E. GRENIER: Couches limites de systèmes paraboliques, Journées Equations aux Dérivées Partielles de Saint-Jean-de-Monts, 1996, exp. n° 8.

[13] E. GRENIER: Boundary layers for parabolic regularizations of totally characteristic quasilinear parabolic equations, J. Math. Pures Appl., 76, (1997), p. 965-990.

[14] E. GRENIER, O. GUÈS: Boundary layers for viscous perturbations of noncharacteristic quasilinear hyperbolic problems, Journal of Diff. Equa. 143, (1998), 110-146.

[15] E. GRENIER, N. MASMOUDI: Ekman layers of rotating fluids, the case of well prepared initial data, Comm. Partial Diff. Equa., 22, (1997), n° 5-6, 953-975.

[16] O. GUÈS: Couches limites pour des problèmes mixtes hyperboliques, séminaire E.D.P. de l'Ecole Polytechnique, 1993-94, exp $n^o 17$.

[17] O. GUÈS: Perturbations visqueuses de problèmes mixtes hyperboliques et couches limites, Ann. Inst. Fourier, Grenoble, (1995), 973-1006.

[18] O. GUÈS: Viscous boundary layers and high frequency oscillations, in IMA Volume 91, "Singularities and Oscillations", p. 61-77, J. Rauch and M. Taylor Editors, Springer-Verlag, (1997).

[19] O. GUÈS: Problèmes mixtes hyperboliques quasilinéaires caractéristiques, Comm. in Partial Diff. Equa., 15, (1990), 595-645.

[20] H.-O. KREISS, J. LORENZ: Initial boundary value problems and the Navier-Stokes equations, Pure and Applied Mathematics, Academic Press, London, 1989.

[21] P. D. LAX, R. S. PHILIPS: Local boundary conditions for dissipative symmetric linear differential operators, Comm. Pure Appl. Math., 13, (1960), 427-455.

[22] O. A. LADYZHENSKAYA, V. A. SOLONNIKOV, N. N. URAL'CEVA: Linear and quasilinear equations of parabolic type, Moscou 1967 and Trans. Math. Monographs, vol. 23, Amer. Math. Soc., Providence, RI, 1968.

[23] N. LEVINSON: The first boundary value problem for $\varepsilon\Delta u + A(x,y)u_x + B(x,y)u_y + C(x,y)u = D(x,y)$ for small ε, Ann. Math., 5 (1950), 428-445.

[24] J.-L. LIONS: Perturbations singulières dans les problèmes aux limites et en contrôle optimal, Lectures Notes in Math, N 323, Springer-Verlag, 1973.

[25] A. MAJDA: compressible fluid flow and systems of conservation laws in several space variables, Appl. Math. Sc. 53, Springer-Verlag (1984).

[26] G. MÉTIVIER: Problèmes mixtes non linéaires et stabilité des chocs multidimensionnels, Séminaire Bourbaki, n.671 (1986-87).

[27] J. RAUCH: Boundary value problems as limits of problems in all space, Séminaire Goulaouic-Schwartz, Ecole Polytechnique, exposé n.3,1978.

[28] J. RAUCH, F. MASSEY: Differentiability of solutions to hyperbolic initial-boundary value problems, Trans. Amer. Math. Soc., Vol. 189, (1974), 303-318.

[29] H. SCHICHTING: Boundary layer theory, McGraw-Hill, 1968.

[30] D. SERRE: Systèmes de lois de conservation II, Diderot editeur, Arts et Sciences, 1996.

[31] M. VISHIK, L. LYUSTERNIK: Asymptotic behavior of solutions of linear differential equations with large or quickly changing coefficients and boundary condition, Russian Math. Surveys, 4 (1960), 23-92.

[32] M. VISHIK, L. LYUSTERNIK: Regular degeneration and boundary layers for linear differential equations with small parameter, Amer. Math. Soc. Transl. (2) 20 (1962), 239-264.

O. GUÈS
Laboratoire Dieudonné, UMR CNRS 6621, Université de Nice-Sophia Antipolis, o6108 Nice, cédex 2. France, E-mail: gues@unice.fr

Ordinary Differential Equations and Generalized Functions

ROBERT HERMANN AND MICHAEL OBERGUGGENBERGER

1 Introduction

The purpose of this note is to present a number of fundamental existence results for systems of ordinary differential equations

$$\begin{aligned} \dot{x}(t) &= F(t, x(t)), \\ x(t_0) &= x_0 \end{aligned} \tag{1.1}$$

in the setting of algebras of generalized functions. More precisely, we shall allow the components F_i of the right-hand side to belong to the class of Colombeau tempered generalized functions [6], $i = 1, ..., n$, and we shall obtain Colombeau generalized solutions $x = (x_1, ..., x_n)$ as well as a generalized flow acting on extended state space $\tilde{\mathbb{R}}^n$.

These results are at the core of various recent applications of algebras of generalized functions in analysis, geometry, general relativity, and so on. In particular, applications include

- generalized symmetries and Lie group transformations;
- geodesics in distributional geometries;
- generalized function potentials;
- hybrid control systems;
- nonsmooth calculus of variations;
- stochastic differential equations.

Concerning Lie theory and geodesics we refer to the contributions [12, 22, 24] in this volume as well as [1, 11, 13, 14, 25]. For stochastic differential equations, see [18];

indications on the remaining three areas of application will be given subsequently in Section 4.

In the framework we adhere to, generalized functions u are represented by (classes of) sequences $(u_\varepsilon)_{\varepsilon>0}$ of smooth functions, satisfying asymptotic growth conditions as $\varepsilon \to 0$, but generally being unbounded and possibly divergent in the usual topologies. When applied to the right-hand side in (1.1) this may lead to the breakdown of local existence, as the time T_ε of existence of a solution may go to zero as $\varepsilon \to 0$. Similar to the classical case, this has to be avoided by imposing conditions on $F_\varepsilon(t,x)$, which in our case will involve both Lipschitz or boundedness conditions with respect to the variable x as well as estimates in terms of the parameter ε. The natural Lipschitz or boundedness conditions or structural assumptions will then lead to existence globally in time. This is why we shall deal here with global solutions only. A refined theory of local solutions in this context is open to further research.

The plan of exposition is as follows. In Section 2, we recall the definitions and results needed from the Colombeau theory of generalized functions. Section 3 contains a representative sample of existence and uniqueness results, while Section 4 hints at some applications in the calculus of variations and control theory. Details are worked out in the manuscript [9].

2 Algebras of generalized functions

In collecting the required definitions and properties, we do not motivate the constructions in this section; for this we refer to the many monographs on the subject, e.g., [2, 7, 8, 17, 20]. For convenience of presentation, we shall work with the so-called *simplified* Colombeau algebras on open subsets Ω of $\mathbb{R}^n$. These are factor algebras in the differential algebra $\mathcal{E}[\Omega]$ of all sequences $(u_\varepsilon)_{\varepsilon>0}$ of real valued smooth functions $u_\varepsilon \in \mathcal{C}^\infty(\Omega)$. Let

$$\begin{aligned}\mathcal{E}_M(\Omega) &= \{(u_\varepsilon)_{\varepsilon>0} \in \mathcal{E}[\Omega] : \forall K \subset\subset \Omega\ \forall \alpha \in \mathbb{N}_0^n\ \exists p \geq 0 : \\ &\quad \sup_{x\in K} |\partial^\alpha u_\varepsilon(x)| = \mathcal{O}(\varepsilon^{-p}), \varepsilon \to 0\},\end{aligned}$$

$$\begin{aligned}\mathcal{N}(\Omega) &= \{(u_\varepsilon)_{\varepsilon>0} \in \mathcal{E}[\Omega] : \forall K \subset\subset \Omega\ \forall \alpha \in \mathbb{N}_0^n\ \forall q \geq 0 : \\ &\quad \sup_{x\in K} |\partial^\alpha u_\varepsilon(x)| = \mathcal{O}(\varepsilon^{q}), \varepsilon \to 0\}.\end{aligned}$$

The *Colombeau algebra of generalized functions* is defined as the factor algebra $\mathcal{G}(\Omega) = \mathcal{E}_M(\Omega)/\mathcal{N}(\Omega)$. Next, we let

$$\begin{aligned}\mathcal{E}_\tau(\Omega) &= \{(u_\varepsilon)_{\varepsilon>0} \in \mathcal{E}[\Omega] : \forall\alpha \in \mathbb{N}_0^n\ \exists p \geq 0 : \\ &\qquad \sup_{x\in\Omega}(1+|x|)^{-p}|\partial^\alpha u_\varepsilon(x)| = \mathcal{O}(\varepsilon^{-p}), \varepsilon \to 0\},\end{aligned}$$

$$\begin{aligned}\mathcal{N}_\tau(\Omega) &= \{(u_\varepsilon)_{\varepsilon>0} \in \mathcal{E}[\Omega] : \forall\alpha \in \mathbb{N}_0^n\ \exists p \geq 0\ \forall q \geq 0 : \\ &\qquad \sup_{x\in\Omega}(1+|x|)^{-p}|\partial^\alpha u_\varepsilon(x)| = \mathcal{O}(\varepsilon^{q}), \varepsilon \to 0\}.\end{aligned}$$

The *Colombeau algebra of tempered generalized functions* is $\mathcal{G}_\tau(\Omega) = \mathcal{E}_\tau(\Omega)/\mathcal{N}_\tau(\Omega)$. Let $\rho \in \mathcal{S}(\mathbb{R}^n)$, $\int \rho(x)dx = 1$, $\int x^\alpha \rho(x)dx = 0$ for all $\alpha \in \mathbb{N}_0^n$, $|\alpha| \geq 1$, and put

$$\rho_\varepsilon(x) = \varepsilon^{-n}\rho(x/\varepsilon). \tag{2.1}$$

The map

$$\iota : w \to \text{ class of } (w * \rho_\varepsilon)_{\varepsilon>0} \tag{2.2}$$

defines an imbedding of $\mathcal{S}'(\mathbb{R}^n)$ into $\mathcal{G}_\tau(\mathbb{R}^n)$ which renders

$$\mathcal{O}_C(\mathbb{R}^n) = \{f \in \mathcal{C}^\infty(\mathbb{R}^n) : \exists p \geq 0\, \forall\alpha \in \mathbb{N}_0^n : \sup_{x\in\mathbb{R}^n}(1+|x|^{-p})|\partial^\alpha f(x)| < \infty\}$$

a faithful subalgebra. Thus $\mathcal{G}_\tau(\mathbb{R}^n)$ is a differential algebra containing the space of tempered distributions $\mathcal{S}'(\mathbb{R}^n)$, where the derivatives extend the usual ones on $\mathcal{S}'(\mathbb{R}^n)$ and the multiplication map extends the usual product on $\mathcal{O}_C(\mathbb{R}^n)$. Further, the space of compactly supported distributions $\mathcal{E}'(\Omega)$ is imbedded in the algebra $\mathcal{G}(\Omega)$ by an imbedding analogous to (2.2). Using the fact that $\mathcal{G}(\Omega)$ is a sheaf of algebras, this can be extended to an imbedding of the space of distributions $\mathcal{D}'(\Omega)$ into $\mathcal{G}(\Omega)$ which now renders $\mathcal{C}^\infty(\Omega)$ a faithful subalgebra. Note that the defining bounds in $\mathcal{G}(\Omega)$ are local, while those in $\mathcal{G}_\tau(\Omega)$ measure global growth with respect to the Euclidean norm. We shall also need a mixed version on subsets $\omega\times\Omega \subset \mathbb{R}^m\times\mathbb{R}^n$, say, defined by $\tilde{\mathcal{G}}_\tau(\omega\times\Omega) = \tilde{\mathcal{E}}_\tau(\omega\times\Omega)/\tilde{\mathcal{N}}_\tau(\omega\times\Omega)$, where

$$\begin{aligned}\tilde{\mathcal{E}}_\tau(\omega\times\Omega) &= \{(u_\varepsilon)_{\varepsilon>0} \in \mathcal{E}[\omega\times\Omega] : \forall K \subset\subset \omega\ \forall\alpha \in \mathbb{N}_0^n\ \exists p \geq 0 : \\ &\qquad \sup_{t\in K}\sup_{x\in\Omega}(1+|x|)^{-p}|\partial^\alpha u_\varepsilon(t,x)| = \mathcal{O}(\varepsilon^{-p}), \varepsilon \to 0\},\end{aligned}$$

$$\tilde{\mathcal{N}}_\tau(\omega \times \Omega) = \{(u_\varepsilon)_{\varepsilon>0} \in \mathcal{E}[\omega \times \Omega] : \forall K \subset\subset \omega \; \forall \alpha \in \mathbb{N}_0^n \; \exists p \geq 0 \; \forall q \geq 0 : \sup_{t \in K} \sup_{x \in \Omega} (1+|x|)^{-p} |\partial^\alpha u_\varepsilon(t,x)| = \mathcal{O}(\varepsilon^q), \varepsilon \to 0\}.$$

All algebras introduced above are invariant under composition with polynomially bounded smooth functions. What is more – and this is the principal reason why we need the algebra $\mathcal{G}_\tau$ here – is the following result:

Proposition 2.1 Let $F \in \mathcal{G}_\tau(\mathbb{R}^m), \Omega$ an open subset of $\mathbb{R}^m, u = (u_1, ..., u_m) \in (\mathcal{G}(\Omega))^m, v = (v_1, ..., v_m) \in (\mathcal{G}_\tau(\Omega))^m$. Then $F(u)$ is a well-defined element of $\mathcal{G}(\Omega)$ and $F(v)$ is a well-defined element of $\mathcal{G}_\tau(\Omega)$.

Proof: The composition is defined componentwise on representatives, i. e. $F(u)$ has $F_\varepsilon(u_{1\varepsilon}, ..., u_{m\varepsilon})$ as a representative. One has to check that it depends neither on the choice of representatives for u nor for F, and that it satisfies the defining bounds. All this is rather straightforward from the definitions. □

Next, we shall discuss generalized point values. The Colombeau *generalized real numbers* $\tilde{\mathbb{R}}$ are defined as the generalized functions in zero dimensions $\mathcal{G}(\mathbb{R}^0)$. They actually coincide with the ring of constants in $\mathcal{G}(\Omega)$, respectively $\mathcal{G}_\tau(\Omega)$ when Ω is connected, and are represented by classes of sequences $(r_\varepsilon)_{\varepsilon>0}$ of real numbers. The *limited generalized numbers* $\tilde{\mathbb{R}}_c$ are those for which there is a compact interval $I \subset \mathbb{R}$ and a representative $(r_\varepsilon)_{\varepsilon>0}$ such that $r_\varepsilon \in I$ eventually as $\varepsilon \to 0$.

Given $u \in \mathcal{G}(\mathbb{R}^n)$ and $x \in \tilde{\mathbb{R}}_c^n$, the generalized point value $u(x)$ is well-defined as the class of $(u_\varepsilon(x_\varepsilon))_{\varepsilon>0}$ in $\tilde{\mathbb{R}}$. Similarly, for $u \in \mathcal{G}_\tau(\mathbb{R}^n)$ and $x \in \tilde{\mathbb{R}}^n, u(x) \in \tilde{\mathbb{R}}$ is well-defined (for point values of generalized functions on open sets Ω, see [19]). An important tool is the fact that generalized functions are characterized by their point values. A proof of the following result is in [19]:

Proposition 2.2 Let $u \in \mathcal{G}(\mathbb{R}^n), v \in \mathcal{G}_\tau(\mathbb{R}^n), w \in \tilde{\mathcal{G}}_\tau(\mathbb{R}^m \times \mathbb{R}^n)$. Then
(a) $u = 0$ in $\mathcal{G}(\mathbb{R}^n)$ iff $u(x) = 0$ in $\tilde{\mathbb{R}}$ for all $x \in \tilde{\mathbb{R}}_c^n$;
(b) $v = 0$ in $\mathcal{G}_\tau(\mathbb{R}^n)$ iff $u(x) = 0$ in $\tilde{\mathbb{R}}$ for all $x \in \tilde{\mathbb{R}}^n$;
(c) $w = 0$ in $\tilde{\mathcal{G}}_\tau(\mathbb{R}^m \times \mathbb{R}^n)$ iff $u(t,x) = 0$ in $\tilde{\mathbb{R}}$ for all $t \in \tilde{\mathbb{R}}_c^m, x \in \tilde{\mathbb{R}}^n$. □

We shall need a number of special types of generalized functions as well:

Definition 2.3 (a) A generalized function u in $\tilde{\mathcal{G}}_\tau(\mathbb{R}^m \times \mathbb{R}^n)$ is called *nonnegative* if it has a representative $(u_\varepsilon)_{\varepsilon>0}$ such that $u_\varepsilon(t,x) \geq 0$ for all $(t,x) \in \mathbb{R}^m \times \mathbb{R}^n$ and all $\varepsilon > 0$.

(b) The generalized function u in $\tilde{\mathcal{G}}_\tau(\mathbb{R}^m \times \mathbb{R}^n)$ is called *of L^∞-type*, if it has a representative $(u_\varepsilon)_{\varepsilon>0}$ such that for every $K \subset\subset \mathbb{R}^m$ there is $p \geq 0$ with

$$\sup_{t\in K}\sup_{x\in\mathbb{R}^n} |u_\varepsilon(t,x)| = \mathcal{O}(\varepsilon^{-p})$$

as $\varepsilon \to 0$.

(c) It is called of *L^∞-log-type*, if for some representative

$$\sup_{t\in K}\sup_{x\in\mathbb{R}^n} |u_\varepsilon(t,x)| = \mathcal{O}(|\log\varepsilon|)$$

as $\varepsilon \to 0$.

Note that in both cases this means that the functions $u_\varepsilon(t,x)$ are bounded, uniformly for $t \in K, x \in \mathbb{R}^n$, but with bounds possibly increasing polynomially, respectively logarithmically, as $\varepsilon \to 0$. The corresponding definitions for elements of $\mathcal{G}(\Omega)$ and $\mathcal{G}_\tau(\Omega)$ are given in an analogous way.

Finally, we recall the notion of *association*. An element $u \in \mathcal{G}(\Omega)$ or $\mathcal{G}_\tau(\Omega)$ is said to *admit an associated distribution* $w \in \mathcal{D}'(\Omega)$, if for some (or - equivalently - each) representative, $u_\varepsilon \to w$ in $\mathcal{D}'(\Omega)$ as $\varepsilon \to 0$. If such a distribution exists, it is unique. We then view w as the "shadow" of the generalized function u in the space of distributions. It is what one can observe when letting u act on test functions in the way distributions act.

3 Existence and uniqueness, the flow

In this section we construct generalized solutions to system (1.1). This means that the differential equations are supposed to hold with equality in the algebra $(\mathcal{G}(\mathbb{R}))^n$, while the initial data are interpreted in the sense of point values in $\tilde{\mathbb{R}}^n$. As is common in the theory of ordinary differential equations, the most basic existence and uniqueness result requires a global Lipschitz condition on the nonlinearity F. In our case, F is a generalized function, and the Lipschitz constants on its representatives $(F_\varepsilon)_{\varepsilon>0}$ are allowed to increase as $\varepsilon \to 0$, but in a controlled manner.

Theorem 3.1 Let $F \in (\tilde{\mathcal{G}}_\tau(\mathbb{R}^{1+n}))^n$ such that its gradient $\nabla_x F$ is of L^∞-log-type. Then given $x_0 \in \tilde{\mathbb{R}}^n$, problem (1.1) has a unique solution $x \in (\mathcal{G}(\mathbb{R}))^n$.

Proof: Let $(F_\varepsilon)_{\varepsilon>0}$ be a representative of F with the indicated properties (see Definition 2.3), $(x_{0\varepsilon})_{\varepsilon>0}$ a representative of x_0. From the classical theory of differential equations we have that there is a unique solution $x_\varepsilon \in (\mathcal{C}^\infty(\mathbb{R}))^n$ to

$$\dot{x}_\varepsilon(t) = F_\varepsilon(t, x_\varepsilon(t));\ x_\varepsilon(t_0) = x_{0\varepsilon}\,.$$

We are going to show that $(x_\varepsilon)_{\varepsilon>0}$ belongs to $(\mathcal{E}_M(\mathbb{R}))^n$. Its class in $(\mathcal{G}(\mathbb{R}))^n$ is a solution to (1.1). We have

$$x_\varepsilon(t) = x_{0\varepsilon} + \int_{t_0}^t F_\varepsilon(s,0)\,ds + \int_{t_0}^t \int_0^1 \nabla_x F_\varepsilon(s, \sigma x_\varepsilon(s))\,d\sigma\, x_\varepsilon(s)\,ds\,.$$

On any fixed time interval $|t - t_0| \le T$, the first two terms on the right-hand side are bounded by $C\varepsilon^{-p}$ while the L^∞-norm of $\nabla_x F_\varepsilon$ is bounded by $C|\log\varepsilon|$ for some constants $C, p \ge 0$ and all small $\varepsilon > 0$. By Gronwall's inequality,

$$|x_\varepsilon(t)| \le C\varepsilon^{-p}\exp(-TC|\log\varepsilon|) = C\varepsilon^{-p-TC},\ |t - t_0| \le T\,.$$

Thus x_ε satisfies the required estimate of order zero. The estimate on its first derivative follows immediately from the differential equation and the growth properties of F_ε. Estimates for the higher order derivatives are obtained by successively differentiating the equation. Thus $(x_\varepsilon)_{\varepsilon>0}$ belongs to $(\mathcal{E}_M(\mathbb{R}))^n$ and this concludes the proof of existence of a solution.

To obtain uniqueness, assume there are two solutions x, y with representatives $(x_\varepsilon)_{\varepsilon>0}, (y_\varepsilon)_{\varepsilon>0}$. Then the difference satisfies

$$\begin{aligned} \dot{x}_\varepsilon(t) - \dot{y}_\varepsilon(t) &= F_\varepsilon(t, x_\varepsilon(t)) - F_\varepsilon(t, y_\varepsilon(t)) + n_\varepsilon(t), \\ x_\varepsilon(t_0) - y_\varepsilon(t_0) &= m_\varepsilon \end{aligned}$$

with $(n_\varepsilon)_{\varepsilon>0} \in (\mathcal{N}(\mathbb{R}))^n$, $(m_\varepsilon)_{\varepsilon>0} \in (\mathcal{N}(\mathbb{R}^0))^n$. Similar arguments as above imply that $(x_\varepsilon - y_\varepsilon)_{\varepsilon>0}$ belongs to $(\mathcal{N}(\mathbb{R}))^n$, so $x = y$ in $(\mathcal{G}(\mathbb{R}))^n$. □

Remarks (a) The solution obtained in Theorem 3.1 does not belong to $(\mathcal{G}_\tau(\mathbb{R}))^n$ in general, as is seen from the example $\dot{x}(t) = x(t)\,, x(0) = 1$.

(b) If the condition on $\nabla_x F$ in Theorem 3.1 is violated, both existence and/or uniqueness may fail to hold. Let $\eta \in \tilde{\mathbb{R}}$ be defined by the sequence $(\varepsilon)_{\varepsilon>0}$. Then

$$\dot{x}(t) = \frac{1}{\eta}x(t)\,, x(0) = 1$$

has no solution in $\mathcal{G}(\mathbb{R})$, while the equation

$$\dot{x}(t) = \frac{1}{\eta}x(t)(1 - x(t))\,, x(0) = 0$$

has more than one solution, for example, the classes of $x_\varepsilon(t) = 0$ and of $y_\varepsilon(t) = e^{t/\varepsilon}/(e^{1/\varepsilon} + e^{t/\varepsilon})$; note that $y_\varepsilon(1) = 1/2$, so that $(y_\varepsilon)_{\varepsilon>0} \notin \mathcal{N}(\mathbb{R})$.

(c) Thanks to the construction of the algebras $\tilde{\mathcal{G}}_\tau$ and $\mathcal{G}_\tau$, we have the following consistency result with classical solutions. If in Theorem 3.1 F belongs to $\mathcal{O}_C(\mathbb{R}^{1+n})$ with $\nabla_x F$ bounded, and x_0 is in $\mathbb{R}^n$, then the generalized solution $x \in (\mathcal{G}(\mathbb{R}))^n$ coincides with the classical $\mathcal{C}^\infty$-solution. This follows from the fact that the latter may serve as a representative for the generalized solution.

Next, we study the flow generated by generalized solutions to system (1.1). It will be useful to introduce the flow as a generalized function of $(t, x_0) \in \mathbb{R}^{1+n}$; its group property will hold in the sense of generalized point values.

Theorem 3.2 Under the hypotheses of Theorem 3.1, there is a unique generalized function $\Phi \in (\tilde{\mathcal{G}}_\tau(\mathbb{R}^{1+n}))^n$ such that

$$\begin{aligned} \frac{d}{dt}\Phi &= F(.,\Phi) \text{ in } (\tilde{\mathcal{G}}_\tau(\mathbb{R}^{1+n}))^n \\ \Phi(0,.) &= \mathrm{I} \text{ in } (\mathcal{G}_\tau(\mathbb{R}^n))^n \\ \Phi(t+s,.) &= \Phi(t,\Phi(s,.)) \text{ in } (\tilde{\mathcal{G}}_\tau(\mathbb{R}^{2+n}))^n . \end{aligned}$$

Further, Φ defines an action of the additive group $(\tilde{\mathbb{R}}_c, +)$ on $\tilde{\mathbb{R}}^n$:

$$\begin{aligned} \Phi(0,x_0) &= x_0 \text{ for all } x_0 \in \tilde{\mathbb{R}}^n, \\ \Phi(t+s,x_0) &= \Phi(t,\Phi(s,x_0)) \text{ for all } x_0 \in \tilde{\mathbb{R}}^n, t, s, \in \tilde{\mathbb{R}}^n_c . \end{aligned}$$

Proof: The first part is obtained by extending the proof of Theorem 3.1 so as to include the dependence on the initial point. Having the group action valid as elements of $(\tilde{\mathcal{G}}_\tau(\mathbb{R}^{2+n}))^n$, the pointwise version follows immediately by the characterization of Colombeau generalized functions by their point values (see Proposition 2.5 and Theorem 2.10 in [19]). □

Remark There are two reasons why Φ has been constructed as an element of $(\tilde{\mathcal{G}}_\tau(\mathbb{R}^{1+n}))^n$ and not as a function $\tilde{\mathbb{R}}_c \times \tilde{\mathbb{R}}^n \to \tilde{\mathbb{R}}^n$. First, a differential calculus for functions on $\tilde{\mathbb{R}}^n$ is lacking so far. Second, we can speak about association in $\tilde{\mathcal{G}}_\tau(\mathbb{R}^{1+n})$, this way having a framework for describing the classical "shadow" of a generalized flow, if it exists.

As in the classical case, a bound on the nonlinear function F, in our case by a generalized constant, gives existence of global solutions as well. The proof is along the same lines as above.

Theorem 3.3 Let $F \in (\tilde{\mathcal{G}}_\tau(\mathbb{R}^{1+n}))^n$ be of L^∞-type. Then given $x_0 \in \tilde{\mathbb{R}}^n$, problem (1.1) has a solution $x \in (\mathcal{G}(\mathbb{R}))^n$. □

In the place of Lipschitz and/or boundedness conditions, structural assumptions on the nonlinear function F may lead to existence and uniqueness as well. We present two examples of this situation. First, consider an autonomous system

$$\dot{x} = F(x)\ , x(t_0) = x_0\ . \tag{3.1}$$

We shall say that $F \in (\mathcal{G}_\tau(\mathbb{R}^n))^n$ is *dissipative*, if the generalized function $G(x,y) = \langle x-y, F(x) - F(y)\rangle$ satisfies $G \le 0$ in $\mathcal{G}_\tau(\mathbb{R}^{2n})$ according to Definition 2.3. The Colombeau algebra on a closed half-ray below is defined in a similar fashion as the algebras in Section 2.

Theorem 3.4 Assume that $F \in (\mathcal{G}_\tau(\mathbb{R}^n))^n$ is dissipative, $x_0 \in \tilde{\mathbb{R}}^n$. Then problem (3.1) has a unique forward solution $x \in (\mathcal{G}[t_0,\infty))^n$.

Proof: The a priori estimate

$$\frac{1}{2}\frac{d}{dt}|x_\varepsilon(t)|^2 = \langle x_\varepsilon(t), F_\varepsilon(x_\varepsilon(t))\rangle \le \langle x_{0\varepsilon}, F_\varepsilon(x_{0\varepsilon})\rangle$$

shows that the classical solution $x_\varepsilon(t)$ exists on $[t_0,\infty)$ for each fixed $\varepsilon > 0$ and further satisfies an $\mathcal{O}(\varepsilon^{-p})$ bound as $\varepsilon \to 0$. Similar bounds on the derivatives of $x_\varepsilon(t)$ follow from the equation, thus producing a representative $(x_\varepsilon)_{\varepsilon>0} \in (\mathcal{E}_\tau[t_0,\infty))^n$ of the solution. Uniqueness is shown along the same lines. □

Our second example concerns gradient systems which arise e.g., as Euler-Lagrange-equations in particle mechanics. Consider the Lagrangian

$$L(t,x,\dot{x}) = \frac{1}{2}|\dot{x}|^2 - V(x)\ .$$

The Euler-Lagrange-equations are, with initial data,

$$\begin{aligned} \ddot{x}(t) + \nabla V(x) &= 0\,, \\ x(t_0) = x_0,\ \dot{x}(t_0) &= \dot{x}_0\,. \end{aligned} \tag{3.2}$$

Theorem 3.5 Let $V \in \mathcal{G}_\tau(\mathbb{R}^n)$ be either of L^∞-type or let $V \ge 0$. Further, let $x_0, \dot{x}_0 \in \tilde{\mathbb{R}}^n$. Then problem (3.2) has a solution $x \in (\mathcal{G}(\mathbb{R}))^n$.

Proof: Conservation of energy gives the a priori estimate

$$\frac{d}{dt}\left(\frac{1}{2}|\dot{x}_\varepsilon(t)|^2 + V_\varepsilon(X_\varepsilon(t))\right) = 0$$

which, as in Theorem 3.4, entails global existence of a classical solution $x_\varepsilon(t), t \in \mathbb{R}$ for fixed $\varepsilon > 0$ as well as an $\mathcal{O}(\varepsilon^{-p})$ bound for $x_\varepsilon(t)$ when t varies in compact sets. The rest follows from the differential equation. □

Remarks (a) This solution actually belongs to $(\mathcal{G}_\tau(\mathbb{R}))^n$.

(b) Uniqueness in $(\mathcal{G}(\mathbb{R}))^n$ is obtained in general if D^2V is of L^∞-log-type (Theorem 3.1) or under additional structural assumptions on the potential V, as e.g., in the case of delta-potentials in the next section.

4 Applications

A classical particle in a delta function potential is formally described by the Lagrangian

$$L(t, x, \dot{x}) = \frac{1}{2}|\dot{x}|^2 - \delta(x) .$$

The Euler-Lagrange-equations

$$\begin{aligned} \ddot{x}(t) + \delta'(x(t)) = 0 \,, \\ x(0) = x_0, \ \dot{x}(0) = \dot{x}_0 \end{aligned} \tag{4.1}$$

have a well-defined meaning in the algebra $\mathcal{G}_\tau(\mathbb{R})$, the delta function and its derivative being obtained from the imbedding (2.2) of $\mathcal{S}'(\mathbb{R})$. We have:

Proposition 4.1 Given $x_0, \dot{x}_0 \in \tilde{\mathbb{R}}$, problem (4.1) has a unique solution $x \in \mathcal{G}_\tau(\mathbb{R})$. Further, this solution is associated with the function $t \to \ \mathrm{sign}(x_0)\,|x_0 + \dot{x}_0 t|$.

Proof: A representative of the delta function is the sequence

$$\rho_\varepsilon(x) = \frac{1}{\varepsilon}\rho(x/\varepsilon) \tag{4.2}$$

with the mollifier ρ as in Section 2. As an element of $\mathcal{G}_\tau(\mathbb{R})$, the delta function is clearly of L^∞-type. Thus Theorem 3.5 with the subsequent remark applies, and so a solution exists.

To prove uniqueness we use a scaling technique. If $x \in \mathcal{G}_\tau(\mathbb{R})$ is a solution to (4.1), we define $y \in \mathcal{G}_\tau(\mathbb{R})$ as the class of $y_\varepsilon(t) = \varepsilon^{-1}x_\varepsilon(\varepsilon^{3/2}t)$, where $(x_\varepsilon)_{\varepsilon>0}$ is a representative of x. It is immediate to check that y solves the equation

$$\ddot{y}(t) + \rho'(y(t)) = 0 \ \text{ in } \ \mathcal{G}_\tau(\mathbb{R})$$

with suitably scaled initial data. But this is an equation in $\mathcal{G}_\tau(\mathbb{R})$ involving the classical bounded function ρ' as nonlinearity, so its solutions are unique (use e.g., the methods in [16]). The uniqueness of the solution x to (4.1) follows by observing that the scaling transformation leaves the ideal $\mathcal{N}_\tau(\mathbb{R})$ invariant.

The qualitative behavior follows from studying the generalized flow. The geometry is clear from the fact that the trajectories $x_\varepsilon, \dot{x}_\varepsilon$ lie on the level curves of the function $E(x,\dot{x}) = \frac{1}{2}\dot{x}^2 + \rho_\varepsilon(x)$. The proof of the actual reflection of the solution trajectories at the origin requires some technical estimates (see [9]). □

A picture of the "shadow" of the flow is below, showing pure reflection of the potential barrier, as known in the physics literature (e.g., [21]). We note, however, that regularizations of the delta function more general than (4.2) need not lead to this picture; it is possible that individual trajectories are trapped at the origin forever.

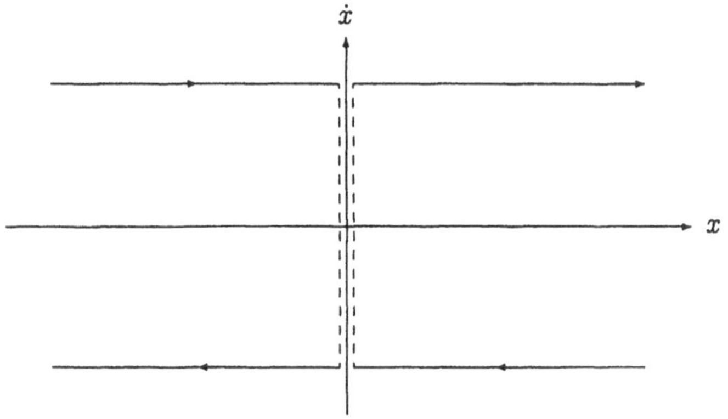

FIGURE: *The flow of* $\ddot{x} + \delta'(x) = 0$.

A further application concerns hybrid control systems. Following [10] we shall focus on affine control systems

$$\dot{x}(t) = V(x(t)) + \sum_{i=1}^{m} u_i(t) W_i(x(t)) \tag{4.3}$$

where in the classical setting $t \to x(t)$ is a smooth curve on a manifold X, describing the state of the system, $\{V, W_1, ..., W_m\}$ are smooth vector fields on $X, t \to (u_1(t), ..., u_m(t))$ is a curve in $\mathbb{R}^m$, defining the controls. Hybrid control theory deals with systems involving both continuous and discrete controls, see e.g., [3, 4, 5]. Many examples of hybrid systems can be constructed using partitions (of state space X). We believe that algebras of generalized functions deliver a suitable framework for hybrid control theory.

To be specific, we assume the Euclidean situation $X = \mathbb{R}^n$ and take open disjoint subsets $\Omega_j, j = 1, ..., k$ with characteristic functions $H_j(x) = 1, x \in \Omega_j, H_j(x) = 0, x \notin \Omega_j$. A hybrid control system can be formally written as

$$\dot{x}(t) = \sum_{j=1}^{k} H_j(x(t)) \left(V_j(x(t)) + \sum_{i=1}^{m} u_i(t) W_{ij}(x(t)) \right) \tag{4.4}$$

where the V_j, W_{ij} are smooth vector fields on $\mathbb{R}^n$. We can view the characteristic functions H_j as elements of $\mathcal{G}_\tau(\mathbb{R}^n)$; then system (4.4) makes perfect sense in the algebra $\mathcal{G}_\tau(\mathbb{R}^n)$. We thus will define a *hybrid control system* more generally as a system of the form (4.3) where the vectorfields V and W_i have coefficients in $\mathcal{G}_\tau(\mathbb{R}^n)$. The controls $u_i(t)$ as well as the components $x_\ell(t)$ of the trajectories are sought in the algebra $\mathcal{G}_\tau(\mathbb{R})$ as well. We note that a coordinate-free formulation is possible by employing generalized functions and generalized vectorfields on manifolds (see [8, 11, 25]).

At this stage, the results of Section 3 apply to produce solutions to the Cauchy problem when the controls are fixed. A central notion is controllability, which in our context reads as follows:

Definition 4.2 System (4.3) is called *controllable*, if given $x_0, x_1 \in \tilde{\mathbb{R}}^n$, there is an input curve $u \in (\mathcal{G}_\tau(\mathbb{R}))^m$ such that the output curve $x \in (\mathcal{G}_\tau(\mathbb{R}))^n$ satisfies $x(0) = 0, x(1) = x_1$.

We conclude this sketch of hybrid control theory in the generalized functions setting by a simple example, enjoying the controllability property above.

Consider the one-dimensional hybrid control system

$$\dot{x}(t) = u(t)(1 + H(x(t))) \tag{4.5}$$

where H is the Heaviside function, the characteristic function of the half-line $(0, \infty)$. We model the Heaviside function as an element of $\mathcal{G}_\tau(\mathbb{R}))$ with representative

$$H_\varepsilon(x) = \int_{-\infty}^{|\log \varepsilon| x} \varphi(\xi)\, d\xi$$

where $\varphi \in \mathcal{D}(\mathbb{R})$ satisfies $\int_{-\infty}^{\infty} \varphi(\xi) d\xi = 1$ and $\varphi \geq 0$. Then $\partial_x H$ is of L^∞-log-type and Theorem 3.1 applies, providing unique solutions to the initial value problem when u is a fixed element of $\mathcal{G}_\tau(\mathbb{R})$. Actually, we can compute the desired control u leading from $x_0 \in \tilde{\mathbb{R}}$ to $x_1 \in \tilde{\mathbb{R}}$ at time $t = 1$ explicitly. Let

$$G_\varepsilon(x) = \int_0^x \frac{1}{1 + H_\varepsilon(\xi)} d\xi .$$

Then $x_\varepsilon(t) = G_\varepsilon^{-1}(G_\varepsilon(x_0) + \int_0^t u_\varepsilon(s)ds)$ so that $u_\varepsilon(s) = G_\varepsilon(x_1) - G_\varepsilon(x_0)$ is the desired control, in fact, a generalized constant. Thus problem (4.5) is controllable in the sense of Definition 4.1.

References

[1] H. Balasin: Geodesics for impulsive gravitational waves and the multiplication of distributions. Class. Quant. Grav. **14** (1997), 455 - 462.

[2] H.A. Biagioni: A Non-linear Theory of Generalized Functions. Lect. Notes Math. Vol. 1421, Springer-Verlag, Berlin 1990.

[3] M. Branicky: Universal computation and other capabilities of hybrid and continuous dynamical systems. Theor. Comput. Sci. **138** (1995), 67 - 100.

[4] M. Branicky: Hybrid dynamical systems, or HDS: The ultimate switching experience. In: S. Morse (Ed.), Control Using Logic-Based Switching. Lect. Notes Control Inf. Sci. Vol. 222, Springer, Berlin 1997, 1 - 12.

[5] P. E. Caines, Y. J. Wei: Hierarchical hybrid control systems. In: S. Morse (Ed.), Control Using Logic-Based Switching. Lect. Notes Control Inf. Sci. Vol. 222, Springer, Berlin 1997, 39 - 48.

[6] J. F. Colombeau: Elementary Introduction to New Generalized Functions. North-Holland Math. Studies Vol. 113, North-Holland, Amsterdam 1985.

[7] J. F. Colombeau: Multiplication of Distributions. Lect. Notes Math. Vol. 1532, Springer-Verlag, Berlin 1992.

[8] R. Hermann: C-O-R Generalized Functions, Current Algebras, and Control. Interdisc. Math. Vol. 30, Math Sci. Press, Brookline 1994.

[9] R. Hermann, M.Oberguggenberger: Generalized functions, calculus of variations, and nonlinear ODEs. Preprint 1997.

[10] V. Jurdjevic: Geometric Control Theory. Studies Adv. Math. Vol. 52, Cambridge Univ. Press, Cambridge 1997.

[11] M. Kunzinger: Lie-Transformation Groups in Colombeau Algebras. Doctoral Thesis, Univ. Vienna 1996.

[12] M. Kunzinger: Lie symmetries of differential equations in a generalized functions setting; this volume.

[13] M. Kunzinger, M. Oberguggenberger: Symmetries of Differential Equations in Colombeau Algebras. In: N. H. Ibragimov, F. M. Mahomed (Eds.), Modern Group Analysis VI, New Age Int. Publ., New Delhi 1997, 9 - 20.

[14] M. Kunzinger, M. Oberguggenberger: Group analysis of differential equations and generalized functions. Preprint No. 533, Erwin-Schrödinger-Institut, Vienna 1997.

[15] M. Nedeljkov, S. Pilipović, D. Scarpalézos: The Linear Theory of Colombeau Generalized Functions. Pitman Research Notes Math. Vol. 385, Longman, Harlow 1998.

[16] M. Oberguggenberger: Generalized solutions to semilinear hyperbolic systems. Monatsh. Math. **103** (1987), 133 - 144.

[17] M. Oberguggenberger: Multiplication of Distributions and Application to Partial Differential Equations, Pitman Research Notes Math. Vol. 259, Longman, Harlow (1992).

[18] M. Oberguggenberger: Generalized functions and stochastic processes. In: E. Bolthausen, M. Dozzi, F. Russo (Eds.), Seminar on Stochastic Analysis, Random Fields and Applications, Birkhäuser, Basel 1995, 215-229.

[19] M. Oberguggenberger, M. Kunzinger: Characterization of Colombeau Generalized Functions by their Pointvalues. Math. Nachr., to appear.

[20] E. E. ROSINGER: Non-Linear Partial Differential Equations. An Algebraic View of Generalized Solutions, North-Holland Math. Studies Vol. 164, North-Holland, Amsterdam 1990.

[21] L. S. SCHULMAN: Applications of the propagator for the delta function potential. In: M. C. GUTZWILLER, A. INOMATA, J. R. KLAUDER, L. STREIT (Eds.), Path Integrals from meV to MeV. World Scientific, Singapore 1986.

[22] R. STEINBAUER: Distributional description of impulsive gravitational waves; this volume.

[23] R. STEINBAUER: Geodesics and geodesic deviation for impulsive gravitational waves. J. Math. Phys. **39** (1998), 2201 - 2212.

[24] J. A. VICKERS: Non-linear generalised functions in general relativity, this volume.

[25] J. A. VICKERS, J. P. WILSON: A nonlinear theory of tensor distributions. Preprint No. 566, Erwin-Schrödinger-Institut, Vienna 1998.

M. OBERGUGGENBERGER
Universität Innsbruck, Institut für Mathematik und Geometrie, A-6020 Innsbruck, Austria, E-mail: michael@mat1.uibk.ac.at
R. HERMANN
MIT, Cambridge, Massachusetts 02139, USA

Conservation Laws, Delta-Shocks and Singular Shocks

BARBARA LEE KEYFITZ

1 Modeling with Conserved Quantities

The theory of hyperbolic conservation laws was developed to study phenomena in unsteady compressible gas dynamics, especially the propagation of nonlinear waves, and the formation and evolution of shock discontinuities. Comparison with experiments shows that the theory is reasonably successful. Although a rigorous theory of conservation laws encompassing existence, uniqueness, well-posedness, and asymptotic behavior has been achieved only for equations in a single space variable, numerical simulations which have been carried out in higher dimensions also give good correspondence with the three-dimensional world. During the last fifty years, the use of conservation laws has grown to include many kinds of fluid and continuum models.

It is also the case that conservation laws are used as qualitative models. This is indeed part of their pedigree; perhaps the first instance is Burgers' use [2] of the scalar equation which now bears his name to describe the formation and decay of coherent structures in turbulence. The basis for Burgers' model was the competition between the steepening of a wave due to amplitude-dependent propagation speed and its weakening due to dissipation. This is an ingredient of many nonlinear systems, and has been derived using weakly nonlinear geometric optics, [5]. Other well-known models balance the same nonlinear convective term against dispersion (as in the Korteweg-de Vries equation), or against higher-order effects (for example, the Kuramoto-Sivashinsky equation).

Alternatively, it is possible to study examples with a particular mathematical structure, and then to see what phenomena they might describe. This is quite a different approach from the original one. Some of these prototype conservation laws have quite novel solutions. Our focus here is on equations which lead to delta-shocks and to singular shocks; we will give some background on physical situations which may lead to such equations, and we will discuss delta-shocks and singular shocks themselves.

2 Basic systems and their use in physical modelling

Starting with the same basic equations and constitutive relations, one can derive several systems of conservation laws which differ in their choice of conserved quantities. We give some examples in this section.

The equations for isentropic gas dynamics in Eulerian form,

$$\rho_t + (\rho u)_x = 0 \tag{1}$$

$$(\rho u)_t + (\rho u^2 + A\rho^\gamma)_x = 0\,, \tag{2}$$

for an ideal gas with pressure density relation $p = A\rho^\gamma$, express conservation of mass and momentum. In many applications these are precisely the quantities one expects will be conserved. The parameter γ, which is the ratio of specific heats, is always greater than unity, and the limit $\gamma \to 1$ does not make physical sense in most applications. However, it is noteworthy that the solutions of (1, 2) are qualitatively the same even if γ assumes the unrealistic value of unity. On the other hand, setting $A = 0$ amounts to neglecting the pressure altogether; while this can be done formally, it is not surprising that the equations that result, called the "pressureless gas dynamics equations", have very different solutions from (1, 2).

Now, carrying out the differentiation in (2), subtracting (1) and dividing by ρ results in another equation which may be put in conservation form,

$$u_t + (\tfrac{1}{2}u^2 + q(\rho))_x = 0, \tag{3}$$

using the *specific enthalpy*

$$q(\rho) = \begin{cases} A\frac{\gamma}{\gamma-1}(\rho^{\gamma-1} - 1)\,, & \gamma > 1 \\ A\log\rho\,, & \gamma = 1 \end{cases}\,. \tag{4}$$

Equation (3) represents "conservation of velocity". Although velocity is not normally a conserved quantity, this equation is used both by Shandarin and Zeldovich

[17] and by Rosenau [14] in place of (2). (Indeed, Burgers' original model, while not a description of compressible gas dynamics, is an evolution equation for a velocity variable.) Specifically, Shandarin and Zeldovich use (3) in the case $A = 0$, along with (1), to model the emergence of velocity and density inhomogeneities in a model of the universe which ignores gravitation. (See Section II.B.3, 'Sticky dust', in [17]; although viscosity is mentioned, it is set to zero before solving. For a complete solution of this system, see [15].) When $A = 0$ the systems (1, 2) and (1, 3) have only a single characteristic speed, $\lambda = u$, and are formally of parabolic, not hyperbolic, type. (The sole characteristic is linearly degenerate; in the terminology of conservation laws, $r \cdot \nabla\lambda \equiv 0$, where r is the eigenvector of the Jacobian matrix of the system. However, the equations still admit shock formation.) The variable ρ appears only to linear power in either system, and the solution of the Cauchy problem, for bounded measurable data, is not of classical type: a unique solution exists, at least for some initial data, but the expression for ρ contains a Dirac δ-function. A measure-theoretic justification of solutions of this form can be found in Bouchut's paper [1]; see also E, Rykov and Sinai [15].

We note that the appearance of measures as solutions for problems with L^∞ data could be predicted from the explicit solution $\rho(x,t) = \rho_0(x) - tu_0'(x)$, $u(x,t) = u_0(x)$ of the system

$$\rho_t + u_x = 0, \qquad u_t = 0$$

obtained by linearizing (1, 2) or (1, 3) about $\rho = 1$, $u = 0$. The qualitative nature of the solutions of (1, 2) and (1, 3) is the same for either choice of conserved quantity, though solving the system (1, 2) gives rise to more straightforward analysis in [1]. Zheng [20] makes some further observations about parabolically degenerate nonlinear systems.

Shandarin and Zeldovich [17] consider also a model with gravitationally interacting particles, which can be written, in simplified notation [15], as

$$\begin{aligned} \rho_t + (\rho u)_x &= 0 \\ u_t + (\tfrac{1}{2}u^2)_x &= -g_x \\ g_{xx} &= G\rho, \end{aligned} \tag{5}$$

(G is a gravitational constant). This system is solved by reducing it to

$$h_t + uh_x = 0$$

$$u_t + (u^2/2)_x = -h$$

(with $h = g_x$), which is hyperbolic, with a double characteristic, but has the disadvantage of not being in conservation form. By using Volpert's theory of BV spaces, a weak solution is found to this system [15]. The trivial pressure term again has the effect of eliminating acoustic waves.

A system with a form similar to (5) was advanced by Rosenau [14] to exhibit threshold phenomena in the velocity evolution of ion-acoustic waves. Rosenau's system uses the enthalpy function $q(\rho)$ from equation (4):

$$\begin{aligned} n_t + (nu)_x &= 0 \\ u_t + (\tfrac{1}{2}u^2)_x &= -q_x \\ \beta q_{xx} &= \rho - n\,. \end{aligned} \tag{6}$$

In this model, ρ measures the (charge) density of electrons, n the ion density and β is a small parameter, the normalized Debye length. The analogy with (5) breaks down in solving (6), as will be discussed later.

This selection of equations is not meant to be exhaustive. We close the section by drawing attention to the facts that in numerous models an equation represents conservation of velocity, and in some reductions the equations are not all in conservation form.

3 Delta-shock waves

As mentioned in the preceding section, some systems with a double eigenvalue, whether or not they have an eigenvector deficiency, exhibit solutions which contain δ-functions or other measure-valued functions, even for bounded initial data. In this section, we discuss a related concept.

The term 'delta-shock waves' was coined by Tan and Zhang in a paper which solved the Riemann problem for a model system of conservation laws in which hyperbolicity fails (because of eigenvalue coincidence and an incomplete set of eigenvectors) on a curve in state space. A prototype is the system

$$u_t + (u^2)_x = 0, \qquad v_t + (uv)_x = 0 \tag{7}$$

first discovered in the Ph.D. dissertation of Dennis Korchinski [11], who solved the Riemann problem using δ-functions and provided some justification for the solutions.

A measure-theoretic justification for delta-shocks, along the lines of Bouchut, could be given. In [18], Tan, Zhang and Zheng give a justification based on approximation by viscous equations. The model on which most of their work has focused is a two-dimensional form of the system:

$$u_t + (u^2)_x + (uv)_y = 0 \tag{8}$$

$$v_t + (uv)_x + (v^2)_y = 0, \tag{9}$$

which is formally obtained from the two-dimensional compressible gas dynamics equations by first setting the pressure to zero and ignoring the density equation and then assuming the density to be constant. Thus it is also a sort of zero-pressure gas dynamics, but differs from the Shandarin-Zeldovich model in being hyperbolic for most states. In (7), one of the characteristic speeds is genuinely nonlinear and the other linearly degenerate for all states. Because the δ-functions occur only in the linear family, they can be dealt with by essentially the same means as the degenerate equations of Shandarin and Zeldovich, E, Rykov and Sinai, and Bouchut. The analysis of these models is still incomplete, though, and may be worth pursuing. A feature of systems such as (7) which makes them similar to systems with eigenvector deficiency in an open set is that the eigenvector at points on the coincidence locus in (7) is tangent to the locus. That is, the coincidence locus is a characteristic variable, and this could create a mechanism for resonance. By contrast, some nonstrictly hyperbolic systems with a parabolic degeneracy along a curve which is crossed transversally by the eigenvectors have classical Riemann solutions which do not contain delta-shocks [7, 8]. We return to this later.

4 The effect of conserving different quantities

We now return to Rosenau's model equations. Following Rosenau, we write $n = \rho + \beta q_{xx}$, invert (4), taking $A = 1$, to obtain $\rho = \rho(q)$ and eliminate n from the first equation of (6) to obtain

$$q_t + uq_x + (\gamma + (\gamma - 1)q)u_x = -\frac{\beta}{\rho'}\Big(q_{xxt} + (uq_{xx})_x\Big) \tag{10}$$

when $\gamma \geq 1$. The term multiplied by β is treated by Rosenau as a perturbation and approximated. The order of differentiation makes this a dispersive rather than dissipative term. However, from the viewpoint of conservation laws we now see that, up to a term that is $\mathcal{O}(\beta)$, n in (6) is the same as ρ in (1, 2) or (1, 3) so the system now derived, consisting of (10) and the second equation of (6), could equally well have been found by adding dispersion to (1, 3). It is interesting that precisely this system, written in the form

$$u_t + uu_x = \sigma_x, \qquad \sigma_t + u\sigma_x = k^2 u_x \tag{11}$$

(so σ, which represents elastic stress, takes the place of $-q$) was extensively studied by Colombeau and Le Roux and others as a prototype nonconservative system (see [4, 13] and the references there). We will return to the question of nonconservative systems.

Since (10) is not in conservation form, Rosenau replaces it, when $\gamma = 1$, by a system of conservation laws, while commenting that the physical interpretation is unclear. The system conserves u (the velocity) and $v = \frac{1}{2}u^2 - q = \frac{1}{2}u^2 - \log\rho$, a convex function of velocity and density. After some algebraic manipulation one obtains

$$u_t + (u^2 - v)_x \;=\; 0\,, \tag{12}$$

$$v_t + (\tfrac{1}{3}u^3 - u)_x \;=\; 0\,. \tag{13}$$

(I have omitted the dispersive term which, after some further approximation, appears as βu_{xxt} in the first equation.)

When $\gamma > 1$, it is possible to obtain an analogous pair of conservation laws; although n and u in Rosenau's model represent density and velocity of ions, they also admit an interpretation as density and velocity in an ordinary γ-law gas. The analogue of (12, 13) when $\gamma > 1$ also conserves u and $v = (1 - \frac{1}{2}\gamma)u^2 - q$:

$$u_t + \left(\tfrac{3-\gamma}{2}u^2 - v\right)_x \;=\; 0 \tag{14}$$

$$v_t + \left(\tfrac{(3\gamma-5)(\gamma-2)}{6}u^3 - \gamma u + (\gamma - 1)uv\right)_x \;=\; 0\,. \tag{15}$$

Again, this represents conservation of velocity and a convex function of u and ρ (when $\gamma = 2$ the second conserved quantity is the enthalpy itself).

The second conserved quantity is an "entropy" function in the terminology of conservation laws (a convex function of the variables which is conserved for classical solutions). It differs from the specific energy by a factor ρ, just as velocity differs from momentum.

5 The use of generalized distributions for nonconservative systems and singular shocks

We can think of (1, 2) with $A = 1$ and $\gamma = 1$, (11) and (12, 13) as three different ways of representing the same equation. Although they appear in the context of different, specific physical models (compressible flow, elasticity, and plasmas, respectively), physics is not offered as a justification for the different forms of the equations. (To be more precise, the elastic model is a simplification of a system which lacks a conservation principle; the plasma model, as we saw, simply uses Burgers' equation for the velocity.)

It is known that such systems have inequivalent weak solutions, even though they have the same classical solutions. A standard example, to fix ideas, is the pair

$$u_t + (\tfrac{1}{2}u^2)_x = 0\,, \qquad (\tfrac{1}{2}u^2)_t + (\tfrac{1}{3}u^3)_x = 0$$

for which the shock joining the states $u = 1$ and $u = 0$ has speed $s = \frac{1}{2}$ and $s = \frac{2}{3}$ respectively. In applications to conservation laws of the theory of generalized distributions, one states that these equations hold only in the sense of association, [3, 4, 13]. While, in principle, one should use the correct form of the equation, in fact, as we have seen, equations used in qualitative modelling are transformed rather freely.

As we shall now demonstrate, different forms of the system may have qualitatively different solutions. In fact, neither (11) nor (12, 13) can be solved using classical conservation law theory, and neither has been completely solved. However, a theory for Riemann solutions and shock admissibility exists.

In the case of equations in nonconservative form, a theory was developed by Vol'pert [19] and extended by Le Floch and others (see [12]), based on geometric measure theory. At the same time, Colombeau, Le Roux and others (see [3, 4]) have shown that Colombeau's calculus of generalized distributions can be used to give a direct definition of products such as $u\sigma_x$ appearing in (11). Oberguggenberger [13] has reconciled the two approaches. Either approach includes an explicit recognition that the nonconservative form is incomplete as given, and that further conditions, perhaps originating in physics but expressed as mathematical constraints, must be supplied. This is true for the original system also. However, the original system in conservation form will supply a shock speed given left and right states, or will

determine that no such solution exists. The nonconservative system requires further information in order to construct shocks.
The third form of the equations, system (12, 13), is exactly the model analysed by Keyfitz and Kranzer [10], as an example of a strictly hyperbolic, genuinely nonlinear system for which the Riemann problem has no classical solution when the left and right states are not close together. We found candidates for nonclassical solutions, although the sense in which they satisfy the equation is still unclear. These solutions are also singular; in fact, v contains a δ-function, like the delta-shocks mentioned earlier, and equation (12) is satisfied in the same sense as the equations of Section 3. However, the function u cannot be constructed as a measure, and requires a more refined space. We exhibited an explicit representation, as a space of weighted measures, in [10]. Another model for the solution space is the class $\mathcal{G}_s$ of generalized distributions [3]. Generalized distributions were used in [10] to supply a template for the singular solutions, which were also shown to converge in a Banach space. Unlike the delta-shocks of Section 3, the singular shocks which are needed to solve (12, 13) are truly nonlinear objects which cannot be defined in the context of classical distribution theory.
Comparing the systems (1, 2) with $\gamma = 1$ and (12, 13), we see that both systems are in conservation form, and are equivalent for smooth solutions, yet one has classical shocks and the other does not. One might conjecture that this is due to the unphysical choice $\gamma = 1$. However, such is not the case. As we have shown, the system (14, 15) is a smooth transformation of (1, 2) when $\gamma > 1$, and we have the following result.

PROPOSITION 1 *In the system (14, 15) for* $1 < \gamma < \frac{5}{3}$, *the set of states* (u_R, v_R) *that can be connected to a given state* (u_L, v_L) *by means of a shock forms a compact set.*

This proposition is easily proved by calculating the Hugoniot locus for shocks. Further investigation of this system shows that its structure is similar to that of the model system (12, 13) for which the Riemann problem was solved using singular shocks in [9]. Hence waves of a similar type will exist. There is an important difference, however: equations (14) and (15) represent a physical system only if $v \leq (1 - \frac{1}{2}\gamma)u^2 + \frac{\gamma}{\gamma - 1}$, corresponding to nonnegative densities. If we attempted to construct singular shocks along the lines of [10], they would be limits of approximate solutions, constructed by the method of [9] for example, which contain nonphysical states. (This is not the case when $\gamma = 1$). Thus it would appear that the mathematical solutions which could be constructed in this case do not have a physical interpretation.

6 Conclusions

We have examined a number of model conservation law systems which do not have classical solutions. Since, in one space dimension, strictly hyperbolic systems always have solutions for sufficiently small data, the difficulties arise for one of two reasons: failure of strict hyperbolicity or problems associated with large data. In a number of examples where strict hyperbolicity fails, specifically cases with a parabolic degeneracy on an open set in state space or on a submanifold, the solutions to initial value problems with bounded data contain measure-valued functions. Construction of solutions in such cases requires careful identification of a suitable class of test functions and application of the Radon-Nikodym theorem. In all the cases of this type which have appeared so far, these solutions are essentially linear and do not require multiplication of distributions.

However, in working with conservation law models, one finds two types of problems which go beyond this, and cannot be handled without some type of nonlinear theory. Both have arisen in problems showing a loss of strict hyperbolicity, but this is probably a coincidence. The first type, discussed only briefly in this paper, consists of quasilinear systems which are not in conservation form. Discontinuities will arise from smooth data in such equations, and continuing the solution beyond the time when shocks form requires new theory. The calculus of generalized distributions has proved to be a valuable technique in this problem.

A second class of problem involves some examples of conservation law systems which do not possess classical weak solutions in the large. The example for which the Riemann problem has been extensively studied shows that any candidate for a weak solution will lie in quite a complicated function space. The theory of generalized distributions provides a useful setting for weak solutions, and, in fact, gave one representation for the solution in [10]. We have no convincing explanation for why some forms of a conservation law system have singular shocks while others do not. Zheng and others have suggested that some notion of "lack of hyperbolicity in the large" may be associated with such systems. However, all forms of the equation have the same characteristic structure, and some do not show the same difficulties with large data. Some time ago, Herbert Kranzer and I found a condition, which we refer to as "opposite variation", which guarantees existence of Riemann solutions in the large for a system of two conservation laws [6]. The condition states that the two characteristic speeds should vary in opposite directions along the eigenvector corresponding to one of them. Different forms of a system may have different nonlinear properties; however, as it happens, the system (1, 2) with $\gamma = 1$ does not have the opposite variation property, and yet it still has global Riemann solutions. Nonetheless, some refinement of this concept may operate here. In a later paper which studied some examples of nonstrictly hyperbolic systems with a parabolic degeneracy on a submanifold [8], Kranzer and I noted that the examples we could solve all exhibited opposite variation of the characteristic variables across the parabolic line, and that a prototype example with the complementary property, "same variation", did not have even a local solution. Schaeffer, Schecter and Shearer [16] later solved the Riemann problem for this system, also using singular shocks. As we noted in Section 3, when a characteristic variable coincides with the parabolic line one may expect delta-shocks to figure in the solution. Thus one is tempted to conjecture that singular shocks are a manifestation of a sort of "nonlinear resonance". At the moment, though, we have no clear physical interpretation of the condition on characteristics, and no explanation of why it does not always lead to singular solutions. These are interesting open questions.

In both problems involving nonconservative quasilinear equations and problems where there are no bounded shock solutions, the nonlinear theory of generalized distributions has played a role.

A physical interpretation of singular shocks has not been found, nor is it even clear in what sense the weak solutions satisfy the equation. However, as we have shown in this paper, systems with singular shocks may be closely related (by a nonlinear change of variables) to common, classical systems. Thus, we conclude that conservation laws systems are quite sensitive to the choice of conserved variable, and choices not motivated by physical principles may lead to singular and possibly nonphysical solutions.

7 Acknowledgments

This research was supported in part by the Department of Energy under Grant DE-FG03-94ER25222. It is a pleasure to thank the Erwin Schrödinger Institute in Vienna for their hospitality and the opportunity to visit and pursue this research during the program on *Nonlinear Theory of Generalized Functions* organized by Michael Oberguggenberger.

References

[1] F. BOUCHUT: On zero pressure gas dynamics, in *Advances in Kinetic Theory and Computing, Selected Papers*, Volume 22 of *Advances in Mathematics for Applied Sciences*, pages 171–190, World Scientific 1994.

[2] J. M. BURGERS: A mathematical model illustrating the theory of turbulence, *Advances in Applied Mathematics*, 1:171–199, 1948.

[3] J.-F. COLOMBEAU: Multiplication of distributions, *Bulletin of the American Mathematical Society*, 23:251–268, 1990.

[4] J. F. COLOMBEAU AND A. Y. LE ROUX: Multiplications of distributions in elasticity and hydrodynamics, *Journal of Mathematical Physics*, 29:315–319, 1988.

[5] J. K. HUNTER AND J. B. KELLER: Weakly nonlinear high frequency waves, *Communications on Pure and Applied Mathematics*, XXXVI:547–569, 1983.

[6] B. L. KEYFITZ AND H. C. KRANZER: Existence and uniqueness of entropy solutions to the Riemann problem for hyperbolic systems of two nonlinear conservation laws, *Journal of Differential Equations*, 27:444–476, 1978.

[7] B. L. KEYFITZ AND H. C. KRANZER: A system of hyperbolic conservation laws arising in elasticity theory, *Archive for Rational Mechanics and Analysis*, 72:219–241, 1980.

[8] B. L. KEYFITZ AND H. C. KRANZER: The Riemann problem for a class of hyperbolic conservation laws exhibiting a parabolic degeneracy, *Journal of Differential Equations*, 47:35–65, 1983.

[9] B. L. KEYFITZ AND H. C. KRANZER: A viscous approximation to a system of conservation laws with no classical Riemann solution, in C. Carasso et al., (eds), *Nonlinear Hyperbolic Problems*, pages 185–197. Springer, Lecture Notes in Mathematics 1402, Berlin 1989.

[10] B. L. KEYFITZ AND H. C. KRANZER: Spaces of weighted measures for conservation laws with singular shock solutions, *Journal of Differential Equations*, 118:420–451, 1995.

[11] D. J. KORCHINSKI: *Solution of a Riemann Problem for a* 2×2 *System of Conservation Laws Possessing No Classical Weak Solution*, Ph.D. thesis, Adelphi University, Garden City, New York 1977.

[12] P. LE FLOCH: Entropy weak solutions to nonlinear hyperbolic systems under nonconservative form, *Communications in Partial Differential Equations*, 13:669–727, 1988.

[13] M. OBERGUGGENBERGER: Case study of a nonlinear, nonconservative, nonstrictly hyperbolic system, *Nonlinear Analysis, Theory, Methods and Applications*, 19:53–79, 1992.

[14] P. ROSENAU: Evolution and breaking of ion-acoustic waves, *Physics of Fluids*, 31:1317–1319, 1988.

[15] W. E, Y. G. RYKOV, AND Y. G. SINAI: Generalized variational principles, global weak solutions and behavior with random initial data for systems of conservation laws arising in adhesion particle dynamics, *preprint*.

[16] D. G. SCHAEFFER, S. SCHECTER, AND M. SHEARER: Nonstrictly hyperbolic conservation laws with a parabolic line, *Journal of Differential Equations*, 103:94–126, 1993.

[17] S. F. SHANDARIN AND Y. B. ZELDOVICH: The large-scale structure of the universe: Turbulence, intermittency, structures in a self-gravitating medium, *Reviews of Modern Physics*, 61:185–220, 1989.

[18] D. TAN, T. ZHANG, AND Y.-X. ZHENG: Delta-shock waves as limits of vanishing viscosity for hyperbolic systems of conservation laws, *Journal of Differential Equations*, 112:1–32, 1994.

[19] A. I. VOL'PERT: The spaces BV and quasilinear equations, *Mathematics of the USSR Sbornik*, 2(73):225–267, 1967.

[20] Y. ZHENG: Systems of conservation laws with incomplete sets of eigenvectors everywhere, in *A Volume Dedicated to Professor Xiaqi Ding's 70th Birthday*. to appear, 1998.

BARBARA LEE KEYFITZ
University of Houston, Department of Mathematics, 4800 Calhoun Street, Houston, Texas 77204-3476 , E-mail: keyfitz@uh.edu

Nonlinear Singular Schrödinger-Type Equations

HORST LANGE, MARKUS POPPENBERG, AND HOLGER TEISMANN

1 Some classes of singular nonlinear Schrödinger equations

In this note we present an overview on some classes of nonlinear Schrödinger equations which contain linear and nonlinear terms of various singular types. In this section we give a list of specific problems of this type from mathematical physics including some which can only live in the framework of 'Generalized Functions'. The appearance of singularities on the following systems are manifold; these could either be inherent in the nonlinearity (inasmuch the solution function could appear in the denominator of a certain rational expression), or the system may contain a singular function (e.g., a singular linear or nonlinear potential); or the nonlinearity could be quasi-linear in such a way that non-positive coefficients of the linear principal part appear when the solution function (wave function) is 'frozen' as a multiplier of the linear part; e.g., the 'ellipticity' or 'skew-symmetry' (when multiplying the equation with $-i$) is not valid any more as in the usual Schrödinger equations.
A list of specific examples of the type mentioned above could be as follows:
(i.) **Quasilinear Schrödinger** equations of type

$$iu_t = -\Delta u + V(x)u + f(|u|^2)u + \epsilon\Delta h(|u|^2)h'(|u|^2)u \tag{1}$$

where f and h are real functions on $\mathbb{R}^+; u = u(t,x)$ is the complex wave function; $V(x)$ is a given 'linear' potential (which may itself have singularities in the space variable x). Special cases of (1) are $h(s) = s$ where (1) plays a role in the theory of superfluid films [17] and in certain models of nonlinear ferromagnets in crystal lattices ([1], [16], [18], [26], [29]); in the case $h(s) = s^{1/2}$ system (1) appears in nonlinear quantum theory (see [13], 'quantum potential equation'), whereas for

$h(s) = (1-s)^{1/2}$, (1) describes models of the Heisenberg ferromagnet ([34], [36]); see also references [4], [5], [19], [20], [21], [25]. The mathematical problems which arise in connection with equations of type (1) are due to the fact that the highest (second) order derivatives in (1) have coefficients (depending nonlinearly on the wave function) which destroy the self-adjointness of the Laplacian (or the skew-symmetry of $-i\Delta$ when multiplying the equation by $-i$ to obtain the usual evolution system) or the principal linear part; also, the lower order terms cannot be treated anymore as just some unimportant perturbations of the main part. In Section 2 we apply a Nash-Moser technique developed by M. Poppenberg to prove local well-posedness of the Cauchy problem for (1) in the case $h(s) = s$ (see [23]).

(ii.) Equations of type

$$iu_t = -\Delta u + V(x)u + R(u, \overline{u}, \partial u, \overline{\partial u}, \ldots)u \tag{2}$$

with a rational function R of u and its complex conjugate $\overline{u}$ (and spatial derivatives of these functions) play a fundamental role in **nonlinear quantum mechanics**; they have been derived by Doebner and Goldin et al. (see [8], [9], [10], [38]); R may look e.g., as $iD[\Delta + |\nabla u|^2/|u|^2](D = \text{const})$ which apparently is singular at places where u vanishes.

In Section 3 we explain a treatment of the Cauchy problem for (2) for a large class of nonlinearities of Doebner-Goldin type by various transformation and semi-group methods (see [38]).

(iii.) A nonlinear Schrödinger equation of type

$$iu_t = -D(t)\Delta u + V(x)u + f(|u|^2)u \tag{3}$$

may describe pulse propagation in a nonlinear fiber in quantum optics where the 'dispersion coefficient' D is a function of t with a finite or infinite number of arbitrarily large jumps at certain times ('**comblike dispersion**'); e.g., $D(t)$ could be a delta-function of type $D(t) = \sum_{n\in\mathbf{Z}} \lambda_n \delta(t - a_n)$ (see [7]). The singularity of (3) could either be treated in the framework of 'classical' theory of nonlinear PDEs with discontinuous coefficients or better in the realm of 'Generalized Solutions' lying e.g., in Colombeau algebras; the last approach would take into account the microscopic structure of the singularities of the equation and its solutions.

(iv.) Singular nonlinear Schrödinger systems from the theory of **quantum chaos** (see [2], [15], [35]) may have the form

$$iu_t = -\Delta u + V(x)\delta_T(t)u + f(|u|^2)u \tag{4}$$

where $\delta_T(t)$ is a 'periodic' δ-function of the type $\sum_{n\in\mathbf{Z}} \lambda_n \delta(t-nT)$ (with a fixed 'period' T); (4) is also known as a model for the 'kicked quantum rotator'; furthermore, it plays a role in the theory of wave propagation in disordered media in nonlinear quantum optics with distributed 'gain and loss' of amplification (see [11], [28]). Here also the notion of a classical or distributional solution is not adequate because of the singularities in the equation; a reasonable approach is the use of 'Generalized Solutions' lying in certain Colombeau algebras (see [24]).

(v.) An even stronger singularity appears in the **Kronig-Penney evolutionary nonlinear Schrödinger equation**

$$iu_t = -\Delta u + V(x)u + \delta_T(x)f(|u|^2)u \tag{5}$$

where the singular term is a part of the nonlinearity; (5) describes wave propagation in quasi-periodic crystal lattices where the medium has some additional nonlinear properties; the 'spatially periodic' δ-function in (5) represents small nonlinear dielectric regions which are 'periodically' embedded in a linear medium; the sequence (λ_n) in the definition of δ_T from (iv.) is usually to be taken as a Fibonacci sequence (see [14]). It is quite clear that solutions to (5) (as in the cases (iii.) and (iv.)) cannot be looked for in classes of distributions, only an approach using Colombeau algebras (or a similar one) seems to be natural (see [22]).

2 Nash-Moser methods for the solution of quasi-linear Schrödinger equations

The purpose of this section is to apply an implicit function theorem of Nash-Moser type for Fréchet spaces (cf. [32]) to prove the local well-posedness for some nonlinear Schrödinger type equation. This method has been proposed in [33] in the case of nonlinear parabolic evolution equations and is applied in [23] to nonlinear Schrödinger equations. As basic function space we take the Fréchet space $H^\infty = H^\infty(\mathbb{R}^n)$, the intersection of all Sobolev spaces $H^k(\mathbb{R}^n)$, equipped with its natural norms

$$||u||_k = (\sum_{|\alpha|\leq k} ||\partial^\alpha u||^2_{L^2(\mathbb{R}^n)})^{1/2} \quad , \quad u \in H^\infty \quad , \quad k = 0, 1, 2, \ldots$$

We consider the Cauchy problem for the quasilinear Schrödinger type equation

$$\left\{ \begin{array}{lcl} iu_t & = & -\Delta u \ + \ \kappa(\Delta|u|^2)\cdot u \ , \quad t \in [-a,a] \\ u(0) & = & \phi \end{array} \right. \tag{6}$$

where κ is a real constant and the initial value $\phi \in H^\infty$ is given. We are looking for a solution $u \in C^1([-a,a], H^\infty)$ for some suitable small $a > 0$. We notice that

$$\mathcal{F}(u) = -\Delta u + \kappa(\Delta|u|^2)\cdot u$$

gives rise to a nonlinear smooth map $\mathcal{F} : H^\infty \to H^\infty$ between Fréchet spaces where H^∞ is considered as a real Fréchet space consisting of complex valued functions; for the notation of differentiability in Fréchet spaces we refer to Hamilton [12]. The first directional derivative of $\mathcal{F}$ is given by

$$\mathcal{F}'(u)z \ = \ -\Delta z + 2\kappa \mathrm{Re}(\Delta(\overline{u}z))\cdot u + \kappa(\Delta|u|^2)\cdot z \ , \quad u, z \in H^\infty .$$

The first step to solve (6) is a transformation in the time variable. For a fixed $a > 0$ we put $t = as$ and write $J = [-1,1]$. Then problem (6) is equivalent to

$$\left\{ \begin{array}{lcl} iu_t & = & a\mathcal{F}(u) \ , \quad t \in J \\ u(0) & = & \phi \end{array} \right. \tag{7}$$

Since the interval $J = [-1,1]$ is fixed this problem can be written by means of a mapping Φ involving the unknown parameter a as an argument. We define

$$\Phi(u,a,\phi) \ = \ (iu_t - a\mathcal{F}(u), u(0) - \phi)$$

and get a smooth map $\Phi : C^1(J, H^\infty) \times \mathrm{I\!R} \times H^\infty \to C(J, H^\infty) \times H^\infty$ between Fréchet spaces. Notice that problem (7) is equivalent to the equation

$$\Phi(u,a,\phi) \ = \ (0,0) \ . \tag{8}$$

We want to solve this equation by an implicit function theorem. We first observe that for a fixed initial value ϕ_0 we already have $\Phi(\phi_0, 0, \phi_0) = (0,0)$. In order to show that (8) is solved by $u = \Psi(a,\phi)$ for all (a,ϕ) near $(0,\phi_0)$ and some suitable smooth map Ψ we have to consider the partial derivative $\Phi_u(u,a,\phi)$; if we have found Ψ then we can fix some suitable small $a > 0$ and the inverse transformation in the time variable gives the desired solution of (6) smoothly depending on the initial value ϕ.

It is well known that implicit function theorems in Fréchet spaces of so called Nash-Moser type (cf. [12], [27], [32]) require more technical assumptions than the corresponding ones in Banach spaces. We shall apply the following result proved in [32], 4.2.

Theorem 2.1 *Let E, F, G be Fréchet spaces having the smoothing property $(\mathrm{S}_\Omega)_\mathrm{t}$ and (DN). Let $U \subset E, V \subset F$ be open and $x_0 \in U, y_0 \in V$. Let $f : U \times V \to G$ be a C^2-map, $f = f(x,y)$, and $f(x_0, y_0) = 0$. Assume that the partial derivative $f_y(w) : F \to G$ is bijective for any $w \in U \times V$ and that there is a fixed $d \geq 0$ such that for any k there is a constant $c_k > 0$ such that the following estimates hold:*

$$\begin{array}{llll} (1) & |f'(w)x|_k & \leq c_k\,[w;x]_{d,k} & , \; w \in U \times V, x \in E \times F. \\ (2) & |f_y(w)^{-1}z|_k & \leq c_k\,[w;z]_{d,k} & , \; w \in U \times V, z \in G. \\ (3) & |f''(w)\{x,x\}|_k & \leq c_k\,[w;x,x]_{d,k} & , \; w \in U \times V, x \in E \times F. \end{array}$$

Then there exist open neighbourhoods $U_0 \subset U$ of x_0, $V_0 \subset V$ of y_0 and a C^2-map $h : U_0 \to V_0$ such that $\{(x,y) \in U_0 \times V_0 : f(x,y) = 0\} = \{(x, h(x)) : x \in U_0\}$. In addition, if f is C^n then h is C^n as well, $2 \leq n \leq \infty$.

The smoothing property $(\mathrm{S}_\Omega)_\mathrm{t}$ is defined in [31], property (DN) is taken from [40]. For the purposes of this paper it is enough to notice that all Fréchet spaces appearing in the definition of the map Φ satisfy both properties $(\mathrm{S}_\Omega)_\mathrm{t}$ and (DN) (cf. [33]). The expressions $[w;x]_{m,k}$ and $[w;x,x]_{m,k}$ are weighted multiseminorms forming a part of a general symbolic calculus defined in [32] which has turned out to be useful when dealing with estimates required by Nash-Moser implicit function theorems. For instance, if E, F are Fréchet spaces both equipped with increasing sequences of seminorms $\|\ \|_0 \leq \|\ \|_1 \leq \ldots$, then for $w \in E, x \in F$ and any integers $m, k \geq 0$ the expression $[w;x]_{m,k}$ is defined by

$$[w;x]_{m,k} = \sup\{\|w\|_{m+i_1} \cdots \|w\|_{m+i_r} \|x\|_{m+j} : 0 \leq r \leq k, 0 \leq i_1 + \ldots i_r + j \leq k\}.$$

An application of Theorem 2.1 gives the following result which is proved in a much more general form for nonlinear evolution equations of arbitrary type in [33], 4.4.

Theorem 2.2 *Let $\phi_0 \in H^\infty$ and let W be an open neighbourhood of ϕ_0 in $C^1(J, H^\infty)$. Assume that there exist $\delta > 0, b \geq 0$ and $c_k > 0$ such that problem*

$$\left\{ \begin{array}{lcl} iz_t(t) & = & a\mathcal{F}'(u(t))z(t) + f(t) \;,\; t \in J \\ z(0) & = & g \end{array} \right. \tag{9}$$

has for any $0 < a < \delta$ *and* $u \in W$ *and for every* $f \in C(J, H^\infty)$ *and* $g \in H^\infty$ *a unique solution* $z \in C^1(J, H^\infty)$ *satisfying for all* k *the estimates*

$$\|z\|_k \;\le\; c_k[u;(f,g)]_{b,k}\,. \tag{10}$$

Then there exist $a > 0$ *and an open neighbourhood* U *of* ϕ_0 *in* H^∞ *such that the Cauchy problem* (6) *has for any* $\phi \in U$ *a unique solution* $u \in C^1([-a,a], H^\infty)$. *Moreover, the solution map* $\Psi : U \to C^1([-a,a], H^\infty)$ *is a* C^1*-map.*

For the solution of our particular nonlinear Schrödinger type equation (6) it thus only remains to solve the linear inhomogeneous equation (9) for all data a, u, f, g and to prove the required estimates (10). We have to solve for all small $a > 0$ and $u \in W$ and every $f \in C(J, H^\infty), g \in H^\infty$ the linear problem

$$\begin{cases} iz_t &= -a\Delta z \;+\; a\kappa\Big\{2u\mathrm{Re}\Delta(\overline{u}z) + (\Delta|u|^2)z\Big\} + f(t)\;,\; t \in J \\ z(0) &= g \end{cases} \tag{11}$$

With $z = z_1 + iz_2 = \binom{z_1}{z_2}$, $u = u_1 + iu_2 = \binom{u_1}{u_2}$, $f = f_1 + if_2 = \binom{f_1}{f_2}$, $g = g_1 + ig_2 = \binom{g_1}{g_2}$ we must solve the linear real (2×2)-system

$$\begin{cases} z_t &= a\Big\{M_u\Delta z \;+\; G_u\nabla z \;+\; H_u z\Big\} \;+\; \binom{f_2(t)}{-f_1(t)}\;,\; t \in J \\ z(0) &= \binom{g_1}{g_2} \end{cases} \tag{12}$$

where

$$M_u \;=\; \begin{pmatrix} 2\kappa u_1u_2 & 2\kappa u_2^2 - 1 \\ 1 - 2\kappa u_1^2 & -2\kappa u_1u_2 \end{pmatrix} \;,\quad G_u \;=\; 4\kappa\begin{pmatrix} u_2\nabla u_1 & u_2\nabla u_2 \\ -u_1\nabla u_1 & -u_1\nabla u_2 \end{pmatrix}$$

$$H_u \;=\; \kappa\begin{pmatrix} 2u_2\Delta u_1 & 2u_2\Delta u_2 + \Delta|u|^2 \\ -2u_1\Delta u_1 - \Delta|u|^2 & -2u_1\Delta u_2 \end{pmatrix}.$$

For the solution of (12) we restrict ourselves to the case $n = 1$. In a first step the well-posedness in L^2 of problem (12) is proved by providing a corresponding evolution operator using the theory of semigroups and evolution systems (cf. [37]). A main difficulty is caused by the fact that the corresponding generating operators fail to be dissipative (even for nice functions u). This obstacle is overcome by a transformation procedure introducing a time-dependent metric as follows:

We assume that $2\kappa|\phi_0(x)|^2 \le \delta < 1$ for all x and some $\delta > 0$; this means no restriction if $\kappa < 0$. We then may assume – changing δ if necessary – that $2\kappa|u(t,x)|^2 \le$

$\delta < 1$ for all $u \in W$ and all t, x. Notice that $\det M_u = 1 - 2\kappa|u|^2 \geq 1 - \delta > 0$. Let $N_u = N_u(t,x)$ denote the unique solution of the ordinary initial value problem $(N_u)_x = -\frac{1}{2}M_u^{-1}G_uN_u$ with $N_u(t,0) = \begin{pmatrix} 1 & 0 \\ 0 & 1 \end{pmatrix}$. With the transformation $\tilde{z} = N_u^{-1}z$ problem (12) is equivalent to

$$\left\{ \begin{array}{rcl} \tilde{z}_t & = & aN_u^{-1}M_uN_u\Delta\tilde{z} \ + \ L_u\tilde{z} \ + \ N_u^{-1}\binom{f_2(t)}{-f_1(t)} \\ \tilde{z}(0) & = & N_{u(0)}^{-1}\binom{g_1}{g_2} \end{array} \right. \tag{13}$$

where the first-order space derivatives vanish and L_u is given by

$$L_u \ = \ a\{N_u^{-1}M_u\Delta N_u \ + \ N_u^{-1}(G_u \cdot \nabla N_u) + N_u^{-1}H_uN_u\} + (\partial_t N_u^{-1})N_u \, .$$

The linear operator $B_u = B_u(t) = N_u^{-1}M_uN_u\Delta$ with domain H^2 is shown to be skew adjoint (i.e., $B_u^* = -B_u$) in the Hilbert space $(L^2, < \ , \ >_t)$ where

$$< f, g >_t = \int (W_u(t,x)f(x)) \cdot g(x)\, dx \ , \ W_u = IN_u^{-1}M_u^{-1}N_u \ , \ I = \begin{pmatrix} 0 & -1 \\ 1 & 0 \end{pmatrix} .$$

Therefore the operators $B_u(t)$ are the infinitesimal generators of strongly continuous groups of bounded linear operators in L^2 which are stable in the sense of [37], 4.3. This is enough for the unique solvability in L^2 of the inhomogeneous problem (13). In a second main step of the proof the well-posedness of problem (13) is shown in H^k for any k and the estimates (10) required by the Nash-Moser technique are derived. For the more technical details of the underlying construction we refer to [23]. Finally, the following main result (cf. [23], Theorem 4.6.) is obtained.

Theorem 2.3 *Let $\kappa \in \mathbb{R}$, $n = 1$ and $\phi_0 \in H^\infty$ so that $2\kappa|\phi(x)|^2 < 1$ for all $x \in \mathbb{R}$. Then there exist $a > 0$ and an open neighbourhood U of ϕ_0 in H^∞ such that problem (6) admits for any $\phi \in U$ a unique solution $u \in C^1([-a,a], H^\infty)$. Moreover, the induced solution map $U \to C^1([-a,a], H^\infty)$ is a C^1-map between Fréchet spaces.*

The carefully proved estimates for the higher order derivatives of the solutions of the linearized problem (12) immediately imply a result on almost global solvability as an application of an inverse function theorem of Nash-Moser type (cf. [23], 4.7.).

Theorem 2.4 *Let $\kappa \in \mathbb{R}$ and $n = 1$. Then for any (large) $a > 0$ there exists an integer $s \geq 0$ and a number $\epsilon > 0$ so that for any $\phi \in H^\infty$ satisfying $||\phi||_s \leq \epsilon$ there is a unique solution $u \in C^1([-a,a], H^\infty)$ of problem (6).*

3 Doebner-Goldin equations

Origin of the equations. In [8] H.-D. Doebner and G.A. Goldin derived the following nonlinear Schrödinger equation, now known as Doebner-Goldin equation

$$i\partial_t u + \tfrac{1}{2}\Delta u = iD\Big(\Delta u + \frac{|\nabla u|^2}{|u|^2}u\Big) \tag{DGE}$$

($u : I \times \mathbb{R}^n \longrightarrow \mathbb{C}$, $(t,x) \mapsto u(t,x)$, denotes the unknown function ('wave function'), $i = \sqrt{-1}$ the imaginary unit and D a real constant) and proposed it as a model for nonlinear quantum mechanics[1]. Later they generalized (DGE) to a seven-parameter family of equations [9, 10]

$$i\partial_t u + \tfrac{1}{2}\,\Delta u = i\tfrac{D}{2}R_2[u]\,u + D'\textstyle\sum_{j=1}^{5} c_j R_j[u]\,u. \tag{GDGE}$$

Here D and D' are real constants, $c_1, \dots, c_5$ are real system parameters and the nonlinear functionals $R_j[.]$ are given by

$$R_1[u] := \frac{\nabla\cdot j}{\rho},\ R_2[u] := \frac{\Delta\rho}{\rho},\ R_3[u] := \frac{(j)^2}{\rho^2},\ R_4[u] := \frac{j\cdot\nabla\rho}{\rho^2},\ R_5[u] := \frac{(\nabla\rho)^2}{\rho^2}$$

where $\rho := |u|^2$ and $j := \mathrm{Im}(\bar{u}\nabla u)$.[2] The starting point for the derivation of (GDGE) is a description of the *kinematics* of quantum systems based on the representation theory of a certain infinite dimensional Lie algebra. The introduction of *dynamics* into the framework of this 'Quantum Borel Kinematics' leads to a Fokker-Planck type equation

$$\partial_t\rho + \nabla\cdot j = D\Delta\rho \tag{FPE}$$

for the probability density ρ and the current density j. The family (GDGE) is a subclass of the class of evolution equations $i\partial_t u + \frac{1}{2}\Delta u = R[u]u$ whose solutions u fulfill (FPE)[3].

---

[1]In this note we consider the free equation only, i.e., we assume that there is no external potential. Furthermore, Planck's constant and the particle mass are set to one.

[2]Easy (formal) calculations show that (GDGE) reduces to (DGE) by setting $D'c_1 = -D$, $c_2 = \ldots = c_5 = 0$ and that (GDGE) can be rewritten by introducing new real parameters ν_1, ν_2, $\mu_1, \dots, \mu_5$ such that only the nonlinear functionals $R_j[.]$ appear: $i\partial_t u = i\sum_{j=1}^{2}\nu_j R_j[u]\,u + \sum_{j=1}^{5}\mu_j R_j[u]\,u$.

[3]Since this condition determines the imaginary part of $R[.]$ only, one imposes further restrictions on $R[.]$, based on physical considerations, which eventually lead to (GDGE).

Cauchy problem. Since (GDGE) is expected to be a fundamental equation for a nonlinear quantum theory, there should be a satisfactory existence theory for the initial value problem (Cauchy problem) for (GDGE). Moreover, this existence theory should be set up in an 'L^2-environment' [4]. Hence, we shall study the following problem:

$$i\partial_t u + \tfrac{1}{2}\,\Delta u = i\tfrac{D}{2}R_2[u]\,u + D'\sum_{j=1}^{5} c_j R_j[u]\,u, \quad u(0) = \phi \in L^2, \quad u \in C(I, L^2).$$

In fact, it is exactly the L^2 condition that causes the main mathematical difficulty in dealing with (GDGE)[5].

Logarithmic structure. In order to make sure that the singular nonlinear functionals $R_j[.]$ are well-defined we look for solutions without zeros. This leads to the ansatz $u = e^v$ associated with a certain logarithmic character of (GDGE). By this we mean the fact that the equation for v is *semilinear*, i.e., contains only first-order space derivatives in a nonlinear way, whereas the original equation (GDGE) is in general[6] *fully nonlinear*. The system for for $r := \mathrm{Re}(v)$ and $s := \mathrm{Im}(v)$ reads as

$$\left.\begin{array}{rcl} \partial_t r - 2\nu_2\Delta r - \nu_1\Delta s &=& \mathrm{Im}(a_2)(\nabla r)^2 + \mathrm{Im}(a_1)(\nabla r)\cdot(\nabla s) \\ \partial_t s + 2\mu_2\Delta r + \mu_1\Delta s &=& -\mathrm{Re}(a_2)(\nabla r)^2 - \mu_3(\nabla s)^2 - \mathrm{Re}(a_1)(\nabla r)\cdot(\nabla s), \end{array}\right\} \quad (14)$$

where $a_1 := 2(i\nu_1 + \mu_1 + \mu_4)$, $a_2 := 4(i\nu_2 + \mu_2 + \mu_5)$. For this system symmetry properties, classification by nonlinear gauge transformations, special solutions for subclasses, etc. have been investigated by several authors (see [10, 30] and the references therein). The main problem is to fulfill the L^2 condition stated above. For instance, if we construct solutions v of (14) by means of L^p methods ($v(t,.) \in L^p$ for every t) the function $u = e^v$ cannot be an L^2 solution of (GDGE) (although it is, by construction, a solution). This is a consequence of the following lemma.

Lemma 3.1 *The set* $\{e^v \in L^2 \mid v \in L^p\}$ *is empty for every* $p \in [1, \infty]$.

[4]This is also necessary to guarantee the consistency with the representation theoretical origin of the equation.

[5]It should be emphasized that we are considering the whole space problem here, i.e., the spatial variable x varies in $\mathbb{R}^n$. The difficulty in constructing an L^2 theory for (GDGE) would probably be less severe if we considered a domain with finite measure.

[6]This means: without any restrictions for the system parameters.

Subsequently we present two different approaches to (GDGE), each of them providing a (partly) successful treatment of its singularities.

First method.

The equation for ∇v. By differentiating (14) with respect to x we get a semilinear evolutionary system for the $2n$ real-valued component functions of the gradient $V := \nabla v$ of v. This system can easily be read from (14) and it can be given the form of the abstract evolution equation

$$\partial_t V + \mathcal{A}V = \nabla F(V), \quad V(0) = \Psi, \tag{15}$$

where Ψ is given by $\frac{1}{\phi}\nabla\phi$ (a suitable class of initial data ϕ is $U[L^p]$ defined below). Under a certain parabolicity assumption for the operator $\partial_t + \mathcal{A}$ equation (15) is solvable in the L^p sense. For reasons which will be explained later we need a slightly more general existence and uniqueness theory involving the Lorentz spaces $L^{n,\infty}$.[7]

Theorem 3.1 *Assume that the determinant and the trace of the matrix* $\begin{pmatrix} 2\nu_2 & \nu_1 \\ -2\mu_2 & -\mu_1 \end{pmatrix}$ *are positive numbers.*

(i) *For any $\Psi \in L^p$ with $p > n$ there is a unique maximal solution $V \in C(I, L^p)$ of (15), where $I := [\,0, T(\Psi)\,[$.*

(ii) *There is a $\delta > 0$, depending on the system parameters only, such that for any $\Psi \in L^{n,\infty}$ with $\|\Psi\|_{L^{n,\infty}} < \delta$ there exists a unique solution $V \in L^\infty(0,\infty; L^{n,\infty}) \cap C(\,]\,0,\infty\,[\,, L^{n,\infty})$ of (15).*

Remark. (a) In (ii) the solution V is continuous as a function with values in $L^{p_0} + L^{p_1}$ (with suitable p_0, p_1). Therefore the assertion $V(0) = \Psi$ makes sense.
(b) The solutions are in fact C^∞ functions on $I \times \mathbb{R}^n$, I being $]\,0, T(\Psi)\,[$ or $]\,0,\infty\,[$.
(c) In (ii) the estimates $\|V(t)\|_{L^p} \le Ct^{-(1/2-n/2p)}$ and $\|\nabla V(t)\|_{L^p} \le Ct^{-1/2-(1/2-n/2p)}$ hold for every $t > 0$ and p with $2p > n$.

The proof of the theorem follows the proofs given in the literature on the Navier-Stokes system (see e. g. [6] and the literature therein).

Solutions for (GDGE). For the rest of this note, let the parabolicity assumption of the theorem be satisfied. For $\Psi \in L^p$ with $p > n$ let V be the solution of (15)

[7] A measurable function $f : \mathbb{R}^n \longrightarrow \mathbb{C}$ belongs by definition to $L^{p,\infty}$ if it satisfies $m_f(t) := \mu(\{x \mid |f(x)| > t\}) \le Ct^{-p}$ for all $t > 0$; $\|f\|_{L^{p,\infty}} := \sup_{t>0} t^{1/p} m_f(t)$ is a quasi-norm for $L^{p,\infty}$.

guaranteed by the theorem. If Ψ is the gradient of a function ψ it can be proved that there is a unique function v such that $\nabla v = V$ and $v(0) = \psi$. Moreover, v is a solution of (14) and therefore $u := e^v$ a solution of (GDGE) such that $u(0) = e^\psi$. Finally, $e^\psi \in L^2$ implies $u \in C([0, T(\Psi)[, L^2)$. These considerations suggest to define the set

$$U[X] := \{e^v \in L^2 \mid v \in L^1_{loc}, \quad \nabla v \in X\},$$

where $X \subset L^1_{loc}$ is a Banach space.

Theorem 3.2 *For any $\phi \in U[L^p]$ with $p > n$ there is a unique maximal solution $u \in C([0, T[, L^2)$ of (GDGE) such that $u(0) = \phi$ and $u(t, .) \in U[L^p]$ for all t.*

The space $U[X]$. The point left open is the determination of the "size" of the set $U[X]$ which depends on the choice of the space X.

Lemma 3.2 *Let $p \in\ ]1, \infty]$.*

(i) *$U[L^p] = \emptyset$ if $p \leq n$ ($p < 2$ if $n = 2$).*

(ii) *$U[L^p \cap L^{n,\infty}]$ is a dense subset of L^2 if $p > n$.*

Remark. (a) The lemma explains our interest in an existence and uniqueness theory for (15) that allows initial data Ψ taken from $L^{n,\infty}$.
(b) Assertion (ii) remains true if $U[L^p \cap L^{n,\infty}]$ is replaced by $U[L^p \cap L^{n,\infty}] \cap H^\infty$.

On a difficulty regarding global existence. Since (15) is globally solvable for initial data with small $L^{n,\infty}$ quasi-norms it is enticing to think that (GDGE), too, is globally solvable under suitable conditions on the initial data and the system parameter. Unfortunately, according to the following lemma, such conditions are by no means easy to find; it seems to be more likely that they do not exist[8].

Lemma 3.3 *For any $n \geq 3$ there is a constant $\sigma_n > 0$ such that for every $(u, V) \in L^2 \times L^{n,\infty}$, $u \neq 0$, satisfying $\nabla u - uV = 0$ the estimate $\|V\|_{L^{n,\infty}} \geq \sigma_n$ holds.*

Second method. The idea is to remove the singularities of (GDGE) by replacing the singular transformation $V = \frac{1}{u}\nabla u$ by the implicit relation $\nabla u - uV = 0$.

Doebner-Goldin system. To begin with, let us remark that there are (bilinear)

[8]In order to apply Theorem (ii) the estimate $\delta < \sigma_n$ would be needed, where σ_n is the constant appearing in the lemma. Notice that every $u = e^v \in U[L^{n,\infty}]$ satisfies equation (17) with $V = \nabla v$.

expressions $\mathbf{R}_j[V]$ such that $R_j[u] = \mathbf{R}_j[V]$ ($j \in \{1, \dots, 5\}$) if $V = \frac{1}{u}\nabla u$; set $\mathbf{R}[V] := i\frac{D}{2}\mathbf{R}_2[V] + D' \sum_{j=1}^{5} c_j \mathbf{R}_j[V]$. If u is a solution of (GDGE) we get

$$i\partial_t u + \tfrac{1}{2}\,\Delta u = \mathbf{R}[V]u. \tag{16}$$

Definition. (i) We call (u, V) a *DG-pair* if u is an L^2 solution of (16) and V is an L^p solution of (15).

(ii) u is called a *solution of (GDGE)* if there is a function V such that (u, V) is a DG-pair satisfying

$$\nabla u - uV = 0. \tag{17}$$

Existence of DG-pairs. The system (15), (16) is partially decoupled because u does not appear in (15). In order to construct a DG-pair (u, V) one first constructs a solution V of (15) by applying Theorem 3.1. Since V is fixed now, (16) is essentially a linear Schrödinger equation with a "potential" $\mathbf{R}[V]$ which can have a **singular time-dependence** of the type $\|\mathbf{R}[V]\|_{L^p} \leq Ct^{-(1-n/2p)}$. The existence theory for linear Schrödinger equations with time-dependent potentials developed by Yajima [41, 42] can be extended such that potentials of this singular type are allowed. This extension is achieved by generalizing Strichartz estimates to certain Lorentz spaces [39].

Theorem 3.3 *Let X be $L^{n,\infty}$ or L^p for some $p > n$. For any $(\phi, \Psi) \in L^2 \times X$ there is a unique DG-pair (u, V) defined on a maximal interval $[\,0, T(\Psi)\,[$.*

Solutions of (GDGE). The key result in the existence theory for (GDGE) is the following "persistence property".[9]

Theorem 3.4 *There is a constant $\delta > 0$ such that for every $(\phi, \Psi) \in H^1 \times L^{n,\infty}$ with $\|\Psi\|_{L^{n,\infty}} < \delta$ and every DG-pair (u, V) with $(u(0), V(0)) = (\phi, \Psi)$ the following holds: $\nabla\phi - \phi\Psi = 0$ implies $\nabla u(t) - u(t)V(t) = 0$ for every $t > 0$.*[10]

If we combine Theorems 3.3 and 3.4 we obtain the solvability of (GDGE) for initial data ϕ taken from the set $\{\phi \in H^1 \mid \text{there is a unique } \Psi \in L^{n,\infty} \text{ s.t. } \nabla\phi - \phi\Psi = 0\}$ which contains $U[L^{n,\infty}] \cap H^1$ (dense subset of L^2 by Lemma 3.2).

[9] We state the result in the special case $\Psi \in L^{n,\infty}$ which is not covered by the first method.
[10] The existence interval of (u, V) is $[\,0, \infty\,[$.

References

[1] BASS, F.G., NASANOV, N.N.: Nonlinear electromagnetic spin waves. Physics Reports **189** (1990), 165-223.

[2] BENVENUTO, F., CASATI, G., PIKOVSKY, A.S., SHEPELYANSKY, D.L.: Manifestations of classical and quantum chaos in nonlinear wave propagation. Physical Review A **44** (1991), R3423-R3426.

[3] DE BOUARD, A., HAYASHI, N., SAUT, J-C.: Global existence of small solutions to a relativistic nonlinear Schrödinger equation. Comm. Math. Phys. **189** (1997), 73-105.

[4] BRÜLL, L., LANGE, H.: Stationary and solitary wave type solutions of singular nonlinear Schrödinger equations. Math. Methods Appl. Sciences **8** (1986), 559-575.

[5] BRÜLL, L., LANGE, H.: Solitary waves for quasilinear Schrödinger equations. Expositiones Math. **4** (1986), 279-288.

[6] CANNONE, M.: Ondelettes, Paraproduits et Navier-Stokes, Diderot Editeur, 1995.

[7] CHERNIKOV, S.V., TAYLOR, J.R.: Comblike dispersion-profiled fiber for soliton pulse train generation. Optics Letters **19** (1994), 539-541.

[8] DOEBNER, H.-D., GOLDIN, G.A.: On a general nonlinear Schrödinger equation admitting diffusion currents. Phys. Letters A **162** (1992), 397-401.

[9] DOEBNER, H.-D., GOLDIN, G.A.: Properties of nonlinear Schrödinger equations associated with diffeomorphism group representations. J. Phys. A **27** (1994), 1771-1780.

[10] DOEBNER, H.-D., GOLDIN, G.A.:, Introducing nonlinear gauge transformations in a family of nonlinear Schrödinger equations. Phys. Review A **54** (1996), 3764-3771.

[11] ELGIN, J.N., KELLY, J.M.: Spectral modulation and growth of resonant modes associated with periodically amplified solitons. Optics Letters **18** (1993), 787-789.

[12] HAMILTON, R.S.: The inverse function theorem of Nash and Moser. Bull. Amer. Math. Soc. **7** (1982), 65-222.

[13] HASSE, R.S.: A general method for the solution of nonlinear soliton and kink Schrödinger equations. Z. Physik B (1980), 83-87.

[14] HENNIG, D., TSIRONIS, G.P., MOLINA, M.I., GABRIEL, H.: A nonlinear quasiperiodic Kronig-Penney model. Physics Letters A **190** (1994), 259-263.

[15] IZRAILEV, F.M., SHEPELYANSKY, D.L.: Quantum resonance for a rotor in a nonlinear periodic field. Soviet Phys. Dokl. **24** (1979), 996-998.

[16] KOSEVICH, A.M., IVANOV, B.A., KOVALEV, A.S.: Magnetic solitons. Physics Reports **194** (1990), 117-238.

[17] KURIHARA, S.: Large-amplitude quasi-solitons in superfluid films. J. Phys. Soc. Japan **50** (1981), 3262-3267.

[18] LAEDKE, E.W., SPATSCHEK, K.H., STENFLO, L.: Evolution theorem for a class of perturbed envelope soliton solutions. J. Math. Phys. **24** (1983), 2764-2769.

[19] LANGE, H.: Derivative nonlinear Schrödinger equations. Lecture Notes Pure Applied Mathematics **155** (1994), 307-318.

[20] LANGE, H.: Nonexistence of solutions to singular nonlinear Schrödinger equations. Mathematische Modellierung. Ed.: A. Kempf (Paris), Hamburg 1986, 114-124.

[21] LANGE, H.: Asymptotical behavior of solutions to quasilinear Schrödinger equations. In: Proc. Trends in Applications of Mathematics to Mechanics, Wassenaar 1987. Eds.: J.F. Besseling, W. Eckhaus, Berlin 1988, 98-107.

[22] LANGE, H., OBERGUGGENBERGER, M.: On nonlinear Schrödinger equations with a nonlinear δ-function term. In preparation, Math. Institut, Univ. Köln, 1998.

[23] LANGE, H., POPPENBERG, M., TEISMANN, H.: Nash-Moser methods for the solution of quasilinear Schrödinger equations, preprint 1997.

[24] LANGE, H., POPPENBERG, M., TEISMANN, H.: On a nonlinear Schrödinger equation with a periodically driven δ-function-like potential. In preparation, Math. Institut, Univ. Köln, 1998.

[25] LANGE, H., TOOMIRE, B., ZWEIFEL, P.F.: Time-dependent dissipation in nonlinear Schrödinger systems. J. Math. Phys. **36** (1995), 1274-1283.

[26] LITVAK, A.G., SERGEEV, A.M.: One dimensional collapse of plasma waves. JETP Letters **27** (1978), 517-520.

[27] LOJASIEWICZ JR., S., ZEHNDER, E.: An inverse function theorem in Fréchet spaces. J. Funct. Anal. **33** (1979), 165-174.

[28] MALOMED, B.A.: Strong periodic amplification of solitons in a lossy optical fiber: analytical results. J. Optical Soc. Amer. B **11** (1994), 1261-1266.

[29] NAKAMURA, A.: Damping and modification of exciton solitary waves. J. Phys. Soc. Japan **42** (1977), 1824-1835.

[30] NATTERMANN, P.: Dynamics in Borel quantization: Nonlinear Schrödinger equations vs. master equations. Inaugural-Dissertation, Math.-Nat. Fakultät, Technische Universität Clausthal, 1997.

[31] POPPENBERG, M.: A smoothing property for Fréchet spaces. J. Funct. Anal. **142** (1996), 193-210.

[32] POPPENBERG, M.: An inverse function theorem for Fréchet spaces satisfying a smoothing property and (DN). To appear in Math. Nachr.

[33] POPPENBERG, M.: Smooth solutions for a class of nonlinear parabolic evolution equations, preprint 1997.

[34] QUISPEL, G.R.W., CAPEL, H.W.: Equation of motion for the Heisenberg spin chain. Physica **110** A (1982), 41-80.

[35] SHEPELYANSKY, D.L.: Delocalization of quantum chaos by weak nonlinearity. Phys. Review Letters **70** (1993), 1787-1790.

[36] TAKENO, S., HOMMA, S.: Classical planar Heisenberg ferromagnet, complex scalar fields and nonlinear excitations. Progr. Theoret. Physics **65** (1981), 172-189.

[37] TANABE, H.: Equations of Evolution. London, Melbourne, San Francisco, Pitman 1979.

[38] TEISMANN, H.: Das Cauchy-Problem für die Doebner-Goldin-Gleichung. Inaugural-Dissertation, Math.-Nat. Fakultät, Univ. Köln, 1997.

[39] TEISMANN, H.: Schrödinger equations with singular time-dependent potentials, preprint 1998.

[40] VOGT, D.: Charakterisierung der Unterräume von s. Math. Z. **155** (1977), 109-117.

[41] YAJIMA, K.: Existence of solutions for Schrödinger evolution equations. Comm. Math. Phys. **110** (1987), 415-426.

[42] YAJIMA, K.: Schrödinger evolution equations with magnetic fields. J. d'Analyse Math. **56** (1991), 29-76.

H. LANGE
Mathematisches Institut, Universität zu Köln, Weyertal 86–90, 50931 Köln, E-mail: lange@mi.uni-koeln.de
M. POPPENBERG
Fachbereich Mathematik, Universität Dortmund, D-44221 Dortmund, Germany, E-mail: poppenberg@math.uni-dortmund.de
H. TEISMANN
Department of Mathematics and Statistics, University of Victoria, P.O. BOX 3045, Victoria, B.C., Canada, E-mail: teismann@math.uvic.ca

Non-Analytic Solutions of Nonlinear Wave Models

YI A. LI, PETER J. OLVER, AND PHILIP ROSENAU

1 Introduction

Historically, the mathematical study of physical phenomena has proceeded in three phases. The introductory phase relies on linear models, and is a necessary prerequisite to the deeper understanding of the actual nonlinear regime. Thus, for instance, the unidirectional linear wave equation $u_t + u_x = 0$ provides the prototype of linear hyperbolic waves. Any initial data propagates completely unchanged with speed $c = 1$. (For simplicity, we adopt units in which all physical parameters are 1 throughout this survey.) Restricting our attention to conservative systems, the next interesting case is the dispersive wave equation $u_t + u_x + u_{xxx} = 0$. Since different wave numbers propagate at different speeds, localized initial data will disperse in time, albeit with conservation of energy. In certain physical regimes, these simple linear models provide the satisfactory explanations of the physically observed wave propagation.

Inspired by the dramatic observations of J. Scott Russell, [26], Boussinesq initiated a systematic derivation of nonlinear wave models that could describe the weakly nonlinear regime, which is the second of the three historical phases. The most celebrated of Boussinesq's models is the equation

$$u_t + u_x + (u^2)_x + u_{xxx} = 0, \tag{1}$$

which initially appears in both [4], eq. (30), p. 77 and [5], eqs. (283, 291). This well-studied equation, now known as Korteweg-de Vries (KdV) equation in honor of its rediscovery, [14], over two decades later, models the weakly nonlinear regime in which nonlinear effects are balanced by dispersion. As recognized by Boussinesq, its solitary wave solutions are described by analytic sech^2 functions — that have exponentially decaying tails. The discovery of the inverse scattering granted to the KdV a status of a paradigm, for it was the first equation shown to be completely integrable, and so a complete analysis can be effected, [9]. Its solitary waves are nowadays referred to as "solitons" since they interact cleanly, except for a phase shift. It admits two distinct, compatible Hamiltonian structures, and hence an infinite hierarchy of local higher order symmetries and associated conservation laws, [20]. Moreover, it is now understood that the KdV and its integrable cousins are not just esoteric equations describing particular phenomena, but ubiquitous descriptions of a weakly nonlinear regime wherein linear dispersion counterbalances the governing nonlinearity to form steadily propagating patterns.

Numerical studies indicate that much of the basic weakly nonlinear phenomenology associated with the KdV and other integrable models is shared with many non-integrable models in the same physical regime. A good example is the BBM or regularized long wave model, [2],

$$u_t + u_x + uu_x - u_{xxt} = 0. \tag{2}$$

Like the KdV, (2) also admits analytic solitary wave solutions, and localized initial data is observed to break up into a finite collection of solitary waves, [12]. Unlike the KdV, (2) has solitary waves of both positive and negative elevation; the positive waves interact almost cleanly — there is a very small but noticeable dispersive effect, [3, 19]. (On the other hand, collisions between positive and negative waves are quite dramatically inelastic, [25].) We are still a long way from being able to establish rigorously these numerical observations yet, one might venture that the basic phenomenology of wave mechanics in the weakly nonlinear regime is fairly well understood.

The final, and most difficult, phase in the modeling process is the transition to fully nonlinear models. Linear and weakly nonlinear models will only predict the physical behavior of weak disturbances — indeed, they are typically derived by rescaling the full physical system to include one or more small parameters, expanding in a power series, then truncating to eliminate some or all of the higher order terms. However, in a genuinely nonlinear regime, nonlinearity plays a dominant role rather than being a higher order correction, and so will introduce new complications, new phenomena, and require new analytical techniques. Weakly nonlinear models cannot describe the fully nonlinear, real world phenomena such as wave breaking, shocks, waves of maximal height, [27], large amplitude disturbances, and so on. The problem of how one derives realistic nonlinear models for such highly nonlinear effects is of fundamental importance but has not as yet received the attention it deserves.

One type of fully nonlinear phenomena — shock waves in gas dynamics and similar first order hyperbolic systems — already has a long history. The simplest case is the inviscid Burgers' — or dispersionless KdV — or Riemann equation $u_t + (u^2)_x = 0$, obtained by replacing the linear convective term in the unidirectional linear model by a fully nonlinear version. The hodograph transformation $(x, t, u) \mapsto (u, t, x)$ will linearize the equation, but the analytic solutions to the transformed equation revert to multiply-valued solutions of the Riemann equation. For the gas dynamics and continuum mechanics applications, a physical wave cannot be represented by a multiply-valued solution, and one must replace it with a discontinuous, nonanalytic shock wave, the placement of the shock governed by a suitable entropy condition, [29]. On the other hand, if one uses this equation to model surface waves, then multiply-valued solutions are physically acceptable — the nonphysical solutions are when the curve crosses itself — and the weak characterization is no longer applicable. Therefore, *the analytical characterization of acceptable solutions depends on the physics underlying the model.*

Just as linear dispersion dramatically alters the hyperbolic traveling wave, so one expects nonlinear dispersion to be capable of causing deep qualitative changes in the nature of dispersive phenomena. Mimicking the passage from the linear hyperbolic equation to the nonlinear Riemann equation, one is naturally led to introduce a simple nonlinearly dispersive model of the form

$$u_t \pm (u^2)_x + (u^2)_{xxx} = 0, \tag{3$\pm$}$$

known as the $K(2,2)$ equation. The $\pm$ sign plays an important role in the structure, and we shall refer to (3+) as the focusing model, while (3−) is the defocusing version. Unlike the KdV model, (1), the $K(2,2)$ equation will model dispersive waves in the fully nonlinear regime. It is one of a large class of simple nonlinearly dispersive wave models, incorporating different powers of u in the nonlinear terms, first proposed in [22, 24]. We shall use (3) as our initial paradigm for the structures that may appear in the fully nonlinear regime, and then indicate additional, stranger denizens, of the nonlinear *terra incognita*.

At this point, it is important to distinguish between a fully nonlinear model, in which the highest order terms in the partial differential equation are nonlinear (typically quasilinear), and the points in phase space where the nonlinearity is actually manifested. These points will be determined by the complete degeneration of the symbol of the partial differential equation. For example, the fully nonlinear $K(2,2)$ equation has degenerate symbol when $u = 0$, and these will be the points of genuine nonlinearity in this case.

A crucial difference between the linear or weakly nonlinear models, versus the fully nonlinear models like (3) is that the latter admit *nonanalytic solutions*. However, unlike the abrupt discontinuities of shock waves in the first order hyperbolic regime, these discontinuities are more subtle, typically occurring at higher order derivatives; moreover, standard existence theorems, e.g., Cauchy–Kovalevskaya, indicate that one can only expect such singularities at the points of genuine nonlinearity in the phase space. This is in contrast to the nonanalytic shock solutions of nonlinear hyperbolic models, which can occur at any point in the phase space, but whose implementation requires additional physical assumptions, e.g., entropy conditions. The first observed manifestations of nonanalyticity were compactons — solitary wave solutions of compact support, [24], — and peakons — solitary wave solutions having peaks or corners, [6]. Both types of solutions appear in the basic $K(2,2)$ equation, as well as other nonlinear models of this type.

How does one characterize such nonclassical solutions? Motivated by the entropy characterization of shock waves, the first idea that might come to mind is to view them as weak solutions in the traditional sense, [29]. Although this is a valid formulation, and typically does serve to distinguish them, both compactons and peakons are solutions in a considerably stronger (although not quite classical) sense. In fact, they are piecewise analytic, satisfying the equation in a classical sense away from the points of genuine nonlinearity; moreover, each term (or, possibly, certain combinations of terms) in the equation is well-defined at the singularities. They remain solutions even when subjected to nonlinear changes of variables — a process that has disastrous consequences for the weak formulation. Consequently, one does not need to introduce an artificial (or physical) entropy condition to prescribe the type of singularity. In [17, 18], we proposed the term *pseudo-classical solution* to distinguish such non-analytic solutions.

Yet another characterization is based on the observation that these solutions can be realized as the limits of classical, analytic solutions. Note that (3) is genuinely nonlinear when the undisturbed (asymptotic) height of the solution is $\lim_{|x|\to\infty} u = 0$, but becomes weakly nonlinear at nonzero amplitudes. Therefore, the solutions with $\lim_{|x|\to\infty} u = \alpha \neq 0$ are analytic, and, in favorable situations, will converge to a solution as $\alpha \to 0$. Alternatively, one can replace u by $u + \alpha$ in (3) — the parameter α then measures linear dispersion. The convergence of analytic solitary wave solutions under vanishing linear dispersion or asymptotic height distinguishes those non-analytic solitary wave solutions which are genuine pseudo-classical solutions.

Since the analysis of general solutions is much too difficult, we shall now concentrate on the traveling wave solutions to the equations under consideration. Substituting the basic ansatz $u = \phi(x - ct)$ for a traveling wave of speed c reduces the partial differential equation in x, t to an ordinary differential equation, which can then be analyzed by dynamical systems methods. The classical, analytic traveling wave solutions are given by classical analytic orbits to the dynamical system: solitary wave solutions correspond to homoclinic orbits at a single equilibrium point; periodic waves come from periodic orbits; while heteroclinic orbits connecting two equilibrium points yield analytic kink solutions. Away from singular points of the dynamical system one can apply standard existence and analytic dependence results to prove that the corresponding solutions are analytic. Therefore, nonanalytic behavior will only arise at the singular points, which are precisely the points of genuine nonlinearity of the original partial differential equation. The admission of nonanalytic traveling wave solutions indicates that classical dynamical systems analysis must be significantly broadened to explain how a classical analytic orbit can give rise to a nonanalytic solution!

In the case of the $K(2,2)$ model (3), the traveling wave solutions satisfy a three-dimensional dynamical system

$$-c\phi' \pm (\phi^2)' + (\phi^2)''' = 0. \qquad (4\pm)$$

Each point $(\alpha, 0, 0)$ lying on the ϕ-axis in the (ϕ, ϕ', ϕ'')–phase space provides a fixed point for the dynamical system (4) representing a constant (undisturbed) solution of height α. The equation (4) can be integrated twice by standard methods, leading to

$$\phi^2(\phi')^2 = \mp\tfrac{1}{4}\phi^4 + \tfrac{1}{3}c\phi^3 + a\phi^2 + b \qquad \text{or} \qquad (\phi')^2 = \frac{b}{\phi^2} + a + \frac{c}{3}\phi \mp \frac{1}{4}\phi^2. \qquad (5)$$

where a, b are the integration constants. If $b \neq 0$ then the potential on the right-hand side of (5) has a pole of order 2 at $\phi = 0$; on the other hand, if $b = 0$ then the pole disappears — nevertheless, $\phi = 0$ remains a singular point of the original form of the equation, and one still anticipates that nonanalytic behavior can occur when an orbit crosses the singular axis in the phase plane. Thus, even when the singular pole is no longer present owing to a cancellation of factors, its nonanalytic legacy will remain. Note particularly that the singular points of the dynamical system arise from the points where the symbol of the original partial differential equation (3) degenerates.

For simplicity, we only consider the case $c > 0$ of positive wave speed. In the focusing case (4+), for each $0 < \alpha < \frac{c}{2}$, there is a homoclinic orbit at the fixed point $(\alpha, 0, 0)$ that represents an analytic solitary wave solution with asymptotic height α. Since each homoclinic orbit can be obtained as the limit of periodic orbits, the solitary wave solution is the limit of periodic traveling wave solutions as the period becomes infinite. When $\alpha = 0$, the origin is a singular point for the dynamical system (4), which implies that the associated homoclinic orbit is traversed in finite time. Therefore, the corresponding traveling wave solution is no longer analytic, but has compact support. In the present case, the dynamical system can be integrated explicitly, leading to the *compacton*

$$\phi_c(x - ct) = \begin{cases} \dfrac{4c}{3}\cos^2\dfrac{x - ct}{4}, & |x - ct| \le 2\pi, \\ 0, & |x - ct| > 2\pi, \end{cases} \tag{6}$$

which can be realized as the limit of solitary waves solutions as their asymptotic height $\alpha \to 0$. (See [17, 18] for the basic method of proof of such results.) In the defocusing case (4−), the fixed point $(\alpha, 0, 0)$ has a homoclinic orbit and corresponding analytic solitary wave solution only when $\alpha > -\frac{2c}{3}$. At the limiting point $\alpha = -\frac{2c}{3}$ the homoclinic orbit passes through the singularity, and the result is a nonanalytic traveling wave that has a corner at its crest $\phi = 0$, known as a *peakon*, which can also be realized as a limit of analytic solitary wave solutions. There is also a closed-form formula for the peakon:

$$\phi_p(x - ct) = \frac{2c}{3}\left(e^{-\frac{|x-ct|}{2}} - 1\right). \tag{7}$$

The peakon solution corresponds to the particular case of (5) where the integration constant $b = 0$, and so the pole is canceled out, but, as discussed above, the $\phi = 0$ singularity remains and permits the non-smooth corner.

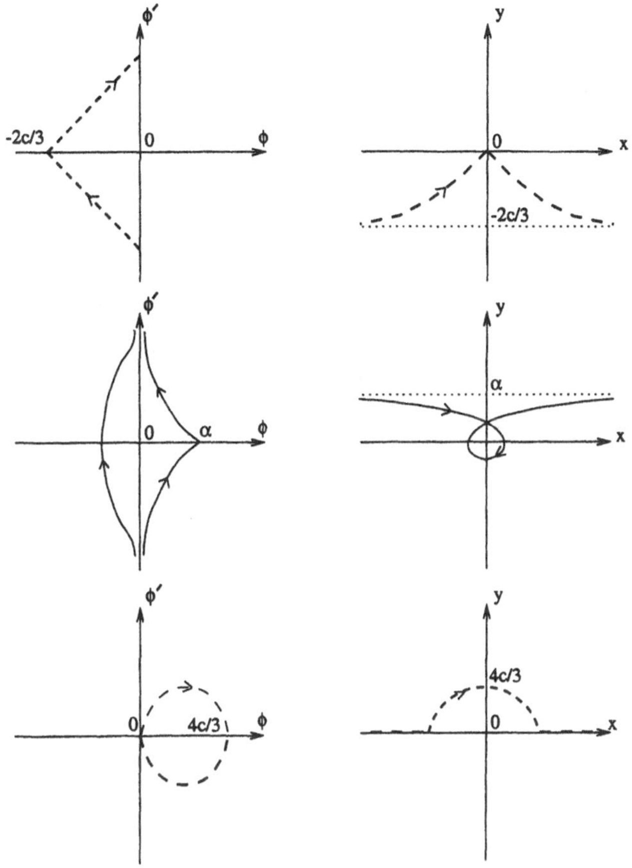

Fig. 1. *Trajectories and graphs of the peakon, loopon and compacton solutions of the $K(2,2)$ equation.*

When $\alpha > 0$, the solutions that pass through the singular axis $\phi = 0$ take the form of multiply-valued looped solitary waves — *loopons*. Loopons first appeared in an integrable model of [28], which, in contrast to the $K(2,2)$ equation, required a changeable sign in the equation. One can justify such solutions either by skewing the coordinate axes, replacing x by $\tilde{x} = x + k\phi$, or by parametrizing the graph by arc length instead of the horizontal coordinate x. As with the Riemann equation, their admissibility as physical solutions depends on the physical system being modeled; unlike breaking surface waves, it seems unlikely that these could emerge from standard initial data. We shall defer the detailed analysis of these loopons until a subsequent publication.

In Figure 1, we display the phase-plane portrait and the corresponding solution for the peakon, compacton and loopon solutions. Note that the phase plane axes are labeled by ϕ, ϕ' and the curves represent the integral curves of the dynamical system (5) with appropriate integration constants a, b. The reader should compare these with the more standard phase portrait for the KdV equation — in that case, changing the integration constants merely translates the picture; in contrast, here the change of integration constants fundamentally alters the potential, allowing new families of solutions to emerge.

The $K(2,2)$ example already indicates the possibility of significantly more interesting phenomena in the fully nonlinear regime. An even richer example is provided by the $K(3,3)$ equation

$$u_t \pm (u^3)_x + (u^3)_{xxx} = 0. \tag{8$\pm$}$$

The resulting dynamical system for its traveling wave solutions $u = \phi(x - ct)$ is twice integrated, with constants a, b, yielding

$$(\phi')^2 = \frac{b}{\phi^4} + \frac{a}{\phi} + \frac{c}{6} \mp \frac{1}{9}\phi^2. \tag{9}$$

The occurrence of two different types of poles at the singularity $\phi = 0$ already demonstrates that the $K(3,3)$ equation has, potentially, a much richer structure. Let us indicate the types of solutions associated with different fixed points $(\phi, \phi') = (\alpha, 0)$ in the phase space.

Consider first the focusing case (8+). For positive wave speeds $c > 0$, if $0 < \alpha < \sqrt{\frac{c}{3}}$, there is an analytic solitary wave of elevation having asymptotic height α. As $\alpha \to 0$ these solitary waves converge to a compacton

$$\phi_c(x - ct) = \begin{cases} \sqrt{\dfrac{3c}{2}} \cos \dfrac{x - ct}{3}, & |x - ct| \leq \dfrac{3\pi}{2}, \\ 0, & |x - ct| > \dfrac{3\pi}{2} \end{cases}. \tag{10}$$

Even though (10) has discontinuous first derivatives, its cube is C^2, and so ϕ_c is a pseudo-classical (and hence weak) solution to (8+). In addition, when $0 < \alpha < \sqrt{\frac{c}{3}}$, the homoclinic "tails" have the singular ϕ' axis as an asymptote, and hence there is a corresponding solution which attains an infinite slope in a finite time. But, as indicated in the phase portrait in Figure 2, the two points of infinite slope can be connected through an analytic solution. The resulting nonanalytic solution forms a symmetric wave of depression that is analytic except for two points of infinite slope, called a *tipon*, [23]. In Figure 2, the solid curve in the phase portrait is the homoclinic orbit that gives rise to the analytic solitary wave, while the dashed curve indicates the homoclinic tails and connecting orbit that provide the nonanalytic tipon.

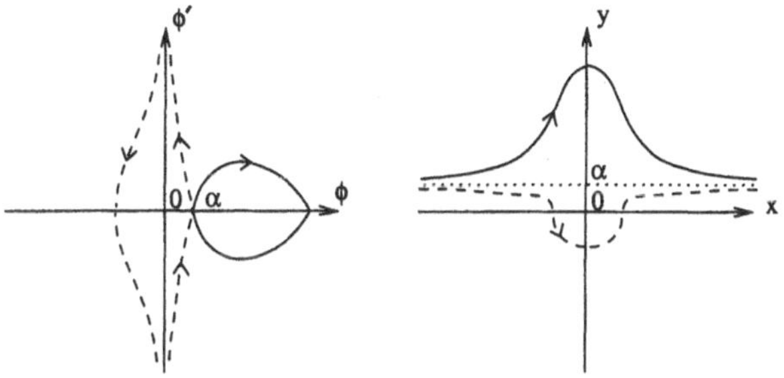

Fig. 2. *Trajectories and graphs of the analytic and tipon solitary wave solutions of the* $K(3,3)$ *equation.*

Further analysis leads to the following sequence of solutions as we move through the fixed points $(\alpha, 0, 0)$:

(a) $\alpha = 0$ — a compacton.

(b) $0 < \alpha < \sqrt{\frac{c}{3}}$ — analytic solitary waves of elevation and solitary tipons of depression. As $\alpha \to 0$ the solitary waves converge to the compacton, while the amplitudes of the solitary tipons decrease to zero. On the other hand, when $\alpha \to \sqrt{\frac{c}{3}}$, the analytic solitary waves shrink to zero while solitary tipons produce a solitary tipon.

(c) $\alpha = \sqrt{\frac{c}{3}}$ — there is no analytic solitary wave, but a solitary tipon of depression remains.

(d) $\sqrt{\frac{c}{3}} < \alpha < \sqrt{c}$ — a family of analytic periodic waves and a second family of *periodic tipons*, which are even, periodic pseudo-classical solutions having isolated points of infinite slope. As the period of the waves increases to ∞, they converge, respectively, to the solitary wave and the solitary tipon at the "complementary" fixed point $\widetilde{\alpha} = \frac{1}{2}(-\alpha + \sqrt{4c - \alpha^2})$. As $\alpha \to \sqrt{\frac{c}{3}}$ from the right, the analytic periodic waves converge to zero while the periodic tipons converge to the solitary tipon. As $\alpha \to \sqrt{c}$, the analytic periodic waves converge to the compacton at $\alpha = 0$, whereas the periodic tipons shrink to zero.

(e) $\alpha = \sqrt{c}$ — analytic periodic waves, which converge to the compacton at $\alpha = 0$ as the period becomes infinite.

(f) $\alpha > \sqrt{c}$ — a family of periodic waves that begin as analytic, then, at a critical period converge to a *periodic cuspon* (also called a billiard solution because the cusps are all upside down touching the real axis) and then become periodic tipons.

The defocusing case (8−) is also interesting. For $c > 0$ there are no periodic or solitary traveling waves at all. For $c < 0$, the following sequence of solutions successively emerges as we move through the fixed points $(\alpha, 0, 0)$:

(a) $0 < \alpha < \sqrt{-\frac{c}{3}}$ — analytic periodic waves.

(b) $\alpha = \sqrt{-\frac{c}{3}}$ — no solitary or periodic waves.

(c) $\sqrt{-\frac{c}{3}} < \alpha < \sqrt{-\frac{c}{2}}$ — a unique analytic solitary wave of depression for each α. As $\alpha \to \sqrt{-\frac{c}{2}}$, the solitary waves converge to a *cuspon*, which is a wave of depression having an infinite slope at its crest.

(d) $\alpha = \sqrt{-\frac{c}{2}}$ — the homoclinic orbit reaches the singularity $\phi = 0$ and becomes a cuspon solution.

(e) $\sqrt{-\frac{c}{2}} < \alpha < \sqrt{-c}$ — the cuspon mutates into a tipon (of depression) whose width becomes larger and larger. (In the $K(3,3)$ equation, we obtain tipons instead of the $K(2,2)$ loops because the associated potential has a pole of order 4 here.)

(f) $\alpha = \sqrt{-c}$ — two *tipon kinks*, each analytic except for a single point of infinite slope.

$\alpha > \sqrt{-c}$ — no solitary or periodic waves.

Unlike the KdV equation, the general $K(m,n)$ equations are not integrable. For example, when $m = n$ they admit only four local conservation laws [1]. Moreover, a zero-mass ripple is left after a collision of, say, two compactons, although their shape remains the same after they collide, [24]. This robustness cannot be explained in terms known to us from the conventional soliton theory. There are, however, a wide variety of integrable nonlinearly dispersive systems, and admit analytic multi-soliton solutions. Historically, the first work in this direction was by Wadati *et. al.*, [28], on cusped and looped solitons. Presently, the most well-studied is the Camassa-Holm equation

$$u_t + u_x - \nu u_{xxt} + 3(u^2)_x - \nu(u^2)_{xxx} + \nu(u_x^2)_x = 0, \tag{11}$$

which was derived as a model for ocean dynamics, [6, 7], and shown to be integrable

[1]This has not, in fact, been rigorously established, but seems almost certain.

by inverse scattering; see also [1, 8]. The biHamiltonian structure of (11) had, in fact, been first written down (albeit with an error in one of the coefficients) by Fuchssteiner, [11], (5.3). The equation for traveling waves $u = \phi(x - ct)$ becomes, upon two integrations,

$$\nu\left(\frac{c}{2} - \phi\right)(\phi')^2 = \left(\frac{c}{2} - \phi\right)\left[\left(\phi + \tfrac{1}{4}\right)^2 + a\right] + b, \tag{12}$$

where a, b are the integration constants. In contrast to the dynamical system (5), the location of the singularity is not universal, but depends on the wave speed; this is reflected in the fact that the characteristic equation $\xi^2\tau + 2u\xi^3 = 0$ for the original partial differential equation (11) has solution-dependent degeneracies, and so the points of genuine nonlinearity are solution-dependent. It is worth re-emphasizing that both cases are fundamentally different from what occurs in the formation of shock waves or more general weak solutions, whose discontinuities are not at all associated with points of genuine nonlinearity. In the focusing case $\nu = +1$, the solitons are *peakons*,

$$\phi_p(x - ct) = \left(\frac{c}{2} + \frac{1}{4}\right)e^{-|x-ct|} - \frac{1}{4}, \tag{13}$$

which can be realized as limits of analytic solitary wave solutions associated with fixed points $(\alpha, 0)$ in the phase plane as $\alpha \to -\frac{1}{4}$ from the right. When $\alpha < -\frac{1}{4}$ or $\alpha > \frac{1}{2}c$, the solutions are cuspons. As $\alpha \to -\frac{1}{4}$ from the left, the cusps suddenly open up into a peak in the limit, so the solitary waves, peakon and cuspons form a single family of solutions. In the defocusing case $\nu = -1$, the singular point $\alpha = \frac{c}{2}$ gives rise to a compacton

$$\phi_c(x - ct) = \begin{cases} \frac{1}{2}c - \left(c + \frac{1}{2}\right)\cos^2\dfrac{x - ct}{2}, & |x - ct| \le \pi, \\ 0, & |x - ct| \ge \pi, \end{cases} \tag{14}$$

which is a limit of analytic solitary wave solutions as the asymptotic height goes to zero. However, in this case, (11) is not well-posed. See [17, 18] for details.

The Camassa-Holm equation (11) can be associated to the Korteweg-deVries equation (1) by a process called "triHamiltonian duality", which was applied in [10, 21] to systematically additional nonlinearly dispersive, integrable systems associated with many other classical weakly nonlinear soliton equations. A particularly interesting example is the generalized Boussinesq system

$$\begin{aligned} v_t + 2w_t - 2v_{xt} &= w_x - v_{xx} + \left(\tfrac{1}{2}v^2 + vw\right)_x - \left(\tfrac{1}{2}v^2\right)_{xx}, \\ w_t + 2w_{xt} &= w_{xx} + \left(vw + \tfrac{1}{2}w^2 + vw_x\right)_x, \end{aligned} \tag{15}$$

that forms the integrable dual to a Boussinesq system arising in shallow water theory that was studied by Whitham, [29], and solved by inverse scattering methods in [13, 15]. Let us summarize the types of solutions found in [16] for the case $c < -1$, focusing on novel features not seen in the preceding examples.

The dynamical system for traveling waves $v = \phi(x - ct)$, $w = \psi(x - ct)$ is

$$\begin{aligned} -c\phi' - 2c\psi' + 2c\phi'' &= \psi' - \phi'' + \left(\tfrac{1}{2}\phi^2 + \phi\psi - \phi\phi'\right)', \\ -c\psi' - 2c\psi'' &= \psi'' + \left(\phi\psi + \tfrac{1}{2}\psi^2 + \phi\psi'\right)'. \end{aligned} \tag{16}$$

The fixed points of (16) all lie in the (ϕ, ψ)–plane; moreover, the line $\phi = -2c - 1$ consists of singular points. Let Γ denote the triangle formed by the three lines $\phi = -2c - 1$, $\psi = 0$ and $\psi = 1 - \phi$. Let Ω be the inscribed ellipse defined by the equation

$$\phi^2 + \phi\psi + \psi^2 + 2c\phi - \psi + c^2 = 0.$$

The curves Γ and Ω divide the (ϕ, ψ)–plane into regions containing fixed points with different properties:

(a) inside the ellipse Ω — periodic traveling wave solutions,

(b) between Ω and Γ — solitary wave solutions,

(c) on the line segments $\psi = 0$ or $\psi = 1 - \phi$ of Γ — two kink solutions.

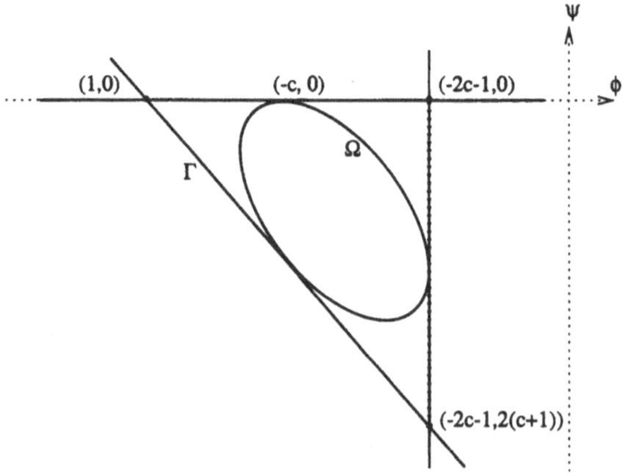

Fig. 3. *The phase plane for the Boussinesq system.*

The most interesting point is the vertex $V_* = (1, 0)$ of Γ, which corresponds to an infinite family of different nonanalytic solutions! Each of these solutions can be realized as a limit of analytic solitary wave solutions associated with points on the interior of Γ, but the limiting solution depends on how the interior points converge to V_*. More specifically, let $\mathcal{P}$ be a curve inside Γ (but outside Ω) which terminates at V_*. The solitary wave solutions (ϕ_a, ψ_a) corresponding to $a \in \mathcal{P}$ will converge to a nonanalytic traveling wave (ϕ_*, ψ_*) as $a \to V_*$, but this solution *depends on the asymptotics of the curve $\mathcal{P}$ at V_**. For example, if $\mathcal{P}$ is not tangent to Γ as it approaches V_*, then the solitary wave solutions converge to a peakon of the form

$$\begin{aligned} \phi_p(\xi) &= 1 - 2(c+1)e^{-|\xi|/2}, \\ \psi_p(\xi) &= (c+1)e^{-|\xi|/2}(\text{sign}(\xi) + 1), \end{aligned} \qquad \text{where} \qquad \xi = x - ct. \tag{17}$$

On the other hand, for the path defined by the equation $\psi(\psi+\phi-1)+e^{-2\lambda/(\phi-1)} = 0$, which tangent to the sides of Γ, the solitary wave solutions converge to a solution that we call a *mesaon* owing to its flat top; the width $2\mu = -2\lambda/(c+1)$ of the top is governed by the parameter λ; in the zero-width limit $\mu \to 0$, the mesaon becomes a peakon. There are two different versions, depending on the branch of the path chosen by the quadratic equation (17); one of these is

$$(\phi_m, \psi_m)(\xi) = \begin{cases} \left(1 - 2(c+1)e^{(\mu+\xi)/2}, 0\right), & \xi < -\mu, \\ (-2c-1, 0) & |\xi| < \mu, \\ \left(1 - 2(c+1)e^{(\mu-\xi)/2}, 2(c+1)e^{(\mu-\xi)/2}\right), & \xi > \mu. \end{cases} \tag{18}$$

Note that the mesaon (18) is constant at the fixed point $(-2c-1, 0)$ of the dynamical system, which *lies on the line of singularities*. (The other mesaon is constant at the second fixed point $(-2c-1, 2c+2)$ on the singular line.) The occurrence of fixed points which are also singularities is the key property that allows the peakon to open up into a flat-topped mesaon of arbitrary width. Indeed, in contrast, the line of singularities of the Camassa-Holm dynamical system (12) contains no embedded fixed points, and this prevents the Camassa-Holm peakons from opening up.

If our path $\mathcal{P}$ coincides with a side of the triangle, the associated analytic kink solutions converge to yet another nonanalytic solution at V_*, called a *rampon* or *semi-compact kink*, of which an example is

$$(\phi_r, \psi_r)(\xi) = \begin{cases} (-2c-1, 0), & \xi < -2\log(-2(c+1)), \\ \left(e^{-\xi/2} + 1, -e^{-\xi/2}\right), & \xi > -2\log(-2(c+1)). \end{cases} \tag{19}$$

Finally, points on the singular line $\phi = -2c - 1$ admit compactons which are found as limits of solitary waves; an example is

$$\phi_c(\xi) = \begin{cases} 2\sqrt{-b(b-2c-2)}\,\cosh\frac{\xi}{2} + 1, & |\xi| < \mu, \\ -2c-1, & |\xi| > \mu, \end{cases} \tag{20}$$

where $\mu = 2\cosh^{-1}\frac{-c-1}{\sqrt{-b(b-2c-2)}}$ determines the support of the compacton.

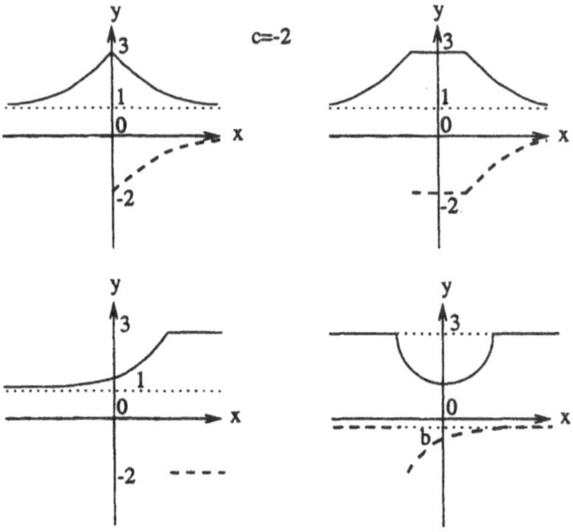

Fig. 4. *Graphs of the peakon, mesaon, rampon and compacton solutions of the dual Boussinesq system.*

Figure 4 shows the different types of nonanalytic solitary wave solutions admitted by (15). From left to right, they show a peakon, a mesaon, a rampon and a compacton. The solid line is $v = \phi$, while the dashed line shows $w = \psi$. Note particularly that w is not continuous — it jumps back to zero at a certain point in the first three cases. Nevertheless, the solution is still quasi-classical and does not require any sort of entropy condition to fix the position of the discontinuity. It is clearly beyond the capacity of the classical dynamical systems theory to explain this phenomenon, and so passage to the fully nonlinear regime introduces new and unexpected phenomena even in such well plowed fields as ordinary dynamical systems!

These examples provide concrete evidence that both the integrable and nonintegrable, nonlinearly dispersive examples treated to date are merely the tip of the proverbial iceberg. Even though, the nature of the nonanalyticity changes in each case, these differences are technical, showing merely the different facets of nonlinearity. All in all, we have not even begun to understand the nature of the nonlinear interactions that govern these equations.

References

[1] BEALS, R., SATTINGER, D.H., SZMIGIELSKI, J., Acoustic scattering and the extended Korteweg-deVries hierarchy, preprint, University of Minnesota, 1998.

[2] BENJAMIN, T.B., BONA, J.L., MAHONEY, J.J., Model equations for long waves in nonlinear dispersive systems, Phil. Trans. Roy. Soc. London A **272** (1972), 47-78.

[3] BONA, J.L., PRITCHARD, W.G., SCOTT, L.R., Solitary-wave interaction, Phys. Fluids **23** (1980), 438-441.

[4] BOUSSINESQ, J., Théorie des ondes et des remous qui se propagent le long d'un canal rectangulaire horizontal, en communiquant au liquide contenu dans ce canal des vitesses sensiblement pareilles de la surface au fond, J. Math. Pures Appl. **17** (2) (1872), 55-108.

[5] BOUSSINESQ, J., Essai sur la théorie des eaux courants, Mém. Acad. Sci. Inst. Nat. France, **23** (1) (1877), 1-680

[6] CAMASSA, R., HOLM, D.D., An integrable shallow water equation with peaked solitons, Phys. Rev. Lett. **71** (1993), 1661-1664.

[7] CAMASSA, R., HOLM, D.D., HYMAN, J.M., A new integrable shallow water equation, Adv. Appl. Mech. **31** (1994), 1-33.

[8] CONSTANTIN, A., MCKEAN, H., The integrability of a shallow water equation with periodic boundary conditions, preprint, Courant Institute, 1997.

[9] DRAZIN, P.G., JOHNSON, R.S., Solitons: An Introduction, Cambridge University Press, Cambridge, 1989.

[10] FOKAS, A.S., OLVER, P.J., ROSENAU, P., A plethora of integrable bi-Hamiltonian equations, Algebraic Aspects of Integrable Systems: In Memory of Irene Dorfman, A.S. Fokas and I.M. Gel'fand, eds., Progress in Nonlinear Differential Equations vol. **26**, Birkhäuser, Boston, 1996, 93-101.

[11] FUCHSSTEINER, B., The Lie algebra structure of nonlinear evolution equations admitting infinite dimensional abelian symmetry groups, Prog. Theor. Phys. **65** (1981) 861-876.

[12] GARDNER, L.R.T., GARDNER, G.A., Solitary waves of the regularised long wave equation, J. Comp. Phys. **91** (1990), 441–459.

[13] KAUP, D.J., A higher-order water-wave equation and the method for solving it, Prog. Theor. Physics **54** (1975), 396-408.

[14] KORTEWEG, D.J., DE VRIES, G., On the change of form of long waves advancing in a rectangular channel, and on a new type of long stationary waves, Phil. Mag.; (5) **39** (1895), 422-443.

[15] KUPERSHMIDT, B.A., Mathematics of dispersive waves, Comm. Math. Phys.; **99** (1985), 51-73.

[16] LI, Y.A., Weak solutions of a generalized Boussinesq system, J. Dyn. Diff. Eq., to appear.

[17] LI, Y.A., OLVER, P.J., Convergence of solitary-wave solutions in a perturbed bi-Hamiltonian dynamical system. I. Compactons and peakons, Discrete Cont. Dyn. Syst. **3** (1997), 419–432.

[18] LI, Y.A., OLVER, P.J., Convergence of solitary-wave solutions in a perturbed bi-Hamiltonian dynamical system. II. Complex analytic behavior and convergence to non-analytic solutions; Discrete Cont. Dyn. Syst. **4** (1998), 159.

[19] MAKHANKOV, V.G., Dynamics of classical solitons (in non-integrable systems), Phys. Rep. **35** (1978), 1-128.

[20] OLVER, P.J., Applications of Lie Groups to Differential Equations, Second Edition, Graduate Texts in Mathematics, vol. **107**, Springer-Verlag, New York, 1993.

[21] OLVER, P.J., ROSENAU, P., Tri-Hamiltonian duality between solitons and solitary-wave solutions having compact support, Phys. Rev. E; **53** (1996), 1900-1906.

[22] ROSENAU, P., Nonlinear dispersion and compact structures, Phys. Rev. Lett. **73** (1994), 1737-1740.

[23] ROSENAU, P., On nonanalytic waves formed by a nonlinear dispersion, Phys. Lett. A **230** (1997), 305-318.

[24] ROSENAU, P., HYMAN, J.M., Compactons: solitons with finite wavelength, Phys. Rev. Lett. **70** (1993), 564-567.

[25] SANTARELLI, A.R., Numerical analysis of the regularized long-wave equation: anelastic collision of solitary waves; Il Nuovo Cim. **46B** (1978), 179-188.

[26] SCOTT RUSSELL, J., Report on Water Waves, British Assoc. Report, 1844.

[27] TOLAND, J.F., On the existence of a wave of greatest height and Stokes' conjecture, Proc. Roy. Soc. London A **363** (1978), 469-485.

[28] WADATI, M., ICHIKAWA, Y.H., SHIMIZU, T., Cusp soliton of a new integrable nonlinear evolution equation, Prog. Theor. Phys. **64** (1980), 1959–1967.

[29] WHITHAM, G.B., Linear and Nonlinear Waves, John Wiley & Sons, New York, 1974.

YI A. LI
School of Mathematics, University of Minnesota, Minneapolis, MN 55455, E-mail: yili@math.umn.edu
PETER J. OLVER
School of Mathematics, University of Minnesota, Minneapolis, MN 55455, E-mail: olver@ima.umn.edu
PHILIP ROSENAU
School of Mathematical Sciences, Tel Aviv University, 69978 Tel Aviv, Israel, E-mail: rosenau@math.tau.ac.il

The Dirichlet Problem and Compact Operators in Colombeau Theory

DIMITRIS SCARPALEZOS

1 Introduction

Colombeau generalized functions are well adapted to nonlinear problems with "rough" data because their products always make sense. In the classical smooth case many existence results are obtained using the Leray-Schauder compact mapping theorem for Banach spaces. In order to use such a theorem in the frame of Colombeau theory we first have to carry out an alternative (but equivalent) construction of the space of Colombeau generalized functions. In this construction, for any given value of a parameter we deal with objects of some Banach space (depending on the parameter); thus we can use theorems adapted to Banach spaces. As an illustration of this idea we study a nonlinear elliptic divergence type Dirichlet problem. As we extensively use results from [2] this work is partly a "parametric extension"-minded commentary of classical results on nonlinear elliptic Dirichlet problems. More general constructions and results as well as the detailed proofs will appear in a common paper with S. Pilipović. The main part of this work has been completed during our stay at the Erwin Schrödinger Institute in Vienna.

2 An alternative construction of $\mathcal{G}(\overline{\Omega})$

Let Ω be a bounded open set of $\mathbb{R}^n$, $\overline{\Omega}$ its closure. Then $\mathcal{G}(\overline{\Omega})$ is usually defined as the quotient space of $\mathcal{E}_M(\overline{\Omega})$ by its subspace $\mathcal{N}(\overline{\Omega})$ where

$$\mathcal{E}_M(\overline{\Omega}) = \{(f_\varepsilon) \in \mathcal{C}^\infty(\overline{\Omega})^{]0,1]} \text{ s.t. } \forall k \in \mathbb{N}\ \exists \alpha_k \in \mathbb{R} \text{ s.t. } ||f_\varepsilon||_{k,\Omega} = o(\varepsilon^{\alpha_k})\}$$

and

$$\mathcal{N}(\overline{\Omega}) = \{(f_\varepsilon) \in \mathcal{C}^\infty(\overline{\Omega})^{]0,1]} \text{ s.t. } \forall k \in \mathbb{N}\ \forall \beta \in \mathbb{R} :\ ||f_\varepsilon||_{k,\Omega} = o(\varepsilon^\beta)\}.$$

Recall that by definition $\mathcal{C}^\infty(\overline{\Omega})$ is the space of all $\mathcal{C}^\infty$ functions on Ω all whose derivatives are continuous up to the boundary and

$$||f||_{k,\Omega} \stackrel{def}{=} \sup_{x\in\Omega, |\alpha|\leq k} (|f^{(\alpha)}(x)|).$$

The least upper bound of all α_k as in the definition of $\mathcal{E}_M(\overline{\Omega})$ will be called the k-th valuation. The set of all such "valuations" defines an ultrametric topology called the "sharp topology". In order to be able to use the Leray-Schauder theorem we will give an alternative construction of $\mathcal{G}(\overline{\Omega})$ based on Banach space valued nets. More generally, given any Fréchet space $(E, (\mu_n))$ (μ_n being an increasing sequence of seminorms), we define its Colombeau extension $\mathcal{G}[E]$ in the following way:

Definition 1 *The Colombeau extension $\mathcal{G}[E]$ of E is defined as the quotient space of the space $\mathcal{E}_M[E]$ by the subspace $\mathcal{N}[E]$ where*

$$\mathcal{E}_M[E] = \{(f_\varepsilon) \in E^{]0,1]} \text{ s.t. } \forall n \in \mathbb{N}\ \exists \alpha_n \in \mathbb{R} \text{ s.t. } \mu_n(f_\varepsilon) = o(\varepsilon^{\alpha_n})\},$$

$$\mathcal{N}[E] = \{(h_\varepsilon) \in E^{]0,1]} \text{ s.t. } \forall n \in \mathbb{N}\ \forall \beta \in \mathbb{R} :\ \mu_n(f_\varepsilon) = o(\varepsilon^\beta)\},$$

$$\mathcal{G}[E] \stackrel{def}{=} \mathcal{E}_M[E]/\mathcal{N}[E].$$

If we consider the least upper bound v_n of all α_n as in the definition of $\mathcal{E}_M[E]$ we obtain the "n-th valuation". The sequence of these valuations defines an ultrametric uniform structure and a topology on $\mathcal{G}[E]$ which are called the "sharp uniform structure" and the "sharp topology", respectively. (For more details on Colombeau extensions and sharp topologies see [1].)

Clearly the above construction is not adapted to Banach space techniques because it is based on Fréchet space valued nets. Thus an alternative construction is necessary. Here we will only deal with the case where all seminorms are norms $||\ ||_n$.

Definition 2 *Let $\bar{E}_n$ be the completion of E for the n-th norm $||\ ||_n$. (Clearly, $\bar{E}_n \subseteq \bar{E}_1$.) The set $\bar{\mathcal{E}}_M[E]$ consists of all nets (f_ε) such that*

a) there exists a sequence ε_n in $]0,1]$ decreasing to zero such that for ε in $]0,\varepsilon_n]$, we have $f_\varepsilon \in \bar{E}_n$;

b) $\forall n \in \mathbb{N}\ \exists a_n \in \mathbb{R}$ s.t. $||f_\varepsilon||_n = o(\varepsilon^{a_n})$.

For any given integer n the least upper bound of such a_n will be called the n-th valuation of (f_ε). As before, the collection of such valuations defines the "sharp topology" on $\mathcal{E}_M[E]$. Let us now consider the subset $\bar{\mathcal{N}}[E]$ of all such nets such that

$$\forall n \in \mathbb{N}\ \forall b \in \mathbb{R}: \ ||f_\varepsilon||_n = o(\varepsilon^b).$$

The quotient space $\bar{\mathcal{G}}[E] = \bar{\mathcal{E}}_M[E]/\bar{\mathcal{N}}[E]$ will be called the "Banach Colombeau ((BC)-) extension of E". The class represented by (f_ε) will be denoted by $[f_\varepsilon]$. Clearly, valuations do not depend on the choice of representatives and thus sharp topologies pass to the respective quotient space.

This construction is equivalent to the usual one for the following reason:

Proposition 1 *There is a canonical isomorphism between $\bar{\mathcal{G}}[E]$ and $\mathcal{G}[E]$. This isomorphism is an isometry between the respective "sharp" ultrametric structures.*

Sketch of proof: The argument is based on the observation that for any (BC)-moderate net (f_ε), for any fixed ε there is an element $\bar{f}_\varepsilon \in E$ such that

$$\text{if } \varepsilon \leq \varepsilon_n \text{ then } ||\bar{f}_\varepsilon - f_\varepsilon||_n \leq e^{-\frac{1}{\varepsilon}}.$$

The net of these differences is clearly negligible; thus the (BC)-class represented by the two nets is the same. It is also clear that their valuations (in their respective frames) are the same. Note that if the space E is one of the usual functional spaces $\mathcal{C}^\infty(\Omega)$ (resp. $\mathcal{C}^\infty(\overline{\Omega})$), we just write $\bar{\mathcal{G}}(\Omega)$ (resp. $\bar{\mathcal{G}}(\overline{\Omega})$).

3 Generalized nonlinear elliptic operators of divergence type

Colombeau type constructions are well adapted to model nonlinear problems with "rough" data even when singularities appear in an expression of a nonlinear operator. This can be obtained by considering nets of "smooth" operators. Here we will investigate the case of "rough" elliptic operators of divergence type.

Definition 3 *Let Ω be an open subset of $\mathbb{R}^n$. We say that a net (Q_ε) of operators of divergence type $Q_\varepsilon(u) = a_\varepsilon^{ij}(x, u, D(u))$ is a uniformly tempered elliptic operator net of divergence type iff:*

a) the matrix of coefficients $a_\varepsilon^{ij}(x, r, p)$ is symmetric, positive definite and the coefficients are Lipschitz with respect to r, p;

b) all coefficients are "tempered", i.e.,

$$\forall \alpha \in \mathbb{N}^n \; \exists C_\alpha > 0, \;\; \gamma_\alpha \in \mathbb{R}, \;\; b_\alpha \in \mathbb{R}, \;\; \varepsilon_\alpha \in]0,1] \text{ s.t.}$$
$$\forall \varepsilon < \varepsilon_\alpha : \; \sup_{x\in\Omega} |\partial^\alpha a_\varepsilon^{ij}(x, r, p)| \leq C_\alpha \varepsilon^{\gamma_\alpha}(1 + |r| + |p|)^{b_\alpha};$$

c) there exist nets of positive functions $(\lambda_\varepsilon), (\Lambda_\varepsilon)$ such that

$$\forall \xi \in \mathbb{R}^n : \;\; 0 < \lambda_\varepsilon(x, r, p)|\xi|^2 \leq a_\varepsilon^{ij}(x, r, p)\xi_i\xi_j \leq \Lambda_\varepsilon(x, r, p)|\xi|^2;$$

d) these nets $(\lambda_\varepsilon), (\Lambda_\varepsilon)$ are "moderate" and "tempered" with "moderate" quotient, i.e.,

$$\exists c_1 > 0, \;\; c_2 > 0, \;\; \mu \in \mathbb{R}, \;\; \nu \in \mathbb{R}, \;\; \varepsilon_0 \in]0,1] \text{ s.t. } \forall \varepsilon \in]0, \varepsilon_0] \; \forall (x, r, p) \in \Omega \times \mathbb{R} \times \mathbb{R}^n :$$

$$c_1 \, \varepsilon^\mu (1 + |r| + |p|)^b < \lambda_\varepsilon(x, r, p) < \Lambda_\varepsilon(x, r, p) < c_2 \, \varepsilon^{\mu+\nu}(1 + |r| + |p|)^b.$$

(Here we have used the Einstein summation convention.)

It is easy to verify that such a net defines a continuous operator from $\bar{\mathcal{G}}(\overline{\Omega})$ to itself. We suppose that the boundary of the bounded open set Ω is smooth and satisfies the following slope condition with respect to (ϕ_ε), where $(\phi_\varepsilon) \in \bar{\mathcal{E}}_M(\overline{\Omega}_1)$, Ω_1 being an open set such that $\Omega \subset\subset \Omega_1$.

Slope Condition: Let $\Gamma_\varepsilon = \{(x, z) \mid x \in \partial\Omega, \; z = \phi_\varepsilon(x)\} \subseteq \mathbb{R}^{n+1}, \; \varepsilon \in]0, 1]$. For any $P \in \Gamma_\varepsilon$ there exist hyperplanes $z = (\pi_P^\pm)_\varepsilon(x)$ in $\mathbb{R}^{n+1}$ passing through P such that

$$(\pi_P^-)_\varepsilon(x) \leq \phi_\varepsilon(x) \leq (\pi_P^+)_\varepsilon(x) \qquad (x \in \partial\Omega)$$

and $\| \operatorname{grad}(\pi_P^\pm)_\varepsilon \| \leq C\varepsilon^a$ where C and a are adequate constants. (As an example, one can verify that if Ω is an Euclidian ball, then it always satisfies such a slope condition.)

In our joint work with S. Pilipović we investigated a Dirichlet problem for such an elliptic operator net on a bounded open set whose boundary satisfies the slope condition. More precisely we proved the following proposition:

Proposition 2 *If Ω is a bounded open subset whose boundary satisfies the slope condition with respect to (ϕ_ε) and if Q is an "elliptic operator" defined by a uniformly elliptic operator net Q_ε then there exists a Colombeau generalized function $u = [u_\varepsilon]$ such that*

a) $Q_\varepsilon(u_\varepsilon) = 0$ in Ω for each ε, hence $Q(u) = 0$ in $\bar{\mathcal{G}}(\overline{\Omega})$;

b) $u_\varepsilon = \phi_\varepsilon$ on $\partial\Omega$ for each ε.

Sketch of proof : Let us first note that if $a \in]0,1[$ then the sequence of Hölder norms $\|f\|_{n,a}$ generates the same topology on $\mathcal{C}^\infty(\overline{\Omega})$ as the usual norms $\|f\|_n$. Thus we can suppose that our (BC)-construction will be made on the basis of the sequence of such Hölder norms for some fixed a. We will now adapt the classical proof of existence of a solution in $\mathcal{C}^{2,a}$ as it appears in [2] (Theorem 11.5). For any given $\varepsilon \in]0, \varepsilon_1]$ and any given $v \in \mathcal{C}^{k,a}$ we know that there exists a unique u_ε^v satisfying the equation

$$a_\varepsilon{}^{ij}(x, v, Dv)D_{ij}u_\varepsilon^v = 0 \text{ in } \Omega,$$
$$u_\varepsilon = \phi_\varepsilon \text{ on } \partial\Omega.$$

Let $T_\varepsilon(v)$ be the unique solution of this well-posed elliptic linear Dirichlet problem. Following the classical methods we show that these operators send bounded sets of $\mathcal{C}_{k,a}(\overline{\Omega})$ into $\mathcal{C}_{k+1,a}(\overline{\Omega})$, i.e., onto relatively compact subsets of $\mathcal{C}_{k,a}(\overline{\Omega})$. Thus this net of operators is a net of compact operators and we can use a corollary of the Leray-Schauder theorem as it appears in [2] (Theorem 11.3). Therefore, for any given ε the corresponding compact operator has a fixed point u_ε which is a solution of the nonlinear equation.

In our generalized Banach-Colombeau frame we also have to prove that the net thus obtained satisfies the required bounds. Let us note that if in the linear equation we replace ϕ_ε by $\sigma\phi_\varepsilon$ where σ belongs to]0,1] then we obtain, for any fixed ε and any given σ, a solution $w_{\varepsilon,\sigma,v}$ which (because of Theorem 6.6 and exercise (6.2) in [2]) satisfies the following inequality:

$$\|w_{\varepsilon,\sigma,v}\|_{k+2} \leq C(n, k, \lambda, \Lambda, a)(\|w_{\varepsilon,\sigma,v}\|_0 + \|\phi_\varepsilon\|_{k+2,a}).$$

By the weak maximum principle for elliptic linear operators we can bound $\|w_{\varepsilon,\sigma,v}\|_0$ by $\|\phi_\varepsilon\|_{0,\partial\Omega}$. Inspecting closely the proofs in [2] we can verify that the constant appearing in this inequality is of "polynomial increase" on the data and on Λ/λ. Using our assumptions on our data of nets we can prove that the supremum of $||w_{\varepsilon,\sigma,v}||_{k+2}$ is of "moderate" growth on ε and of "tempered" growth on $||v||_1$.

If u_ε^σ satisfies $a_\varepsilon{}^{ij}(x, u_\varepsilon^\sigma, Du_\varepsilon^\sigma)D_{ij}u_\varepsilon^\sigma = 0$ in Ω, and $u_\varepsilon^\sigma = \sigma\phi_\varepsilon$ on $\partial\Omega$ then its $(k+2, a)$-norm would be of "moderate growth" on ε and of "tempered growth" on its own $||\ ||_1$ norm. Now by the weak maximum principle we only have to prove that its gradient is of tempered behaviour on the data. This is obtained using the slope condition and following the same strategy as in Chapter 11.3 of [2]. In conclusion it is of moderate growth on ε and of tempered growth on the norms of ϕ_ε with constants non-dependent on σ. Using now the Leray-Schauder theorem we can conclude that the net of solutions exists and satisfies the same bounds.

Let us denote by (u_ε^k) the net obtained when we study our problem in the $C^{k,a}$ frame and let us put by definition $u_\varepsilon = u_\varepsilon^k$ for $\varepsilon \in]\varepsilon_{k+1}, \varepsilon_k]$. It is now easy to verify that the class $[u_\varepsilon]$ satisfies the given generalized nonlinear Dirichlet problem.

We saw that adapting definitions we can use (BC)-methods in the frame of Colombeau generalized functions. We also saw that when the same uniform structure of a Fréchet space is given by different sets of seminorms we can freely choose the one which is most convenient in order to use known results valid for some classical case. Here we applied such ideas to a simple divergent type nonlinear problem, but one can see how the same could be done in many other cases where the Leray-Schauder theorem is used.

References

[1] SCARPALEZOS D.: Some remarks on Colombeau generalized functions, to appear in *Integral transforms and special functions.*

[2] GILBARG, TRUDINGER: Elliptic partial differential equations of second order (second edition) *Springer* 1983.

D. SCARPALEZOS

Universite Paris 7, F-75251 Paris Cedex 05, France

Highly Oscillatory Shock Waves

YAGUANG WANG

1 Introduction

The topic of nonlinear geometric optics concerns the formal analysis of small amplitude, high frequency oscillatory waves in nonlinear problems, and the rigorous justification of the asymptotic properties of oscillations. Most of the literature deals with the case of smooth background states. See the article [2] of Joly, Métivier and Rauch for a review of this research. A natural interesting problem is to study the case of singular background states, e.g. shock waves, which leads us to investigate the interaction of singularities and oscillations. The rigorous justification of the asymptotic analysis of a single highly oscillatory shock wave was established recently by Corli [1], Wang [6] and Williams [8].

We are interested in the problem of multiple shock waves being perturbed by high frequency oscillations. As a starting step, this paper is devoted to the case of 2×2 conservation laws in one space variable, from which some important ideas and interesting phenomena are already obtained. For the problem of two weak shocks perturbed by small amplitude, high frequency oscillatory waves, we obtain that the leading profiles of oscillatory shocks are solutions to an integro-differential system with free boundaries, the leading terms of shock fronts do not oscillate, and oscillations only appear in the leading terms of shock speeds. The case of general $m \times m$ systems will be studied by the author in [7].

In the next section, we formulate the problems and state the main results. Then, in §3, we sketch the proof of the main results briefly, which gives our nonlinear geometric optics.

2 Formulation of problems and main results

Suppose that the following 2×2 conservation law in one space variable:

$$Lu = \partial_t u + \partial_x f(u) = 0 \tag{2.1}$$

is strictly hyperbolic with respect to t for each $u \in I\!R^2$, let $\lambda_k(u)$ $(k = 1, 2)$ be the eigenvalue of $A(u) = \nabla f(u)$ with $\lambda_1(u) < \lambda_2(u)$, and $r_k(u)$ ($l_k(u)$ resp.) be the right (left resp.) eigenvector of $A(u)$ with respect to $\lambda_k(u)$.
We assume that the Cauchy problem of (2.1) with piecewise constant initial data:

$$u(0, x) = \begin{cases} u_-, & x < 0 \\ u_+, & x > 0 \end{cases} \tag{2.2}$$

admits two shocks

$$\underline{u} = \begin{cases} u_-, & -\infty < x < \sigma_1 t \\ u_\star, & \sigma_1 t < x < \sigma_2 t \\ u_+, & \sigma_2 t < x < \infty \end{cases} \tag{2.3}$$

with $\sigma_1 < \sigma_2$ and $u_\star$ being constants, and $\underline{u}$ satisfies the Rankine-Hugoniot condition

$$\sigma_1(u_\star - u_-) = f(u_\star) - f(u_-), \quad \sigma_2(u_+ - u_\star) = f(u_+) - f(u_\star) \tag{2.4}$$

and the Lax entropy condition:

$$\begin{cases} \lambda_1^\star < \sigma_1 < \lambda_1^-, & \sigma_1 < \lambda_2^\star \\ \lambda_2^+ < \sigma_2 < \lambda_2^\star, & \sigma_2 > \lambda_1^\star \end{cases} \tag{2.5}$$

where $\lambda_k^\flat = \lambda_k(u_\flat)$ for any $k \in \{1, 2\}$ and $\flat \in \{\pm, \star\}$.
Denote by $r_k^\star = r_k(u_\star)$ and $A_\star = A(u_\star)$, and let

$$\begin{cases} (\sigma_1 I - A_\star)^{-1}(u_\star - u_-) = \alpha r_1^\star + \beta r_2^\star \\ (A_\star - \sigma_2 I)^{-1}(u_+ - u_\star) = \gamma r_1^\star + \delta r_2^\star \end{cases}, \tag{2.6}$$

we suppose that the following stability condition

$$|\beta\gamma| < |\alpha\delta| \tag{2.7}$$

holds.
As shown by Métivier in [4, Appendice A], when $\lambda_1(u)$ and $\lambda_2(u)$ are genuinely nonlinear, and the shocks $\underline{u}$ are weak enough, the stability condition (2.7) is always valid.

Given a small closed neighborhood $\omega \subset \{t=0\}$ of the origin, suppose Ω is the closure of a determinacy domain of ω for the Cauchy problem of (2.1) when $|u-\underline{u}| < \eta$ for a fixed small constant $\eta > 0$. Equip the space $C^k(\Omega)$ of the k-th continuously differentiable functions in Ω with the family of norms

$$\|u\|^{\epsilon}_{k,\Omega} = \sum_{|\alpha|\le k} \epsilon^{|\alpha|} \|\partial^{\alpha} u\|_{L^{\infty}(\Omega)}.$$

We say that a family $u^{\epsilon} \in C^k(\Omega)$ is *bounded in* $C^k_{\epsilon}(\Omega)$ if norms $\|u^{\epsilon}\|^{\epsilon}_{k,\Omega}$ are bounded, and $\phi^{\epsilon} \in C^k([0,T])$ *bounded in* $\tilde{C}^k_{\epsilon}([0,T])$ if $\|d_t\phi^{\epsilon}\|^{\epsilon}_{k-1,[0,T]}$ are bounded for $k \ge 1$. Let $C^0_p(\Omega : I\!R^q)$ be the space of continuous functions in $(t,x) \in \Omega$ with values in the continuous almost periodic functions in $\theta \in I\!R^q$. For $k \in I\!N$, define the space $C^k_p(\Omega : I\!R^q)$ as the set of all $U \in C^0_p(\Omega : I\!R^q)$ whose derivatives $\partial^{\alpha}_{(t,x;\theta)} U$ belong to $C^0_p(\Omega : I\!R^q)$ for any $|\alpha| \le k$.

Consider the following Cauchy problem for small perturbations of $\underline{u}$:

$$\begin{cases} \partial_t U^{\epsilon} + \partial_x f(U^{\epsilon}) = 0, \quad t > 0,\ x \in I\!R \\ U^{\epsilon}(0,x) = \begin{cases} u_+ + \epsilon u^{\epsilon}_{+,0}(x), & x > 0 \\ u_- + \epsilon u^{\epsilon}_{-,0}(x), & x < 0 \end{cases} \end{cases} \tag{2.8}$$

where $\epsilon > 0$ is small enough, and $u^{\epsilon}_{\pm,0} \in C^1(\omega^{\pm})$ with $\omega^+ = \omega \cap \{x > 0\}$ and $\omega^- = \omega \cap \{x < 0\}$, satisfying that there are $U_{\pm,0} \in C^1_p(\omega^{\pm} : I\!R)$ such that

$$\|u^{\epsilon}_{\pm,0}(x) - U_{\pm,0}(x, \frac{x}{\epsilon})\|^{\epsilon}_{1,\omega^{\pm}} = o(1) \tag{2.9}$$

when ϵ goes to zero. Obviously, from (2.9) we have the boundedness of $u^{\epsilon}_{\pm,0}$ in $C^1_{\epsilon}(\omega^{\pm})$.

Inspired by the result of Sablé-Tougeron in [5], we suppose that the problem (2.8) determines two shocks:

$$U^{\epsilon}(t,x) = \begin{cases} u_- + \epsilon u^{\epsilon}_-(t,x), & x < \sigma_1 t + \epsilon\varphi^{\epsilon}_1(t) \\ u_\star + \epsilon u^{\epsilon}_\star(t,x), & \sigma_1 t + \epsilon\varphi^{\epsilon}_1(t) < x < \sigma_2 t + \epsilon\varphi^{\epsilon}_2(t) \\ u_+ + \epsilon u^{\epsilon}_+(t,x), & x > \sigma_2 t + \epsilon\varphi^{\epsilon}_2(t) \end{cases}. \tag{2.10}$$

The purpose of this paper is to study the local existence of U^{ϵ} in the form (2.10) to the problem (2.8), and to analyse the asymptotic properties of $(u^{\epsilon}_{\flat}, \varphi^{\epsilon}_1, \varphi^{\epsilon}_2)$ with respect to ϵ for $\flat \in \{\pm, \star\}$.

At first, let us solve $u^{\epsilon}_{\pm}(t,x)$, and investigate their asymptotics. Denote by $\tilde{u}^{\epsilon}_{\pm,0} \in C^1_{\epsilon}(\omega)$ ($\tilde{U}_{\pm,0} \in C^1_p(\omega : I\!R)$ resp.) the extension of $u^{\epsilon}_{\pm,0}$ ($U_{\pm,0}(x,\theta)$ resp.) in $x \in \omega$

with the asymptotic property

$$\|\tilde{u}^{\epsilon}_{\pm,0}(x) - \tilde{U}_{\pm,0}(x, \frac{x}{\epsilon})\|^{\epsilon}_{1,\omega} = o(1), \quad \text{when } \epsilon \to 0 \tag{2.11}$$

valid.

By employing the result of Joly, Métivier and Rauch [3, Theorem 2.9.1] for the following Cauchy problem:

$$\begin{cases} \partial_t U^{\epsilon}_{\pm} + \partial_x f(U^{\epsilon}_{\pm}) = 0, & t > 0,\ x \in I\!R \\ U^{\epsilon}_{\pm}(0,x) = u_{\pm} + \epsilon \tilde{u}^{\epsilon}_{\pm,0}(x) \end{cases} \tag{2.12}$$

we obtain T, $\epsilon_0 > 0$ such that on $\Omega_T = \Omega \cap \{t \le T\}$, there is a unique solution $U^{\epsilon}_{\pm}(t,x) = u_{\pm} + \epsilon u^{\epsilon}_{\pm}(t,x)$ of (2.12) with $u^{\epsilon}_{\pm}(t,x)$ bounded in $C^1_{\epsilon}(\Omega_T)$ for any $\epsilon \in (0, \epsilon_0]$; moreover, we have the asymptotic expansion:

$$u^{\epsilon}_{\pm}(t,x) = U_{\pm}(t,x; \frac{t}{\epsilon}, \frac{x}{\epsilon}) + o(1) \quad \text{in } C^1_{\epsilon}(\Omega_T) \tag{2.13}$$

when $\epsilon \to 0$, where $U_{\pm} \in C^1_p(\Omega_T : I\!R^2)$ is the unique solution to the initial value problem of an integro-differential system, see Joly et al. [3] for the details.

By using the entropy condition (2.5), we know that $\Omega^-_T = \{(t,x) \in \Omega_T | x < \sigma_1 t + \epsilon\varphi^{\epsilon}_1(t)\}$ and $\Omega^+_T = \{(t,x) \in \Omega_T | x > \sigma_2 t + \epsilon\varphi^{\epsilon}_2(t)\}$ are determinacy domains of ω^- and ω^+, respectively, when $(d_t\varphi^{\epsilon}_1, d_t\varphi^{\epsilon}_2)$ remains bounded, which will be proved later. Thus, the restrictions of $U^{\epsilon}_{\pm}(t,x)$ and $U_{\pm}(t,x;\tau,\theta)$ to $\Omega^{\pm}_T$ are independent of the way of extending $u^{\epsilon}_{\pm,0}$ and $U_{\pm,0}$ in ω, and satisfy the equations (2.8) and the asymptotic property (2.13) in the domains $\Omega^{\pm}_T$.

It remains to study the existence of $(u^{\epsilon}_{\star}, \varphi^{\epsilon}_1, \varphi^{\epsilon}_2)$, which makes (2.10) to be a shock wave solution to the problem (2.8). From the Rankine-Hugoniot condition, we know that $(u^{\epsilon}_{\star}, \varphi^{\epsilon}_1, \varphi^{\epsilon}_2)$ satisfy the following problem on $\Omega^{\star} = \{\sigma_1 t + \epsilon\varphi^{\epsilon}_1(t) < x < \sigma_2 t + \epsilon\varphi^{\epsilon}_2(t)\}$:

$$\begin{cases} \partial_t u^{\epsilon}_{\star} + A(u_{\star} + \epsilon u^{\epsilon}_{\star})\partial_x u^{\epsilon}_{\star} = 0, \quad (t,x) \in \Omega^{\star} \\ (\sigma_1 + \epsilon d_t\varphi^{\epsilon}_1)(\epsilon u^{\epsilon}_{\star} - \epsilon u^{\epsilon}_{-} + u_{\star} - u_{-}) \\ \quad = f(u_{\star} + \epsilon u^{\epsilon}_{\star}) - f(u_{-} + \epsilon u^{\epsilon}_{-}), \quad \text{on } x = \sigma_1 t + \epsilon\varphi^{\epsilon}_1(t) \\ (\sigma_2 + \epsilon d_t\varphi^{\epsilon}_2)(\epsilon u^{\epsilon}_{+} - \epsilon u^{\epsilon}_{\star} + u_{+} - u_{\star}) \\ \quad = f(u_{+} + \epsilon u^{\epsilon}_{+}) - f(u_{\star} + \epsilon u^{\epsilon}_{\star}), \quad \text{on } x = \sigma_2 t + \epsilon\varphi^{\epsilon}_2(t) \\ \varphi^{\epsilon}_1(0) = \varphi^{\epsilon}_2(0) = 0 \end{cases} \tag{2.14}$$

which is a Goursat problem with free boundaries, where $U^{\epsilon}_{\pm}(t,x) = u_{\pm} + \epsilon u^{\epsilon}_{\pm}(t,x)$ is given by (2.12).

As in Métivier [4], by using the transformation

$$\begin{cases} y = t\frac{x-(\sigma_1 t+\epsilon\varphi_1^\epsilon(t))}{(\sigma_2 t+\epsilon\varphi_2^\epsilon(t))-(\sigma_1 t+\epsilon\varphi_1^\epsilon(t))} \\ z = t\frac{\sigma_2 t+\epsilon\varphi_2^\epsilon(t)-x}{(\sigma_2 t+\epsilon\varphi_2^\epsilon(t))-(\sigma_1 t+\epsilon\varphi_1^\epsilon(t))} \end{cases} \tag{2.15}$$

in the problem (2.14), we know that $(u_\star^\epsilon, \varphi_1^\epsilon, \varphi_2^\epsilon)$ satisfy the following Goursat problem in $\Omega^\star = \{y,\ z > 0\}$:

$$\begin{cases} L^\epsilon(u_\star^\epsilon, \varphi_1^\epsilon, \varphi_2^\epsilon)u_\star^\epsilon = 0, & (y,z) \in \Omega^\star \\ \mathcal{F}_1^\epsilon(t, \gamma_1 u_\star^\epsilon, d_t\varphi_1^\epsilon, \varphi_1^\epsilon) = 0, & \text{on } y = 0 \\ \mathcal{F}_2^\epsilon(t, \gamma_2 u_\star^\epsilon, d_t\varphi_2^\epsilon, \varphi_2^\epsilon) = 0, & \text{on } z = 0 \\ \varphi_1^\epsilon(0) = \varphi_2^\epsilon(0) = 0 \end{cases} \tag{2.16}$$

where $\gamma_1 u_\star^\epsilon(t) = u_\star^\epsilon(0,t)$, $\gamma_2 u_\star^\epsilon(t) = u_\star^\epsilon(t,0)$, $\mathcal{F}_1^\epsilon = 0$ and $\mathcal{F}_2^\epsilon = 0$ are the Rankine-Hugoniot conditions given in (2.14), and $L^\epsilon(u_\star^\epsilon, \varphi_1^\epsilon, \varphi_2^\epsilon) = (\sigma_2 - \sigma_1)(M^\epsilon\partial_y + N^\epsilon\partial_z)$ with $M^\epsilon = \frac{\partial y}{\partial t}I + \frac{\partial y}{\partial x}A(u_\star + \epsilon u_\star^\epsilon)$ and $N^\epsilon = I - M^\epsilon$.

We note that M^ϵ and N^ϵ are smooth with respect to $V^\epsilon = (\epsilon u_\star^\epsilon, \epsilon d_t\varphi_1^\epsilon, \epsilon d_t\varphi_2^\epsilon, \epsilon\theta^\epsilon, \epsilon y d_t\theta^\epsilon, \epsilon z d_t\theta^\epsilon)$ near the origin in $\mathbb{R}^7$ with $\theta^\epsilon = (\varphi_2^\epsilon - \varphi_1^\epsilon)/t$, and at the origin $0 \in \mathbb{R}^7$, we have

$$M^\epsilon(0) = \frac{1}{\sigma_2 - \sigma_1}(A_\star - \sigma_1 I), \quad N^\epsilon(0) = \frac{1}{\sigma_2 - \sigma_1}(\sigma_2 I - A_\star).$$

Without loss of generality, we suppose that $A_\star = diag[\lambda_1^\star,\ \lambda_2^\star]$ is diagonal. In this case, the operator L^ϵ frozen at $V^\epsilon = 0$ takes the simple form $L^\epsilon(0) = (\sigma_2 - \sigma_1)P(\partial_y, \partial_z)$, where

$$P(\partial_y, \partial_z) = \begin{pmatrix} \frac{\sigma_2-\lambda_1^\star}{\sigma_2-\sigma_1} & 0 \\ 0 & \frac{\lambda_2^\star-\sigma_1}{\sigma_2-\sigma_1} \end{pmatrix} \left[\begin{pmatrix} -b_1 & 0 \\ 0 & 1 \end{pmatrix} \partial_y + \begin{pmatrix} 1 & 0 \\ 0 & -b_2 \end{pmatrix} \partial_z \right] \tag{2.17}$$

with

$$b_1 = \frac{\lambda_1^\star - \sigma_1}{\lambda_1^\star - \sigma_2} \quad \text{and} \quad b_2 = \frac{\lambda_2^\star - \sigma_2}{\lambda_2^\star - \sigma_1}. \tag{2.18}$$

Let us discuss the problem (2.16) in formal analysis. Suppose that the solutions $(u_\star^\epsilon, \varphi_1^\epsilon, \varphi_2^\epsilon)$ are the form of

$$\begin{cases} u_\star^\epsilon(y,z) = U_\star(y,z;\frac{y}{\epsilon},\frac{z}{\epsilon}) + \epsilon V_\star(y,z;\frac{y}{\epsilon},\frac{z}{\epsilon}) + o(\epsilon) \\ \varphi_k^\epsilon(t) = \varphi_k(t,\frac{t}{\epsilon}) + \epsilon\phi_k(t,\frac{t}{\epsilon}) + o(\epsilon), \ k = 1,2 \end{cases} \tag{2.19}$$

where $(U_\star(y,z;\xi,\eta), V_\star(y,z;\xi,\eta))$ and $(\varphi_k(t,\tau), \phi_k(t,\tau))$ are almost periodic in $(\xi,\eta) \in \mathbb{R}^2$ and $\tau \in \mathbb{R}$, respectively.

By plugging the expression (2.19) into the boundary conditions in (2.16), we deduce

$$\partial_\tau \varphi_1 = \partial_\tau \varphi_2 = 0 \tag{2.20}$$

which means the leading terms (φ_1, φ_2) of $(\varphi_1^\epsilon, \varphi_2^\epsilon)$ do not oscillate.

In order to formulate the problem for $U_\star$, we define $\mathbb{E} = diag[\mathbb{E}_1, \mathbb{E}_2]$, where

$$\begin{cases} \mathbb{E}_1 u(y,z;\xi,\eta) = \lim_{\rho\to\infty} \frac{1}{2\rho}\int_{-\rho}^{\rho} u(y,z;\xi - b_1 s, \eta + s)ds \\ \mathbb{E}_2 u(y,z;\xi,\eta) = \lim_{\rho\to\infty} \frac{1}{2\rho}\int_{-\rho}^{\rho} u(y,z;\xi + s, \eta - b_2 s)ds \end{cases} \tag{2.21}$$

are two mean value operators for any $u \in C_p^0(\Omega : \mathbb{R}^2)$.

By the formal analysis, we obtain that leading terms of $(u_\star^\epsilon, d_t\varphi_1^\epsilon, d_t\varphi_2^\epsilon)$ satisfy the problem:

$$\begin{cases} \mathbb{E}U_\star = U_\star \\ P(\partial_y, \partial_z)U_\star + (dI + hA_\star)(\partial_\xi - \partial_\eta)U_\star + \mathbb{E}B_\star((\partial_\xi - \partial_\eta)U_\star, U_\star) = 0 \\ \chi_1(u_\star - u_-) + (\sigma_1 I - A_\star)\gamma_1 U_\star - (\sigma_1 I - A_-)\mathbf{U}_- = 0, \quad \text{on } y = \xi = 0 \\ \chi_2(u_+ - u_\star) + (\sigma_2 I - A_+)\mathbf{U}_+ - (\sigma_2 I - A_\star)\gamma_2 U_\star = 0, \quad \text{on } z = \eta = 0 \end{cases} \tag{2.22}$$

where $B_\star = \nabla^2 f(u_\star)$, $\chi_k(t,\tau) = \frac{d\varphi_k(t)}{dt} + \frac{\partial\phi_k(t,\tau)}{\partial\tau}$ $(k = 1,2)$, $\gamma_1 U_\star(t,\tau) = U_\star(0,t;0,\tau)$, $\gamma_2 U_\star(t,\tau) = U_\star(t,0;\tau,0)$, $\mathbf{U}_-(t,\tau) = U_-(t,\sigma_1 t;\tau,\sigma_1\tau + \varphi_1(t))$ and $\mathbf{U}_+(t,\tau) = U_+(t,\sigma_2 t;\tau,\sigma_2\tau + \varphi_2(t))$ with $U_\pm$ being given in (2.13), and

$$\begin{cases} d = \frac{(z\sigma_1 + y\sigma_2)(\varphi_2 - \varphi_1)}{(\sigma_2 - \sigma_1)(y+z)^2} - \frac{yd_t\varphi_2 + zd_t\varphi_1}{y+z} \\ h = \frac{\varphi_1 - \varphi_2}{(\sigma_2 - \sigma_1)(y+z)} \end{cases}$$

From (2.22), we know that the level sets of the oscillation phase functions $\phi_1(y,z) = y + b_1 z$ and $\phi_2(y,z) = b_2 y + z$ of $U_\star^1$ and $U_\star^2$ respectively are characteristic lines of $L^\epsilon(0)$.

Acting the mean value operator

$$\mathbb{E}_0 u = \overline{u}(y,z) = \lim_{\rho\to\infty} \frac{1}{(2\rho)^2}\int_{-\rho}^{\rho}\int_{-\rho}^{\rho} u(y,z;\xi,\eta)d\xi d\eta$$

on the problem (2.22), we get that $(\bar{U}_\star, \varphi_1, \varphi_2)$ satisfy the problem:

$$\begin{cases} P(\partial_y, \partial_z)\bar{U}_\star = 0, \quad y,z > 0 \\ d_t\varphi_1(u_\star - u_-) + (\sigma_1 I - A_\star)\gamma_1 \bar{U}_\star - (\sigma_1 I - A_-)\bar{\mathbf{U}}_- = 0, \quad \text{on } y = 0 \\ d_t\varphi_2(u_+ - u_\star) + (\sigma_2 I - A_+)\bar{\mathbf{U}}_+ - (\sigma_2 I - A_\star)\gamma_2 \bar{U}_\star = 0, \quad \text{on } z = 0 \\ \varphi_1(0) = \varphi_2(0) = 0 \end{cases} \tag{2.23}$$

where $\gamma_1\bar{U}_\star(t) = \bar{U}_\star(0,t)$, $\gamma_2\bar{U}_\star(t) = \bar{U}_\star(t,0)$, $\bar{\mathbf{U}}_-(t) = \bar{U}_-(t,\sigma_1 t)$ and $\bar{\mathbf{U}}_+(t) = \bar{U}_+(t,\sigma_2 t)$.

The main results of this paper are stated as follows:

MAIN RESULTS: *Under the assumption (2.9), we have:*
(1) there are $T, \epsilon_0 > 0$ *such that the problem (2.16) has unique solutions* $u_\star^\epsilon$ *and* $(\varphi_1^\epsilon, \varphi_2^\epsilon)$ *bounded in* $C_\epsilon^1(\Omega_T^\star)$ *and* $\tilde{C}_\epsilon^2([0,T])$ *respectively for any* $\epsilon \in (0, \epsilon_0]$*;*
(2) there are unique solutions $U_\star \in C_p^1(\Omega_T^\star : I\!R^2)$, $(\chi_1, \chi_2) \in C_p^1([0,T] : I\!R)$ *and* $(\varphi_1, \varphi_2) \in C^2[0,T]$ *to the problems (2.22) and (2.23);*
(3) for the above solutions $(u_\star^\epsilon, \varphi_1^\epsilon, \varphi_2^\epsilon)$, *we have the following asymptotic properties:*

$$\|u_\star^\epsilon(y,z) - U_\star(y,z;\frac{y}{\epsilon},\frac{z}{\epsilon})\|_{1,\Omega_T^\star}^\epsilon = o(1) \tag{2.24}$$

and

$$\begin{cases} \|d_t\varphi_k^\epsilon(t) - \chi_k(t,\frac{t}{\epsilon})\|_{1,[0,T]}^\epsilon = o(1) \\ \|\varphi_k^\epsilon(t) - \varphi_k(t)\|_{L^\infty[0,T]} = o(1) \end{cases} \tag{2.25}$$

when ϵ *goes to zero for* $k \in \{1,2\}$.

3 Sketch of Proof

The proof of the main results is given in three steps:

STEP 1: We solve the problem (2.16) by the following iteration scheme:

$$\begin{cases} L^\epsilon(u_\star^{\epsilon,\nu}, \varphi_1^{\epsilon,\nu}, \varphi_2^{\epsilon,\nu})u_\star^{\epsilon,\nu+1} = 0, \quad y,z>0 \\ F_{k,(\gamma_k u_\star^{\epsilon,\nu}, d_t\varphi_k^{\epsilon,\nu})}^\epsilon(\gamma_k u_\star^{\epsilon,\nu+1}, d_t\varphi_k^{\epsilon,\nu+1}) = -\mathcal{F}_k^\epsilon(t, \gamma_k u_\star^{\epsilon,\nu}, d_t\varphi_k^{\epsilon,\nu}, \varphi_k^{\epsilon,\nu}) \\ \qquad\qquad + F_{k,(\gamma_k u_\star^{\epsilon,\nu}, d_t\varphi_k^{\epsilon,\nu})}^\epsilon(\gamma_k u_\star^{\epsilon,\nu}, d_t\varphi_k^{\epsilon,\nu}) \\ \varphi_k^{\epsilon,\nu+1}(0) = 0, \quad k=1,2 \end{cases} \tag{3.1}$$

where $(u_\star^{\epsilon,0}, \varphi_1^{\epsilon,0}, \varphi_2^{\epsilon,0}) = 0$, and $F_{k,(\gamma_k u_\star^\epsilon, d_t\varphi_k^\epsilon)}^\epsilon(v,\tau)$ is the Fréchet derivative of $\mathcal{F}_k^\epsilon$ with respect to $(\gamma_k u_\star^\epsilon, d_t\varphi_k^\epsilon)$. That is to say that we use the usual Picard iteration for the equation, and the Newton iteration for the boundary conditions in (2.16).
By studying a linear Goursat problem similar to (3.1), we can obtain

(1) there are $T, \epsilon_0 > 0$ such that the sequences $\{u_\star^{\epsilon,\nu}, d_t\varphi_1^{\epsilon,\nu}, d_t\varphi_2^{\epsilon,\nu}\}_{\nu\ge 0}$ are bounded in $C_\epsilon^1(\Omega_T^\star)$ with $\epsilon \in (0,\epsilon_0]$;

(2) when ν goes to infinity, $u_\star^{\epsilon,\nu}$ ($(\varphi_1^{\epsilon,\nu}, \varphi_2^{\epsilon,\nu})$ resp.) is convergent in $C^0(\Omega_T^\star)$ (resp. $C^1([0,T])$) uniformly in $\epsilon \in (0,\epsilon_0]$;

(3) for any $\epsilon \in (0,\epsilon_0]$ fixed, the family $\{\nabla u_\star^{\epsilon,\nu}, d_t^2\varphi_1^{\epsilon,\nu}, d_t^2\varphi_2^{\epsilon,\nu}\}_{\nu\ge 0}$ is equicontinuous,

from which we have that the convergence of $u_\star^{\epsilon,\nu}$ and $(\varphi_1^{\epsilon,\nu}, \varphi_2^{\epsilon,\nu})$ also holds in $C^1(\Omega_T^\star)$ and $C^2([0,T])$ respectively as $\nu \to \infty$; moreover, their limits $(u_\star^\epsilon, \varphi_1^\epsilon, \varphi_2^\epsilon)$ are unique solutions of (2.16).

STEP 2: Noting that the problem (2.23) is a linear Goursat one, we can immediately obtain $(\varphi_1, \varphi_2) \in C^2([0,T])$.

Solve the nonlinear problem (2.22) by the following iteration scheme:

$$\begin{cases} I\!E_\star U_\star^{\nu+1} = U_\star^{\nu+1} \\ P_\star(\partial_y, \partial_z)U_\star^{\nu+1} + (dI + hA_\star)(\partial_\xi - \partial_\eta)U_\star^{\nu+1} + I\!E_\star B_\star((\partial_\xi - \partial_\eta)U_\star^{\nu+1}, U_\star^\nu) = 0 \\ \chi_1^{\nu+1}(u_\star - u_-) + (\sigma_1 I - A_\star)\gamma_1 U_\star^{\nu+1} - (\sigma_1 I - A_-)\mathbf{U}_- = 0, \quad \text{on } y = \xi = 0 \\ \chi_2^{\nu+1}(u_+ - u_\star) + (\sigma_2 I - A_+)\mathbf{U}_+ - (\sigma_2 I - A_\star)\gamma_2 U_\star^{\nu+1} = 0, \quad \text{on } z = \eta = 0 \end{cases} \tag{3.2}$$

with $(U_\star^0, \chi_1^0, \chi_2^0)$ being identically zero.

By studying a linear problem similar to (3.2), in the same way as in the first step, we can get that there is $T > 0$ such that the sequences $\{U_\star^\nu, \chi_1^\nu, \chi_2^\nu\}_{\nu \geq 0}$ are convergent in $C_p^1(\Omega_T^\star : I\!R^2)$, and the limits $(U_\star, \chi_1, \chi_2)$ are unique solutions to (2.22).

STEP 3: At this step, we explain the idea of the justification of the asymptotic properties (2.24) and (2.25).

From the problems (2.16) and (2.22), we know that the first line of (2.25) can be infered from (2.24). As in [6, I, Theorem 5.2], the second line of (2.25) can be easily obtained from the first line by using

$$d_t\varphi_k(t) = I\!E_0\chi_k(t) = \lim_{\rho\to\infty} \frac{1}{2\rho}\int_{-\rho}^{\rho} \chi_k(t,\tau)d\tau. \tag{3.3}$$

The proof of (2.24) in the L^∞−norm is given by using the technique of simultaneous Picard iteration: for each $\nu \geq 0$, we establish the asymptotics

$$u_\star^{\epsilon,\nu}(y,z) = U_\star^\nu(y,z;\frac{y}{\epsilon},\frac{z}{\epsilon}) + o(1) \quad \text{in} \quad L^\infty(\Omega_T^\star) \tag{3.4}$$

which immediately implies the part of (2.24) in the L^∞−norm by using the uniformity of the L^∞−convergence of $u_\star^{\epsilon,\nu}$ in $\epsilon \in (0, \epsilon_0]$.

The asymptotic property of $\epsilon\nabla_{(y,z)}u_\star^\epsilon$ is obtained by studying the nonlinear problem (2.16) directly.

The detailed demonstration of the main results can be found in [6, II].

ACKNOWLEDGMENTS: The author would like to express his gratitude to Michael Oberguggenberger for his invitation to participate in the workshop "Nonlinear theory of generalized functions" in November 1997 in the ESI, and to the adminstrators of the ESI for their hospitality and the financial support. This research was partially supported by a Lise Meitner postdoctoral fellowship of the Austrian Science Foundation, the NNSF-19501027 of China, and the Shanghai Qimingxing Foundation 97QA14040.

References

[1] Corli, A.: Weakly nonlinear geometric optics for hyperbolic systems of conservation laws with shock waves. *Asymptotic Anal.,* 10(1995), 117-172.

[2] Joly, J.L., Métivier, G. & Rauch, J.: Several recent results in nonlinear geometric optics, in "*Partial Differential Equations and Mathematical Physics*", L.Hörmander & A.Melin Eds., Birkhaüser, Boston 1996, 181-206.

[3] Joly, J.L., Métivier, G. & Rauch, J.: Resonantly one dimensional nonlinear geometric optics. *J. Funct. Anal.,* 114(1993), 106-231.

[4] Métivier, G.: Interaction de deux chocs pour un système de deux lois de conservation, en dimension deux d'espace. *Trans. Amer. Math. Soc.,* 296(1986), 431-479.

[5] Sablé-Tougeron, M.: Stabilité BV de la structure d'une solution de Riemann à deux grands chocs. Preprint, Univ. de Rennes 1, 1995.

[6] Wang, Y.G.: Nonlinear geometric optics for shock waves, I: scalar case; II: system case. *Z. Anal. Anwendungen,* 16(1997), No. 3, 607-619; No. 4, 857-918.

[7] Wang, Y.G.: Multiple highly oscillatory shock waves in 1-d, in preparation.

[8] Williams, M.: Highly oscillatory multidimensional shocks. Preprint, Univ. of North Carolina, 1997.

Y. G. Wang
Department of Applied Mathematics, Shanghai Jiao Tong University, Shanghai 200240, P. R. China, E-mail: ygwang@online.sh.cn

STRUCTURE THEORY

Sharp Topologies on $(\mathcal{C}, \mathcal{E}, \mathcal{P})$-Algebras

ANTOINE DELCROIX AND DIMITRIS SCARPALEZOS

1 Introduction: $(\mathcal{C}, \mathcal{E}, \mathcal{P})$-extensions

1.1 Introduction

In ([5]) J.-A. Marti introduces the notion of $(\mathcal{C}, \mathcal{E}, \mathcal{P})$-extensions of algebras. These algebras are based on a ring $\mathcal{C}$ of generalized numbers and a topological algebra $\mathcal{E}$, the topology of $\mathcal{E}$ being given by a family $\mathcal{P}$ of seminorms. By various choices of $\mathcal{C}$, $\mathcal{E}$ and $\mathcal{P}$, one recovers most of the known constructions of generalized functions (see for example [2], [6], [7]).

In this paper, we introduce topologies on these extensions, or at least on large classes of $(\mathcal{C}, \mathcal{E}, \mathcal{P})$-extensions. Generalizing some results of [4] and [8], we give an *extension principle* which contains sufficient conditions for extending a map between two topological algebras to their respective extensions. This principle is the main tool to study differential equations with generalized data or operators. Some applications and the detailed proofs will be presented in a forthcoming paper.

1.2 Algebraical settings

Notations.

1. $\mathbb{K}$ will stand either for the real or the complex field.
2. For every set $\mathcal{F}$, elements of $\mathcal{F}^{(0,1]}$ will be denoted by $s = (s_\varepsilon)_\varepsilon$.
3. For every subset $B \subset \mathbb{K}^{(0,1]}$, we put $B^+ = \{(r_\varepsilon)_\varepsilon \in B \mid \forall \varepsilon \in (0,1] : \ s_\varepsilon \geq 0\}$.

We shall consider $\mathbb{K}^{(0,1]}$ with its natural structure of a $\mathbb{K}$-algebra. Let A denote a (fixed) subring of $\mathbb{K}^{(0,1]}$ which is *solid* as a ring (with compatible lattice structure) in the following sense:

Definition 1 *A is said to be* solid *if from $(s_\varepsilon)_\varepsilon \in A$ and $|t_\varepsilon| \leq |s_\varepsilon|$ for each $\varepsilon \in (0,1]$ it follows $(t_\varepsilon)_\varepsilon \in A$.*

As an immediate consequence of this, A is even an algebra over $\mathbb{K}$.
Let I be an ideal of A which is solid as well. The quotient ring $\mathcal{C} = A/I$ is then called a *ring of generalized numbers.*

Let $\mathcal{E}$ be a vector space on $\mathbb{K}$ with a topological structure given by a family of seminorms $\mathcal{P} = (p_j)_{j\in J}$ satisfying

$$\forall (j_1, j_2, \ldots, j_n) \in J^n \;\exists k \in J \;\forall f \in \mathcal{E} \;\forall l \in \{1, \ldots, n\} : \quad p_{j_l}(f) \leq p_k(f). \tag{1}$$

We define two subsets of $\mathcal{E}^{(0,1]}$ by

$$\mathcal{H}_A(\mathcal{E}) = \left\{ (u_\varepsilon)_\varepsilon \in \mathcal{E}^{(0,1]} \mid \forall j \in J : \; (p_j(u_\varepsilon))_\varepsilon \in A^+ \right\},$$

$$\mathcal{I}_A(\mathcal{E}) = \left\{ (u_\varepsilon)_\varepsilon \in \mathcal{E}^{(0,1]} \mid \forall j \in J : \; (p_j(u_\varepsilon))_\varepsilon \in I^+ \right\}.$$

Proposition 1 *The following properties hold:*
1. *$\mathcal{H}_A(\mathcal{E})$ is a vector subspace of $\mathcal{E}^{(0,1]}$ and $\mathcal{I}_A(\mathcal{E})$ is a vector subspace of $\mathcal{H}_A(\mathcal{E})$.*
2. *The factor space $\mathcal{H}_A(\mathbb{K})/\mathcal{I}_A(\mathbb{K})$, where the topology on $\mathbb{K}$ is given by the modulus, is equal to the factor space A/I.*

Definition 2 *The vector space $\mathcal{A}_{\mathcal{C}}(\mathcal{E}) = \mathcal{H}_A(\mathcal{E})/\mathcal{I}_A(\mathcal{E})$ is called the* $(\mathcal{C}, \mathcal{E}, \mathcal{P})$-extension of $\mathcal{E}$.

Suppose that in addition, $\mathcal{E}$ is a topological algebra. Then, in particular, the following property holds:

$$\forall i \in J \;\exists j \in J \;\exists k \in J : \qquad p_i(fg) \leq p_j(f) p_k(g). \tag{2}$$

We can show that under this assumption $\mathcal{H}_A(\mathcal{E})$ is a subalgebra of $\mathcal{E}^{(0,1]}$ and $\mathcal{I}_A(\mathcal{E})$ is an ideal of $\mathcal{H}_A(\mathcal{E})$. $\mathcal{A}_{\mathcal{C}}(\mathcal{E})$ is called a $(\mathcal{C}, \mathcal{E}, \mathcal{P})$-*algebra.*

2 The sharp topology on $(\mathcal{C}, \mathcal{E}, \mathcal{P})$-algebras

In order to simplify some notations, we define a relation of partial strict order in $\mathbb{R}^{(0,1]}$ by

$$\forall (a_\varepsilon)_\varepsilon \in \mathbb{R}^{(0,1]} \;\forall (b_\varepsilon)_\varepsilon \in \mathbb{R}^{(0,1]} : \qquad (a_\varepsilon)_\varepsilon \prec (b_\varepsilon)_\varepsilon \Leftrightarrow \exists \eta > 0 \;\forall \varepsilon \in (0, \eta] : \quad a_\varepsilon < b_\varepsilon.$$

Let $\mathcal{C} = A/I$ be a ring of generalized numbers.

Definition 3 *A set $\mathcal{B} \subset (\mathbb{R}^+)^{(0,1]}$ will be called a set of* bounds for A *if the following properties hold:*

1. for all $(a_\varepsilon)_\varepsilon \in A$ there exists $(r_\varepsilon)_\varepsilon \in \mathcal{B}$ such that $(|a_\varepsilon|)_\varepsilon \prec (r_\varepsilon)_\varepsilon$;

2. for all $(r_\varepsilon)_\varepsilon$ and $(s_\varepsilon)_\varepsilon \in \mathcal{B}$, there exists $(t_\varepsilon)_\varepsilon$ and $(t'_\varepsilon)_\varepsilon$ in $\mathcal{B}$ such that

$$(t_\varepsilon)_\varepsilon \prec (\inf(r_\varepsilon, s_\varepsilon))_\varepsilon \qquad (\sup(r_\varepsilon, s_\varepsilon))_\varepsilon \prec (t'_\varepsilon)_\varepsilon;$$

3. for all $(r_\varepsilon)_\varepsilon \in \mathcal{B}$ there exist $(s_\varepsilon)_\varepsilon$ and $(s'_\varepsilon)_\varepsilon$ in $\mathcal{B}$ such that

$$(s_\varepsilon)_\varepsilon \prec (1/2)(r_\varepsilon)_\varepsilon \qquad 2(r_\varepsilon)_\varepsilon \prec (s'_\varepsilon)_\varepsilon;$$

4. for all $(r_\varepsilon)_\varepsilon$ and $(s_\varepsilon)_\varepsilon$ in $\mathcal{B}$ there exists $(t_\varepsilon)_\varepsilon \in \mathcal{B}$ such that $(r_\varepsilon t_\varepsilon)_\varepsilon \prec (s_\varepsilon)_\varepsilon$.

Example. If A is a solid subring of $\mathbb{K}^{(0,1]}$ containing the unit element $(1)_\varepsilon$, the set of positive invertible elements of A is a set of bounds.

We suppose now that for A a family of bounds $\mathcal{B}$ is given. For $a = (a_\varepsilon)_\varepsilon \in A$ and $(r_\varepsilon)_\varepsilon \in \mathcal{B}$, the ball of center a and "radius" $(r_\varepsilon)_\varepsilon$ is the set

$$B(a, r_\varepsilon) = \{(t_\varepsilon)_\varepsilon \in A \mid (|t_\varepsilon - a_\varepsilon|)_\varepsilon \prec (r_\varepsilon)_\varepsilon\}.$$

As A is an additive group, we only have to define the neighborhood filter of 0, denoted by $\mathcal{V}(0)$:

$$\mathcal{V}(0) = \left\{V \subset \mathbb{K}^{(0,1]} \mid \exists (r_\varepsilon)_\varepsilon \in \mathcal{B} \text{ such that } B(0, r_\varepsilon) \subset V\right\}.$$

Lemma 2 *The following properties hold:*

a. for each $(r_\varepsilon)_\varepsilon \in \mathcal{B}$, $B(0, r_\varepsilon) \neq \emptyset$;

b. for each $(r_\varepsilon)_\varepsilon \in \mathcal{B}$ and $(s_\varepsilon)_\varepsilon \in \mathcal{B}$ there exists $(t_\varepsilon)_\varepsilon \in \mathcal{B}$ such that

$$B(0, t_\varepsilon) \subset B(0, r_\varepsilon) \cap B(0, s_\varepsilon);$$

c. for each $U \in \mathcal{V}(0)$, there exists $V \in \mathcal{V}(0)$ such that

$$\forall (x_\varepsilon)_\varepsilon \in V\ \exists W \in \mathcal{V}((x_\varepsilon)_\varepsilon) : \ W \subset U.$$

Assertions *a.*, *b.* and *c.* of Lemma 2 show that $\mathcal{V}(0)$ defines a topology on A, called the sharp topology.

Lemma 3 *The addition and the multiplication on A are continuous for the sharp topology.*

The **Proof** is straightforward.

The ring A is then a topological ring. This topology transfers to the quotient space $\mathcal{C} = A/I$ and will be called the *sharp topology* related to $\mathcal{B}$.

To obtain a nondiscrete topology on $\mathcal{C}$, we shall suppose from now on that $\forall (r_\varepsilon)_\varepsilon \in \mathcal{B} \quad I \subset B(0, r_\varepsilon)$.

Note that this topology is Hausdorff if and only if $I = \bigcap_{V \in \mathcal{V}(0)} V = \bigcap_{(\eta_\varepsilon)_\varepsilon \in \mathcal{B}} B(0, \eta_\varepsilon)$.

Example. For a solid subring A of $\mathbb{K}^{(0,1]}$, put

$$\mathcal{B}' = \left\{ (r_\varepsilon)_\varepsilon \in A^+ \text{ such that } (r_\varepsilon)_\varepsilon \text{ is invertible} \right\}.$$

Then verify that $I' = \bigcap_{(\eta_\varepsilon)_\varepsilon \in \mathcal{B}'} B(0, \eta_\varepsilon)$ is an ideal of A. Hence, $\mathcal{C} = A/I'$ is a ring of generalized numbers, Hausdorff for the sharp topology given by $\mathcal{B}'$.

Let us consider a topological algebra $\left(\mathcal{E}, \mathcal{P} = (p_j)_{j \in J}\right)$ over $\mathbb{K}$ satisfying properties 1 and 2 (notations are the same as in Section 1.2.).

For each $(\eta_\varepsilon)_\varepsilon \in \mathcal{B}$ and each $j \in J$, we set

$$B_j(0, \eta_\varepsilon) = \left\{ (f_\varepsilon)_\varepsilon \in \mathcal{E}^{(0,1]} \mid \exists \varepsilon_0 \in (0,1]\ \forall \varepsilon \in (0, \varepsilon_0] : \ p_j(f_\varepsilon) < \eta_\varepsilon \right\}$$

and define the neighborhood filter of 0 in $\mathcal{E}^{(0,1]}$ by

$$\mathcal{V}(0) = \left\{ V \subset \mathcal{E}^{(0,1]} \mid \exists (\eta_\varepsilon)_\varepsilon \in \mathcal{B}\ \exists j \in J \text{ such that } B_j(0, \eta_\varepsilon) \subset V \right\}. \tag{3}$$

Proposition 4 *With the previous definitions, $\mathcal{H}_A(\mathcal{E})$ is a topological algebra.*

The **proof** of Proposition 4 proceeds like the proof of Lemmas 2 and 3. Properties 1 and 2 of the family $\mathcal{P}$ of seminorms are used in an essential way.

The topology carries over to the quotient algebra $\mathcal{H}_A(\mathcal{E})/\mathcal{I}_A(\mathcal{E})$ and will be called the *sharp topology*. As before, the algebra $\mathcal{A}_\mathcal{C}(\mathcal{E})$ is Hausdorff if and only if

$$\mathcal{I}_A(\mathcal{E}) = \bigcap_{V \in \mathcal{V}(0)} V = \bigcap_{(\eta_\varepsilon)_\varepsilon \in \mathcal{B},\, j \in J} B_j(0, \eta_\varepsilon).$$

One verifies that this occurs if and only if $I = \bigcap_{(\eta_\varepsilon)_\varepsilon \in \mathcal{B}} B(0, \eta_\varepsilon)$.

3 Examples

3.1 Asymptotic algebras

Let $\mathcal{S} = (a_n : (0,1] \to \mathbb{R}_+ \setminus \{0\})_{n\in\mathbb{Z}}$ be a sequence of functions and $\mathcal{E}$ a topological algebra with its uniform structure given by an increasing countable family of seminorms $(p_n)_{n\in\mathbb{N}}$.

We assume that $\mathcal{S}$ is an *asymptotic scale,* that is a_0 is bounded and

1. for each $n \in \mathbb{N}$ (except possibly $n = 0$) $\lim_{\varepsilon\to 0} a_n(\varepsilon) = 0$,
2. $\forall n \in \mathbb{N}: \ a_{n+1}(\varepsilon) = \mathrm{o}(a_n(\varepsilon))$ for $\varepsilon \to 0$,
3. $\forall n \in \mathbb{N}\ \forall m \in \mathbb{N}\ \exists N \in \mathbb{N}: \ a_N(\varepsilon) = \mathrm{O}(a_n(\varepsilon)a_m(\varepsilon))$ for $\varepsilon \to 0$,
4. $\forall n \in \mathbb{N}\setminus\{0\}$: $a_{-n}(\varepsilon) = 1/a_n(\varepsilon)$.

We choose

$$A_S = \left\{(s_\varepsilon)_\varepsilon \in \mathbb{K}^{(0,1]} \mid \exists j \in \mathbb{Z}: \ s_\varepsilon = \mathrm{o}(a_j(\varepsilon)) \text{ for } \varepsilon \to 0\right\},$$

$$I_S = \left\{(s_\varepsilon)_\varepsilon \in \mathbb{K}^{(0,1]} \mid \forall j \in \mathbb{Z}: \ s_\varepsilon = \mathrm{o}(a_j(\varepsilon)) \text{ for } \varepsilon \to 0\right\}.$$

Both of these sets are solid. The algebra $\mathcal{A}_S(\mathcal{E}) = \mathcal{H}_{A_S}(\mathcal{E})/\mathcal{I}_{A_S}(\mathcal{E})$ is called the *asymptotic algebra of scale $\mathcal{S}$ related to $\mathcal{E}$.*

We recall briefly how the sharp topology was introduced in [4] for asymptotic algebras. Let $u = (u_\varepsilon)$ be an element of $\mathcal{H}_{A_S}(\mathcal{E})$ and $n \in \mathbb{N}$ an integer. The *n-valuation of* u, noted $v_n(u)$, is the least upper bound (eventually infinite) of the $q \in \mathbb{Z}$ such that $\mu_n(u_\varepsilon) = \mathrm{O}(a_q(\varepsilon))$. A family δ_n of ultrametric pseudodistances on $\mathcal{H}_{A_S}(\mathcal{E})$ is then defined by

$$\forall n \in \mathbb{N} \quad \forall (u,v) \in (\mathcal{H}_{A_S}(\mathcal{E}))^2: \quad \delta_n(u,v) = \exp(-v_n(u-v)).$$

Clearly, an element u of $\mathcal{H}_{A_S}(\mathcal{E})$ is in $\mathcal{I}_{A_S}(\mathcal{E})$ if and only if, for each n in $\mathbb{N}$, $\delta_n(u,0) = 0$. Consequently, the above definitions carry over to the quotient space $\mathcal{A}_S(\mathcal{E})$ and we obtain valuations and ultrametric pseudodistances on $\mathcal{A}_S(\mathcal{E})$.

On the other hand, it is easy to check that $\mathcal{B} = \{a_n(\varepsilon),\ n \in \mathbb{Z}\setminus\{0\}\}$ is a set of bounds for $\mathcal{A}_S$. The following properties hold:

$$\forall (n,m) \in \mathbb{N}\times(\mathbb{N}\setminus\{0\}): \qquad B_n(0, a_m(\varepsilon)) \subseteq \left\{(u_\varepsilon)_\varepsilon \in \mathcal{H}_{A_S}(\mathcal{E}) \mid \delta_n(u,0) < e^{-m+1}\right\}.$$

$$\forall (n,m) \in \mathbb{N}\times(\mathbb{N}\setminus\{0\}): \qquad B_n(0, a_m(\varepsilon)) \supseteq \left\{(u_\varepsilon)_\varepsilon \in \mathcal{H}_{A_S}(\mathcal{E}) \mid \delta_n(u,0) < e^{-m}\right\}.$$

Hence the topologies given by the valuations and by the bounds $\mathcal{B}$, respectively, have the same base of neighborhoods of zero. Thus the sharp topology introduced in [4] and the topology given by $\mathcal{B}$ are equal on $\mathcal{H}_{A_S}(\mathcal{E})$ and on $\mathcal{A}_S(\mathcal{E})$.

As the ideal $\mathcal{I}_{A_S}(\mathcal{E})$ is equal to the intersection of the neighborhoods of 0, $\mathcal{A}_S(\mathcal{E})$ is Hausdorff. (See [4] and [8] for more properties about sharp topologies).

3.2 Algebras with infra-exponential growth

We take here

- $A = \left\{ (s_\varepsilon)_\varepsilon \in \mathbb{K}^{(0,1]} \mid \forall \sigma > 0 \ \exists \eta > 0 \ \forall \varepsilon \in (0,\eta] : \ |s_\varepsilon| < \exp(\sigma/\varepsilon) \right\}$, the set of nets of infra-exponential growth;
- $I = \left\{ (s_\varepsilon)_\varepsilon \in \mathbb{K}^{(0,1]} \mid \exists \rho > 0 \ \exists \eta > 0 \ \forall \varepsilon \in (0,\eta] : \ |s_\varepsilon| < \exp(-\rho/\varepsilon) \right\}$, the set of nets of exponential decrease.

It is easy to see that I is an ideal of A. The ring $\mathcal{C} = A/I$ occurs in problems linked with the embedding of periodic hyperfunctions in algebras of generalized functions. We remark that here A is naturally defined as an intersection of subsets of $\mathbb{K}^{(0,1]}$ and I as an union, this being exactly the contrary of the asymptotic algebras.
Let us define $\mathcal{B} = \left\{ (r_\varepsilon)_\varepsilon \in (\mathbb{K}^+ \setminus \{0\})^{(0,1]} \mid \ \limsup_{\varepsilon \to 0^+} \varepsilon \ln (r_\varepsilon) = 0 \right\}$.

Lemma 5 *The set $\mathcal{B}$ constitutes a set of bounds for which $I = \bigcap_{(\eta_\varepsilon)_\varepsilon \in \mathcal{B}} B(0, \eta_\varepsilon)$.*

Therefore, a $(\mathcal{C}, \mathcal{E}, \mathcal{P})$-extension based on this choice of the ring $\mathcal{C}$ can be endowed with a Hausdorff topology given by the family $\mathcal{B}$ of bounds.

4 Extension of maps and functorial properties

We investigate functorial properties of the above construction. We are led to look for which mappings from a seminormed space to another can be extended to mappings between the respective extensions of the initial spaces. This is a generalization of an initial idea of J.-F. Colombeau and H. Biagioni ([1]) and this has been developed for asymptotic algebras in [3] and [4].

Let $(\mathcal{C}_1 = A_1/I_1, \ \mathcal{C}_2 = A_2/I_2)$ be a couple of rings of generalized numbers and $\mathcal{B}_1$ and $\mathcal{B}_2$ be respective sets of bounds.

Definition 4 *Let* $g=(g_\varepsilon)_\varepsilon$ *be a net of functions from* $\mathbb{R}$ *to* $\mathbb{R}_+\setminus\{0\}$.

1. *The net* g *is* $(\mathcal{C}_1,\mathcal{C}_2)$-moderate *if for all* $(u_\varepsilon)_\varepsilon\in A_1^+$, $(g_\varepsilon(u_\varepsilon))_\varepsilon$ *is in* A_2^+.
2. *The net* g *is* algebraically $(\mathcal{C}_1,\mathcal{C}_2)$-compatible *if for all* $(u_\varepsilon)_\varepsilon\in I_1^+$, $(g_\varepsilon(u_\varepsilon))_\varepsilon$ *is in* I_2^+.
3. *The net* g *is* topologically $(\mathcal{B}_1,\mathcal{B}_2)$-compatible *if for each* ε *the function* $g_\varepsilon:\mathbb{R}\to\mathbb{R}_+\setminus\{0\}$ *is increasing and if*

$$\forall(a_\varepsilon)_\varepsilon\in\mathcal{B}_2\ \ \forall(s_\varepsilon)_\varepsilon\in\mathcal{B}_2\ \exists(r_\varepsilon)_\varepsilon\in\mathcal{B}_1:\quad(a_\varepsilon g_\varepsilon(r_\varepsilon))_\varepsilon\prec(s_\varepsilon)_\varepsilon.$$

Proposition 6 *The set of moderate nets is stable under addition, ring multiplication and multiplication by positive scalars. It contains the polynomials with coefficients in* A_2^+, *that is the nets of the form*

$$g_\varepsilon(u)=\textstyle\sum_{i\in I}a_\varepsilon^i u^i,\quad I \text{ a finite subset of } \mathbb{N},\ \left(a_\varepsilon^i\right)\in A_2^+.$$

Let $\left(\mathcal{E},\mathcal{P}=(p_j)_{j\in J}\right)$ and $\left(\mathcal{F},\mathcal{Q}=(q_k)_{k\in K}\right)$ be two semi-normed algebras as in section 1.

Definition 5 *Let* Φ *be a map from* $\mathcal{E}$ *to* $\mathcal{F}$.

1. *The map* Φ *is* algebraically $(\mathcal{C}_1,\mathcal{C}_2)$-temperate, *if*
   i. *for all* $k\in K$ *there exists* $j\in J$ *and a* $(\mathcal{C}_1,\mathcal{C}_2)$*-moderate net* g *such that*

$$\forall\varepsilon\in(0,1]\ \ \forall x\in\mathcal{E}:\qquad q_k(\Phi(x))\le g_\varepsilon(p_j(x));$$

   ii. *for all* $k\in K$ *there exists* $(j,q)\in J\times J$, *a* $(\mathcal{C}_1,\mathcal{C}_2)$*-moderate net* g *and a* $(\mathcal{C}_1,\mathcal{C}_2)$*-compatible net* h *such that*

$$\forall\varepsilon\in(0,1]\ \ \forall(x,y)\in\mathcal{E}^2:\qquad q_k(\Phi(x+y)-\Phi(x))\le g_\varepsilon(p_j(x))\,h_\varepsilon(p_q(y)).\quad(4)$$

2. *The map* Φ *is* topologically temperate *with respect to* $(\mathcal{C}_1,\mathcal{C}_2)$ *and* $(\mathcal{B}_1,\mathcal{B}_2)$ *if* Φ *is*

   algebraically $(\mathcal{C}_1,\mathcal{C}_2)$*-temperate and if in (4) the net* h *can be chosen topologically* $(\mathcal{B}_1,\mathcal{B}_2)$*-compatible.*

Theorem 7 *Any algebraically temperate map* Φ *from* $\mathcal{E}$ *to* $\mathcal{F}$ *admits a canonical extension from* $\mathcal{A}_{\mathcal{C}_1}(\mathcal{E})$ *to* $\mathcal{A}_{\mathcal{C}_2}(\mathcal{F})$. *Moreover, if* Φ *is topologically temperate, this extension is continuous for the sharp topologies* $\mathcal{A}_{\mathcal{C}_1}(\mathcal{E})$ *and* $\mathcal{A}_{\mathcal{C}_2}(\mathcal{F})$.

Proof. Let $\Phi : \mathcal{E} \mapsto \mathcal{F}$ be an algebraically temperate map. There is a canonical extension of Φ (also denoted Φ) from $\mathcal{H}_A(\mathcal{E})$ to $\mathcal{F}^{(0,1]}$, given by $\Phi(u) = (\Phi(u_\varepsilon))_\varepsilon$ for all $u = (u_\varepsilon)_\varepsilon \in \mathcal{H}_A(\mathcal{E})$.

1. *For all $u = (u_\varepsilon)_\varepsilon \in \mathcal{H}_{A_1}(\mathcal{E})$, $(\Phi(u_\varepsilon))_\varepsilon$ belongs to $\mathcal{H}_{A_2}(\mathcal{E})$.*

For $k \in K$ there exists $j \in J$ and a moderate net g such that

$$\forall \varepsilon \in (0,1] : \qquad q_k(\Phi(u_\varepsilon)) \leq g_\varepsilon(p_j(u_\varepsilon)).$$

As u is in $\mathcal{H}_{A_1}(\mathcal{E})$, $(p_j(u_\varepsilon))_\varepsilon$ belongs to A_1^+ and, since g is moderate, $(g_\varepsilon(p_j(u_\varepsilon)))_\varepsilon$ to A_2^+. Using the solidity of A_2, we obtain that $(q_k(\Phi(u_\varepsilon)))_\varepsilon$ is in A_2^+ and $(\Phi(u_\varepsilon))_\varepsilon$ is in $\mathcal{H}_{A_2}(\mathcal{E})$.

2. *For all $u = (u_\varepsilon)_\varepsilon \in \mathcal{H}_{A_1}(\mathcal{E})$ and all $\eta = (\eta_\varepsilon)_\varepsilon \in \mathcal{I}_{A_1}(\mathcal{E})$, the equality $[\Phi(u)] = [\Phi(u+\eta)]$ holds, where $[\Phi]$ denote the class of Φ in $\mathcal{A}_{\mathcal{C}_2}(\mathcal{F})$.*

For $k \in K$, there exist $(j,q) \in I \times I$, a moderate net g and an algebraically compatible net h such that

$$\forall \varepsilon \in (0,1] : \qquad q_k(\Phi(u_\varepsilon + \eta_\varepsilon) - \Phi(u_\varepsilon)) \leq g_\varepsilon(p_j(u_\varepsilon))\, h_\varepsilon(p_q(\eta_\varepsilon)). \qquad (5)$$

As in step 1., $(g_\varepsilon(p_j(u_\varepsilon)))_\varepsilon$ belongs to A_2^+. The compatibility of h similarly shows that $(h_\varepsilon(p_q(\eta_\varepsilon)))_\varepsilon$ is in I_2^+, for all $(\eta_\varepsilon)_\varepsilon$ in I_1^+. Then the product on the right-hand side of (5) belongs to I_2^+. Hence, $(\Phi(u_\varepsilon + \eta_\varepsilon) - \Phi(u_\varepsilon))_\varepsilon$ is in $\mathcal{I}_{A_2}(\mathcal{F})$.

We can now define the extension $\widetilde{\Phi}$ from $\mathcal{A}_{\mathcal{C}_1}(\mathcal{E})$ to $\mathcal{A}_{\mathcal{C}_2}(\mathcal{F})$ by $\widetilde{\Phi}(U) = [\Phi(u)]$ where $u = (u_\varepsilon)_\varepsilon$ is any representative of $U \in \mathcal{A}_{\mathcal{C}_1}(\mathcal{E})$.

Suppose now that Φ is topologically temperate. We remark that inequality (5) holds for all $(\eta_\varepsilon)_\varepsilon \in \mathcal{H}_{A_1}(\mathcal{E})$ and put $(a_\varepsilon)_\varepsilon = (g(p_j(u_\varepsilon)))_\varepsilon \in A_2^+$. Let $(s_\varepsilon)_\varepsilon$ be in $\mathcal{B}_2$. As the net h can be chosen topologically compatible, there exists $(r_\varepsilon)_\varepsilon$ in $\mathcal{B}_1$ such that $(a_\varepsilon h_\varepsilon(r_\varepsilon))_\varepsilon \prec (s_\varepsilon)_\varepsilon$. Thus we have

$$\forall (\eta_\varepsilon)_\varepsilon \in \mathcal{H}_{A_1}(\mathcal{E}), \quad (p_q(\eta_\varepsilon))_\varepsilon \prec (r_\varepsilon)_\varepsilon \Longrightarrow (q_k(\Phi(u_\varepsilon + \eta_\varepsilon) - \Phi(u_\varepsilon)))_\varepsilon \leq (s_\varepsilon)_\varepsilon$$

because each h_ε is an increasing map. We finally obtain

$$\begin{gathered}\forall k \in K \ \forall (s_\varepsilon)_\varepsilon \in \mathcal{B}_2 \ \exists q \in I \ \exists (r_\varepsilon)_\varepsilon \in \mathcal{B}_1 \ \forall (\eta_\varepsilon)_\varepsilon \in B_q(u, r_\varepsilon) : \\ (\Phi(u_\varepsilon + \eta_\varepsilon))_\varepsilon \in B_k((\Phi(u_\varepsilon))_\varepsilon, s_\varepsilon).\end{gathered}$$

Hence the canonical extension of Φ is continuous as a mapping from $\mathcal{H}_{A_1}(\mathcal{E})$ to $\mathcal{H}_{A_2}(\mathcal{E})$. This property carries over to the extension of Φ to the respective quotient spaces. ■

Acknowledgments. The authors are grateful to M. Oberguggenberger and to the Erwin Schrödinger Institute for their invitation and support.

References

[1] BIAGIONI H. A., COLOMBEAU J.-F.: New generalized functions and C^∞ functions with values in generalized complex numbers. J. London Math. Soc. 33(2)(1986),167-179.

[2] COLOMBEAU J.-F.: Elementary Introduction to New Generalized Functions, North Holland Math. Studies, n^o113, Amsterdam 1985.

[3] DELCROIX A. AND SCARPALEZOS D.: Asymptotic scales, asymptotic algebras, to appear in Integral Transforms and Special Functions.

[4] DELCROIX A. AND SCARPALEZOS D.: Topology on asymptotic algebras of generalized functions and applications, Submitted to Monatshefte für Mathematik.

[5] MARTI J.-A., NUIRO P. S., VALMORIN V.: A nonlinear Goursat problem, to appear in Integral Transforms and Special Functions.

[6] OBERGUGGENBERGER M.: Nonlinear theories of generalized functions, Proceedings of the international conference on Applications of Nonstandard Analysis, Functional Analysis and Probability Theory. Blaubeuren, 1992.

[7] ROSINGER E. E.: Generalized solutions of nonlinear partial differential equations, North Holland, Math. Studies n^o146, 1987.

[8] SCARPALEZOS D.: Some remarks on the functoriality of Colombeau's construction, on topological and microlocal aspects and applications, to appear in Integral Transforms and Special Functions.

A. DELCROIX
IUFM des Antilles et de la Guyane, Morne Ferret, BP 399, 97159 Pointe à Pitre Cedex, Guadeloupe, France, E-mail: antoine.delcroix@univ-ag.fr
D. SCARPALEZOS
Universite Paris 7, F-75251 Paris Cedex 05, France

$(\mathcal{C},\mathcal{E},\mathcal{P})$-Sheaf Structures and Applications

JEAN-ANDRE MARTI

1 Introduction

Let $\mathcal{C}$ be a ring of generalized numbers, $\mathcal{E}$ a sheaf of algebras on a topological space X and, for each open set Ω of X, $\mathcal{P}(\Omega)$ a family of seminorms on $\mathcal{E}(\Omega)$. Then, we can construct a sheaf $\mathcal{A}$ of $(\mathcal{C}$,$\mathcal{E}$,$\mathcal{P})$-algebras which generalize many previous approaches in the theory of generalized functions [2][3][4][8][15][1]. The sheaf structure of $\mathcal{A}$ permits to define localization and microlocalization processes [9][16][17]. Then we develop the idea to adapt $\mathcal{C}$,$\mathcal{E}$,$\mathcal{P}$ as parameters to each problem to solve [10][13][14][11][12]. Restriction and derivation can be defined and (with product) are good tools to pose and solve many non linear (and linear) problems with irregular data which have no solution in the distributional sense [14][18]. We give here another application. We will show that some perturbation problems

$$\begin{cases} \varphi(\varepsilon)\frac{dX}{dt} = f_\varepsilon(t,X) \\ X(0) = \psi(\varepsilon) \end{cases}$$

can be interpreted as fixed point problems in some algebras whose parameters are adjusted to the perturbations.

2 The sheaves of $(\mathcal{C},\mathcal{E},\mathcal{P})$-algebras

2.1 The algebra structure

a) Suppose given

- a subring A of the ring $\mathbb{K}^{]0,1]}$, ($\mathbb{K} = \mathbb{R}$ or $\mathbb{C}$)|
- $A^+ = \{(r_\varepsilon)_\varepsilon \in A : r_\varepsilon \geq 0\}$
- the following *stability by overestimation property* for A: whenever $(|s_\varepsilon|)_\varepsilon \leq (r_\varepsilon)_\varepsilon$ (that is to say: for each ε we have: $|s_\varepsilon|_\varepsilon \leq r_\varepsilon$) for any pair $((s_\varepsilon)_\varepsilon, (r_\varepsilon)_\varepsilon) \in \mathbb{K}^{]0,1]} \times A^+$,

it follows that we have: $(s_\varepsilon)_\varepsilon \in A$.

• an ideal I_A of A with the same stability by overestimation property

• a sheaf $\mathcal{E}$ of $\mathbb{K}$-algebras on a topological space X, such that, for each open set Ω in X, the algebra $\mathcal{E}(\Omega)$ is endowed with a family $\mathcal{P}(\Omega) = (p_i)_{i\in I(\Omega)}$ of seminorms verifying

$$\forall i \in I(\Omega), \exists (j,k,C) \in I(\Omega) \times I(\Omega) \times \mathbb{R}_+^* : p_i(fg) \leq Cp_j(f)\, p_k(g)$$

• the following property: for two open subsets of X such that $\Omega_1 \subset \Omega_2$, we have $I(\Omega_1) \subset I(\Omega_2)$, and that for every $i \in I(\Omega_1)$ and $u \in \mathcal{E}(\Omega_2)$, we have $p_i(u|_{\Omega_1}) = p_i(u)$.

b) We put

$$\mathcal{H}_{(A,\mathcal{E}_\mathcal{P})}(\Omega) = \left\{ (u_\varepsilon)_\varepsilon \in [\mathcal{E}(\Omega)]^{]0,1]}\, \forall i \in I(\Omega), \forall \varepsilon \in]0,1] : (p_i(u_\varepsilon))_\varepsilon \in A^+ \right\},$$
$$\mathcal{J}_{(I_A,\mathcal{E}_\mathcal{P})}(\Omega) = \left\{ (u_\varepsilon)_\varepsilon \in [\mathcal{E}(\Omega)]^{]0,1]}\, \forall i \in I(\Omega), \forall \varepsilon \in]0,1] : (p_i(u_\varepsilon))_\varepsilon \in I_A^+ \right\}.$$
$$\mathcal{C} = A/I_A$$

It may easily be seen that A^+ is not a subring of A, but is stable under addition and multiplication. The same holds for I_A^+.

Finally, we set

$$A^* = \left\{ (|r_\varepsilon|)_\varepsilon \in \mathbb{R}^{]0,1]}, (r_\varepsilon)_\varepsilon \in A \right\}$$
$$I_A^* = \left\{ (|r_\varepsilon|)_\varepsilon \in \mathbb{R}^{]0,1]}, (r_\varepsilon)_\varepsilon \in I_A \right\}$$

and we suppose that $A^+ = A^*$ and $I_A^+ = I_A^*$.

c) We can show [10], that under the above hypothesis:

Proposition 2.1. *(i)* $\mathcal{H}_{(A,\mathcal{E}_\mathcal{P})}$ *is a sheaf of sub-algebras of the sheaf* $\mathcal{E}^{]0,1]}$ *;*
(ii) $\mathcal{J}_{(I_A,\mathcal{E}_\mathcal{P})}$ *is a sheaf of ideals of* $\mathcal{H}_{(A,\mathcal{E}_\mathcal{P})}$ *;*
(iii) The constant factor sheaf $\mathcal{H}_{(A,\mathbb{K}_{|.|})}/\mathcal{J}_{(I_A,\mathbb{K}_{|.|})}$ *is exactly the factor ring* $\mathcal{C} = A/I_A$.

d) We can now give the following

Definition 2.2. *We call* $(\mathcal{C},\mathcal{E},\mathcal{P})$*-algebra every factor algebra*

$$\mathcal{A} = \mathcal{H}_{(A,\mathcal{E}_\mathcal{P})}/\mathcal{J}_{(I_A,\mathcal{E}_\mathcal{P})}.$$

2.2 Other definitions, operations and processes

By a convenient choice of $\mathcal{C}$, $\mathcal{E}$ and $\mathcal{P}$ as parameters, we can describe many algebras of generalized functions [8] and define some other operations than algebraic ones such as differentiation, restriction, sheaf embedding of $\mathcal{D}'$. The association or weak equality previously defined by Colombeau [3] is a very useful process to study some differential equations in non conservative form, e.g., Burgers' equation.

2.2.1 Overgenerated rings

In view of applications, it is interesting to define rings generated by some given elements. More precisely, let B be the subset of $\mathbb{R}_+^{]0,1]}$ generated algebraically by the elements of the set

$$B_p = \left\{ (r_{n,\varepsilon})_\varepsilon \in \mathbb{R}_+^{]0,1]}, n = 1, 2, ...p. \right\}.$$

Let A be the subset of $\mathbb{K}^{]0,1]}$ defined by

$$A = \left\{ (a_\varepsilon)_\varepsilon \in \mathbb{K}^{]0,1]} \exists \, (b_\varepsilon)_\varepsilon \in B : |a_\varepsilon| \leq b_\varepsilon \right\}.$$

It is easy to see that A is a subring of $\mathbb{K}^{]0,1]}$, with the stability by overestimation property. Then, we say that

Definition 2.3. *A is overgenerated by B_p. And if I_A is some ideal of A with the same stability by overestimation property, we can also say that $\mathcal{C} = A/I_A$ is overgenerated by B_p.*

2.2.2 Relationship with distribution theory

The main advantage of the $(\mathcal{C}, \mathcal{E}, \mathcal{P})$-algebra theory is that we can multiply the distributions in $\mathcal{A}$ as well in the other theories of generalized functions. So we have first to embed $\mathcal{D}'$ in $\mathcal{A}$. If $(\varphi_\varepsilon)_\varepsilon$ is some given family of mollifiers

$$\varphi_\varepsilon(x) = \frac{1}{\varepsilon^n} \varphi\left(\frac{x}{\varepsilon}\right), x \in \mathbb{R}^n, \int \varphi(x)\, dx = 1,$$

we can prove that if one has: $T \in \mathcal{D}'(\mathbb{R}^n)$, the convolution product family $(T * \varphi_\varepsilon)_\varepsilon$ is a family of smooth functions slowly increasing in $\frac{1}{\varepsilon}$. So we can choose the subring A in such a way that it is overgenerated by some subset B_p of $\mathbb{R}_+^{]0,1]}$ containing the family $(\varepsilon)_\varepsilon$.

Example 2.4. If B_p contains only that family, it is easy to prove that we have:

$$A = \left\{ (r_\varepsilon)_\varepsilon \in \mathbb{R}^{]0,1]}, \exists p \in \mathbb{N}, \exists C > 0 : (|r_\varepsilon|)_\varepsilon \leq \left(\frac{C}{\varepsilon^p}\right)_\varepsilon \right\},$$

and we can take:

$$I_A = \left\{ (r_\varepsilon)_\varepsilon \in \mathbb{R}^{]0,1]}, \forall q \in \mathbb{N}, \exists D > 0 : (|r_\varepsilon|)_\varepsilon \leq (D\varepsilon^q)_\varepsilon \right\}.$$

Now, for each open set Ω in X, let us take $\mathcal{E}(\Omega) = \mathrm{C}^\infty(\Omega)$ with the usual $\mathcal{P}(\Omega)$ topology of uniform convergence in all derivatives on compact subsets of Ω. In this case we have:

$$\mathcal{H}_{(A,\mathcal{E}_\mathcal{P})}(\Omega) = \{(u_\varepsilon)_\varepsilon \in [\mathrm{C}^\infty(\Omega)]^{]0,1]} : \forall K \subset\subset \Omega, \forall \alpha \in \mathbb{N}^n, \exists p \in \mathbb{N}, \exists C > 0$$
$$(\sup_{x \in K} |D^\alpha u_\varepsilon(x)|)_\varepsilon \leq (\tfrac{C}{\varepsilon^p})_\varepsilon\},$$

$$\mathcal{J}_{(I_A,\mathcal{E}_\mathcal{P})}(\Omega) = \{(u_\varepsilon)_\varepsilon \in [\mathrm{C}^\infty(\Omega)]^{]0,1]} : \forall K \subset\subset \Omega, \forall \alpha \in \mathbb{N}^n, \forall q \in \mathbb{N}, \exists D > 0$$
$$(\sup_{x \in K} |D^\alpha u_\varepsilon(x)|)_\varepsilon \leq (D\varepsilon^q)_\varepsilon\}.$$

In this case, the algebra $\mathcal{A}(\Omega) = \mathcal{H}_{(A,\mathcal{E}_\mathcal{P})}(\Omega) / \mathcal{J}_{(I_A,\mathcal{E}_\mathcal{P})}(\Omega)$ is exactly Colombeau's simplified algebra, and we can embed $\mathcal{D}'(\mathbb{R}^n)$ into $\mathcal{A}(\mathbb{R}^n)$ by the mapping:

$$T \to cl\,(T * \varphi_\varepsilon)_\varepsilon$$

because $(T * \varphi_\varepsilon)_\varepsilon$ belongs to $\mathcal{H}_{(A,\mathcal{E}_\mathcal{P})}(\mathbb{R}^n)$.

Definition 2.5. *In the same way, with the help of a cutoff function, we can define for each open set Ω in $\mathbb{R}^n$ an embedding of $\mathcal{D}'(\Omega)$ into $\mathcal{A}(\Omega)$ and, finally, a sheaf embedding: $\mathcal{D}' \to \mathcal{A}$.*

This embedding depends on the choice of the mollifier φ_ε and it is easy to define a canonical embedding $\mathrm{C}^\infty \to \mathcal{A}$. But C^k may only be embedded as a subspace of $\mathcal{D}'$.

2.2.3 Restriction

When $\mathcal{E} = \mathrm{C}^\infty$, the restriction to the submanifold

$$\{x = (x_1, x_2, ...x_j...x_n) : x_j = 0\} \subset \mathbb{R}^n$$

of the generalized function $u = cl\,(u_\varepsilon)_\varepsilon \in \mathcal{A}(\mathbb{R}^n)$ is the generalized one:

Definition 2.6.

$$u\,|_{\{x_j=0\}} = cl\,(x \to u_\varepsilon(x_1, x_2, ...x_{j-1}, 0, x_{j+1}, ...x_n))_\varepsilon.$$

That restriction belongs to a subalgebra of $\mathcal{A}(\mathbb{R}^n)$ which is canonically identified with $\mathcal{A}(\mathbb{R}^{n-1})$ and agrees with the usual process met in the literature.

2.2.4 Derivation

When $\mathcal{E} = \mathrm{C}^\infty$, derivation is defined for each $u = cl\,(u_\varepsilon)_\varepsilon \in \mathcal{A}\,(\Omega)$, by:

$$\mathcal{D}^\alpha u = cl(D^\alpha u_\varepsilon)_\varepsilon$$

where D^α is the classical derivation. Naturally the restriction of $\mathcal{D}^\alpha$ to C^∞ or $\mathcal{D}'$ agrees with the classical derivation.

If $\mathcal{E}$ is a sheaf on X, with a "derivation operator", that is to say a sheaf endomorphism D^α such that:

• $D^{\alpha+\beta} = D^\alpha \circ D^\beta = D^\beta \circ D^\alpha$

• $\forall f, g \in \mathcal{E}\,(\Omega)$: $D^\alpha\,(fg) = \sum_{\beta \leq \alpha} \frac{\alpha!}{\beta!(\alpha-\beta)!} D^\beta f D^{\alpha-\beta} g$,

then it is possible to define a sheaf endomorphism $\mathcal{D}^\alpha$ on $\mathcal{A}$ with the same properties as D^α by putting as above, if $u = cl\,(u_\varepsilon)_\varepsilon \in \mathcal{A}\,(\Omega)$,

Definition 2.7. *$\mathcal{D}^\alpha u = cl(D^\alpha u_\varepsilon)_\varepsilon$ under the sole condition:*

$$\textit{If } (u_\varepsilon)_\varepsilon \in \mathcal{J}_{(I_A, \mathcal{E}_\mathcal{P})}\,(\Omega)\,,\ \textit{then: } (D^\alpha u_\varepsilon)_\varepsilon \in \mathcal{J}_{(I_A, \mathcal{E}_\mathcal{P})}.$$

2.2.5 Local analysis

Let us denote by:

- $\mathcal{A}$: a sheaf on X of $\mathbb{K}$-vector spaces (which may moreover be $(\mathcal{C}, \mathcal{E}, \mathcal{P})$ - algebras),
- $\mathcal{B}$: a subsheaf of vector spaces of $\mathcal{A}$,
- Ω: any given open set in X,
- $\mathcal{V}\,(x)$ the set of all open neighbourhoods in Ω of $x \in \Omega$,
- $u = cl\,(u_\varepsilon)_\varepsilon \in \mathcal{A}\,(\Omega)\,,$
- $O_\mathcal{B}^\mathcal{A} u = \{x \in \Omega, \exists V_x \in \mathcal{V}\,(x) : u|_{V_x} \in \mathcal{B}\,(V_x)\}\,.$

Definition 2.8. *The $\mathcal{B}$-support of u, denoted by $\mathcal{B}$-supp u, or $S_\mathcal{B}^\mathcal{A} u$ is defined by*

$$S_\mathcal{B}^\mathcal{A} u = \Omega \setminus O_\mathcal{B}^\mathcal{A} u.$$

Example 2.9. 1)$\mathcal{B} = \{0\}$. Then we have: $\{0\}$-supp u = supp u.
2) $\mathcal{A} = \mathcal{D}'$ (the sheaf of distributions on $X = \mathbb{R}^n$) and $\mathcal{B} = \mathcal{C}^\infty$. Then, for each $u \in \mathcal{D}'(\Omega)$, $S^{\mathcal{D}'}_{\mathcal{C}^\infty} u$ is exactly sing supp u in the Hörmander sense [7].
3) $\mathcal{A} = \mathcal{A}_{(\mathcal{C},\mathcal{E}_\mathcal{P})}$ and $\mathcal{B} = \mathcal{D}'_\mathcal{A}$, with, for each open set Ω in $\mathbb{R}^n$:

$$\mathcal{D}'_\mathcal{A}(\Omega) = \left\{ u = cl\,(u_\varepsilon)_\varepsilon \in \mathcal{A}(\Omega) : \lim_{\varepsilon \to 0} u_\varepsilon \in \mathcal{D}'(\Omega) \right\}.$$

Then, $\mathcal{D}'_\mathcal{A}$ is a subsheaf of vector spaces of $\mathcal{A}$ and we can describe the $\mathcal{D}'$-singularity of $u \in \mathcal{A}(\Omega)$ by means of $\mathcal{D}'_\mathcal{A}$-supp $u = S^\mathcal{A}_{\mathcal{D}'_\mathcal{A}} u$. In that case we have:

$$\operatorname{supp} u \supset \mathcal{C}^\infty\text{-}\operatorname{supp} u \supset \mathcal{D}'_\mathcal{A}\text{-}\operatorname{supp} u.$$

2.2.6 The association process

Let us denote by • Ω any open set in $X = \mathbb{R}^n$
• E a given sheaf of topological $\mathbb{K}^n$ -vector spaces containing $\mathcal{E}$ as a subsheaf.
• $\mathcal{N}^\mathcal{A}_E(\Omega)$ the vector subspace of $\mathcal{H}_{(A,\mathcal{E}_\mathcal{P})}(\Omega)$ such that:

$$\mathcal{N}^\mathcal{A}_E(\Omega) = \{(u_\varepsilon)_\varepsilon \in \mathcal{H}_{(A,\mathcal{E}_\mathcal{P})}(\Omega) : \lim_{\substack{E(\Omega)\\ \varepsilon \to 0}} = 0\}.$$

We make the following assumptions:
-The inclusion: $\mathcal{J}_{(I_A,\mathcal{E}_\mathcal{P})}(\Omega) \subset \mathcal{N}^\mathcal{A}_E(\Omega)$ holds.
-The restriction and limit operations commute in E.
-E is a sheaf of C$^\infty$-modules.
-For each Ω, the topology of $E(\Omega)$ is compatible with its $\mathcal{C}^\infty(\Omega)$-module structure.
Then if we set:

Definition 2.10.

$$E_\mathcal{A}(\Omega) = \{u = cl\,(u_\varepsilon)_\varepsilon \in \mathcal{A}(\Omega) : \lim_{\substack{E(\Omega)\\ \varepsilon \to 0}} = 0\}\},$$
$$O_\mathcal{A}(\Omega) = \{u = cl\,(u_\varepsilon)_\varepsilon \in \mathcal{A}(\Omega) : \lim_{\substack{E(\Omega)\\ \varepsilon \to 0}} = 0\}\},$$

we can show that the functorial maps $\Omega \to O_\mathcal{A}(\Omega)$ and $\Omega \to E_\mathcal{A}(\Omega)$ define two subsheaves $E_\mathcal{A}$ and $O_\mathcal{A}$ according to the following

Proposition 2.11. *$E_\mathcal{A}$ is a subsheaf of vector spaces of $\mathcal{A}$ and $O_\mathcal{A}$ is a subsheaf of $E_\mathcal{A}$.*

We don't give the proof here. The main difficulty is to glue together the bits. This can be achieved with the help of our above hypothesis.

Example 2.12. The above assumptions are satisfied for $E = \mathcal{D}'$. Then $\mathcal{D}'_{\mathcal{A}}$ previously defined in Example 2.9 is a subsheaf of $\mathcal{A}$.

Definition 2.13. *If* $u = cl\,(u_\varepsilon)_\varepsilon$ *belongs to* $E_{\mathcal{A}}(\Omega)$ *with* $\lim\limits_{\substack{E(\Omega)\\ \varepsilon\to 0}} u_\varepsilon = T$ *we write:* $u \underset{E(\Omega)}{\sim} T$ *and we say that* u *and* T *are associated according to* $E(\Omega)$*. If* u *and* v *belong to* $\mathcal{A}(\Omega)$ *with* $u - v \in O_{\mathcal{A}}(\Omega)$*, we write* $u \underset{E(\Omega)}{\approx} v$ *and we say that* u *and* v *are associated according to* $E(\Omega)$*.*

Definition 2.14. *For* $u = cl(u_\varepsilon)_\varepsilon$ *and* $v = cl(v_\varepsilon)_\varepsilon \in \mathcal{A}(\Omega)$ *we denote by*

$$u \underset{\mathcal{E}(\Omega)}{\overset{p}{\approx}} v \quad (resp.\ u \underset{E(\Omega)}{\overset{p}{\approx}} v)$$

the p-$\mathcal{E}$ *(or p-E)-association process defined by: There exists some* $p \in \mathbb{N}$ *such that*

$$\lim_{\substack{E(\Omega)\\ \varepsilon\to 0}} \frac{1}{\varepsilon^p}(u_\varepsilon - v_\varepsilon) = 0 \quad (resp.\ \lim_{\substack{E(\Omega)\\ \varepsilon\to 0}} \frac{1}{\varepsilon^p}(u_\varepsilon - v_\varepsilon) = 0),$$

and by $u \underset{\mathcal{E}(\Omega)}{\approxeq} v$ $(resp.\ u \underset{E(\Omega)}{\approxeq} v)$ *the* $\mathcal{E}$ *(or E)-association process defined by: For each* $p \in \mathbb{N}$ *we have*

$$\lim_{\substack{E(\Omega)\\ \varepsilon\to 0}} \frac{1}{\varepsilon^p}(u_\varepsilon - v_\varepsilon) = 0 \quad (resp. \lim_{\substack{E(\Omega)\\ \varepsilon\to 0}} \frac{1}{\varepsilon^p}(u_\varepsilon - v_\varepsilon) = 0).$$

Remark 1. *If* $\mathcal{A}(\Omega)$ *is the Colombeau simplified algebra, the* $\mathcal{E}$*-association process is equivalent to the equality in* $\mathcal{E}(\Omega)$*. Indeed, if one has* $u \underset{\mathcal{E}(\Omega)}{\approxeq} v$ *then for each* $i \in I(\Omega)$*, and* $p \in \mathbb{N}$*,* $p_i\left(\frac{1}{\varepsilon^p}(u_\varepsilon - v_\varepsilon)\right)$ *is bounded, that is to say:* $(u_\varepsilon - v_\varepsilon)_\varepsilon$ *is in the set:*

$$\left\{(u_\varepsilon)_\varepsilon \in [\mathcal{E}(\Omega)]^{]0,1]}\, \forall i \in I(\Omega), \forall \varepsilon \in]0,1] : (p_i(u_\varepsilon))_\varepsilon \in I_A^+\right\} = \mathcal{J}_{(I_A, \mathcal{E}_{\mathcal{P}})}(\Omega).$$

Then we have $u = v$*, and the converse can easily be proved.*

Remark 2. *If there is a derivation operation* D^α *in* $\mathcal{E}$ *with the same properties as the ones described in this section, and continuous for the E-topology, it is easy to see that the association process is compatible with the derivation in* $\mathcal{A}$ *but not with the multiplication.*

Example 2.15. When taking $E = \mathcal{D}'$, we find again the association process defined by Colombeau (see Remark 1) or Egorov [5] who works in the $(\mathcal{C}, \mathcal{E}, \mathcal{P})$-algebra defined by choosing $\mathcal{C} = \mathbb{C}^{]0,1]}/I_{\mathbb{C}^{]0,1]}}$ where $I_{\mathbb{C}^{]0,1]}}$ is the ring of families of complex numbers $(z_\varepsilon)_\varepsilon$ such that $z_\varepsilon = 0$ if ε is small enough, and where $\mathcal{E}$ is the sheaf of C^∞-complex valued functions.

3 Solving a perturbation problem

Consider the perturbation problem

$$(1)\begin{cases} \varphi(\varepsilon)\frac{dx}{dt} = f_\varepsilon(t, x) \\ x(0) = \psi(\varepsilon) \end{cases}$$

in $C^1(\mathbb{R}_+)$, where φ, ψ, and F with $F(\varepsilon, t, w) = f_\varepsilon(t, w)$ are given functions of the parameter ε, F being continuous in the (t, w) variables.
We can try to find a solution of the equivalent system

$$(2)\begin{cases} \varphi(\varepsilon)\frac{dy}{dt} = f_\varepsilon(t, y + \psi(\varepsilon)) \\ y(0) = 0 \end{cases}$$

as a fixed point of the map $\Phi_\varepsilon : C^0(\mathbb{R}_+) \longrightarrow C^0(\mathbb{R}_+)$ such that

$$(\Phi_\varepsilon(y))(t) = \frac{1}{\varphi(\varepsilon)}\int_0^t f_\varepsilon(s, y(s) + \psi(\varepsilon))\, ds$$

We assume a Lipschitz condition

$$\forall \varepsilon \in]0,1]\, \forall t \in \mathbb{R}_+ \forall (v, w) \in \mathbb{R} \times \mathbb{R} : |f_\varepsilon(t, v) - f_\varepsilon(t, w)| \leq g_\varepsilon(t)\, |v - w|$$

where g_ε is bounded on the compact subsets of $\mathbb{R}$ with $M_{T,\varepsilon} = \sup_{t\in[0,T]} g_\varepsilon(t)$. The idea ([6], [10]) is to replace the standard family $(p_T)_{T\in\mathbb{R}_+}$ of seminorms on $C^0(\mathbb{R}_+)$ with $p_T(x) = \sup_{t\in[0,T]} |x(t)|$ by the equivalent family $(q_{T,\varepsilon})_{T\in\mathbb{R}_+}$, defined for each fixed ε by

$$q_{T,\varepsilon}(x) = \sup_{t\in[0,T]}\left(|x(t)|\, e^{-\frac{M_{T,\varepsilon}}{\varphi(\varepsilon)}t}\right)$$

for which the mapping Φ_ε has a contraction property in (the locally convex semi-normed Hausdorff and complete space) $C^0(\mathbb{R}_+)$. The reason why we change the seminorms is that the first family doesn't induce this contraction property of Φ_ε.

So, by a standard process, we can construct a fixed point z_ε of Φ_ε, which is a solution of (2). The Gronwall's lemma implies uniqueness.
Thus we can summarize all the informations included in the families $(\Phi_\varepsilon)_\varepsilon$ and $(z_\varepsilon)_\varepsilon$ in single objects Φ and z such that:
-z is lying in a convenient $(\mathcal{C}, \mathcal{E}, \mathcal{P})$-algebra $\mathcal{A}(\mathbb{R}_+)$,
-Φ is a mapping from $\mathcal{A}(\mathbb{R}_+)$ to $\mathcal{A}(\mathbb{R}_+)$,
-z is a fixed point of Φ in $\mathcal{A}(\mathbb{R}_+)$.

4 Fixed points in $(\mathcal{C}, \mathcal{E}, \mathcal{P})$-algebras

4.1 The main result

This result is proved in [10]. We suppose here that the algebra $\mathcal{E}(\Omega)$ is complete, for a given open set Ω of X. Let $(\Phi_\varepsilon)_\varepsilon$ be a family of maps from $\mathcal{E}(\Omega)$ into itself such that:

(i) there exists a family $(a_{n,\varepsilon})_{(n,\varepsilon)\in\mathbb{N}\times]0,1]}$ of positive numbers verifying:
for each $i \in I$ there exists $N(i) \in \mathbb{N}$ and $j(i) \in I$ such that, for each $\varepsilon \in]0,1]$ and $x \in \mathcal{E}(\Omega)$

$$p_i(\Phi_\varepsilon(x)) \leq \sum_{n=0}^{N(i)} a_{n,\varepsilon} p_{j(i)}^n(x)$$

(ii) for each $i \in I$ and $\varepsilon \in]0,1]$ there exists a seminorm $q_{i,\varepsilon}$ on $\mathcal{E}(\Omega)$, strictly positive constants $a_{i,\varepsilon}$ and $b_{i,\varepsilon}$, and a constant $k_{i,\varepsilon} \in [0,1[$ such that

$$a_{i,\varepsilon} p_i \leq q_{i,\varepsilon} \leq b_{i,\varepsilon} p_i$$

and for each $(x,y) \in \mathcal{E}(\Omega) \times \mathcal{E}(\Omega)$

$$q_{i,\varepsilon}(\Phi_\varepsilon(x) - \Phi_\varepsilon(y)) \leq k_{i,\varepsilon} q_{i,\varepsilon}(x-y)$$

(iii) for each $i \in I$, the families $\left(\frac{b_{i,\varepsilon}}{a_{i,\varepsilon}}\right)_\varepsilon$, $\left(\frac{1}{1-k_{i,\varepsilon}}\right)_\varepsilon$ and $(a_{n,\varepsilon})_\varepsilon$ for $n = 0, 1, .., N(i)$, are in A^+.

Then, setting $\Phi(u) = cl(\Phi_\varepsilon(u_\varepsilon))_\varepsilon$ for each $u = cl(u_\varepsilon)_\varepsilon \in \mathcal{A}$ one can define a mapping Φ from $\mathcal{A}(\Omega)$ into itself and Φ has a fixed point in $\mathcal{A}(\Omega)$.

Sketch of proof

a) From (i) and (ii) it follows that the map Φ is well defined by

$$\Phi : u = cl\,(u_\varepsilon)_\varepsilon \to cl\,(\Phi\,(u_\varepsilon))_\varepsilon\,.$$

b) From (ii) we conclude that for each ε the map Φ has a fixed point $z_\varepsilon \in \mathcal{E}\,(\Omega)$.
c) Using (i) and (iii) we can prove that we have: $(z_\varepsilon)_\varepsilon \in \mathcal{H}_{(A,\mathcal{E}_{\mathcal{P}})}\,(\Omega)$, that is to say: $z = cl\,(z_\varepsilon)_\varepsilon \in \mathcal{A}\,(\Omega)$.

5 Application to the previous perturbation problem

We additionally assume that $f_\varepsilon\,(t,0) = 0$.
Recall that we have defined the family of seminorms on $\mathcal{E}\,(\Omega) = C^0\,(\mathbb{R}_+)$

$$q_{T,\varepsilon}\,(x) = \sum t \in [0,T] \left(|x\,(t)|\, e^{-\frac{M_{T,\varepsilon}}{\varphi(\varepsilon)}t} \right).$$

Computing the estimates (i), (ii), and (iii) of the main theorem we find here:
• $a_{0,\varepsilon} = \frac{\psi(\varepsilon) T M_{T,\varepsilon}}{\varphi(\varepsilon)}$, $a_{1,\varepsilon} = \frac{T M_{T,\varepsilon}}{\varphi(\varepsilon)}$, $a_{n,\varepsilon} = 0$ for $n > 1$
• $a_{T,\varepsilon} = e^{\frac{M_{T,\varepsilon}}{\varphi(\varepsilon)}}$, and $b_{T,\varepsilon} = 1$
• $k_{T,\varepsilon} = 1 - e^{\frac{M_{T,\varepsilon}}{\varphi(\varepsilon)}}$.
If we have some polynomial estimate of the form

$$M_{T,\varepsilon} \leq \sum_0^N T^n M_\varepsilon^n$$

we see that the fixed point associated with the solution of

$$(1) \left\{ \begin{array}{c} \varphi\,(\varepsilon)\,\frac{dx}{dt} = f_\varepsilon\,(t,x) \\ X\,(0) = \psi\,(\varepsilon) \end{array} \right.$$

is lying in a $(\mathcal{C}$,$\mathcal{E}$,$\mathcal{P})$-algebra $\mathcal{A}\,(\Omega)$ where
• $\mathcal{E} = C^0$
• $X = \mathbb{R}$ and $\Omega = \mathbb{R}_+$
• $\mathcal{P}\,(\Omega) = (p_T)_{T \in \mathbb{R}_+}$ with $p_T\,(x) = \sum t \in [0,T]\,|x\,(t)|$
• $\mathcal{C} = A/I_A$ is overgenerated by the two factors of singular perturbation $(\psi\,(\varepsilon))_\varepsilon$ and $\left(e^{\frac{M_\varepsilon}{\varphi(\varepsilon)}}\right)_\varepsilon$.

Acknowledgment. The author wants to thank professor M. Oberguggenberger for his invitation and the Erwin Schrödinger International Institute for its support.

References

[1] A. B. ANTONEVITCH, YA. V. RADYNO, On a General Method for Constructing Algebras of Generalized Functions, Soviet Math. Dokl. Vol. 43 (1991), N°3.

[2] H.A. BIAGIONI, Generalized Solutions to the Kuramoto-Sivashinsky Equation, Proceedings of the International Conference on Generalized Functions-Linear and nonlinear Problems. Novi Sad, Yugoslavia, 1996.

[3] J.F. COLOMBEAU, Elementary introduction to new generalized functions. North-Holland Math. Studies 113, 1985.

[4] A. DELCROIX, Fonctions généralisées et Analyse non standard, Monatshefte fur Mathematik, 123, p.127-134, 1997.

[5] Y.V. EGOROV, A contribution to the theory of generalized functions, Russian Math. surveys 45:5, p.1-49, 1990.

[6] M. FRIGON, Points fixes dans les espaces de Fréchet. Preprint Université de Montréal, I996.

[7] L. HÖRMANDER, The Analysis of Linear Partial Differential Operators I, Springer-Verlag, 1983.

[8] J.-A. MARTI. Fundamental structures and asymptotic microlocalization in sheaves of generalized functions. To appear in Integral transforms and special functions.

[9] J.-A. MARTI, Analyse locale et microlocale des algèbres de fonctions généralisées. To appear in Annales de la Faculté des Sciences de Toulouse.

[10] J.-A. MARTI, Fixed points in algebras of generalized functions. International Conference ICIMAF. La Habana Cuba, 1997.

[11] J.-A. MARTI, P. NUIRO, Génération d'algèbres différentielles et applications. Proceedings of the 3rd Catalan Days on Applied Mathematics, Lleida, Espagne, p. 139-150, 1996.

[12] J.-A. MARTI, P. NUIRO, Analyse algébrique d'un problème de Dirichlet non linéaire et singulier. Preprint, 1998.

[13] J.-A. MARTI, S.P. NUIRO, V.S. VALMORIN, On a nonlinear Goursat problem, Proceedings of the International Conference on Generalized Functions-Linear and Nonlinear Problems. Novi Sad, Yugoslavia, 1996.

[14] J.-A. MARTI, P. NUIRO, V. VALMORIN, Sur un probleme de Goursat non linéaire à données irrégulières. To appear in Annales de la Faculté des Sciences de Toulouse.

[15] H. A. BIAGIONI, M. OBERGUGGENBERGER, Generalized Solutions to the Korteweg-de Vries and the Regularized Long-Wave Equations, Siam J. Math. Anal. Vol.23, N°4, pp.923-940, 1992.

[16] S. PILIPOVIC, M. NEDELJKOV, Paley-Wiener Type Theorems for Colombeau's Generalized Functions. Journal of Mathematical Analysis and Applications 195, 108-122 (1995).

[17] D. SCARPALEZOS, Some remarks on functoriality of Colombeau's construction on topological and microlocal aspects and applications, Proceedings of the International Conference on Generalized Functions - Linear and Nonlinear Problems. Novi Sad, Yugoslavia, 1996.

[18] V.VALMORIN, A new algebra of periodic generalized functions (V. Valmorin). Journal for Analysis and its Applications. Vol. 15, N°1, p.57-74,1996.

J.-A. MARTI, Département de Mathématiques et Informatique, Université Antilles Guyane, 97159 Pointe à Pitre, Guadeloupe (F.W.I. France).

Local and Microlocal Analysis in the Space of Colombeau Generalized Functions

STEVAN PILIPOVIĆ

1 Introduction

Local and microlocal properties of distributions, ultradistributions and hyperfunctions have been extensively studied in the last three decades. Mainly, the applications were made in the analysis of linear problems while the non-linear problems required constructions of new algebras in which the product and some other non-linear operations are defined. This was the motivation for introducing the space of Colombeau generalized functions $\mathcal{G}$ ([4]). This is an algebra with many useful properties for applications.

In a series of papers, the author together with his collaborators tried to establish a tool of microlocal analysis within Colombeau generalized functions. This paper is an overview of this work.

First, we investigate the hypoellipticity in $\mathcal{G}$. By following the classical method of solving $P(D)g = h$, we found in [20] a family of solutions with the power order growth rate with respect to ε to a family of equations

$$P_\varepsilon(D)G = \sum_{|\alpha|\leq m} a_{\alpha,\varepsilon} D^\alpha G = H_\varepsilon,\ \varepsilon \in (0,1), \tag{1.1}$$

with coefficients which have power order growth rate with respect to ε. More precisely, an equation $P_\varepsilon(D)U_\varepsilon = H_\varepsilon$, where H_ε is an element in Colombeau space $\mathcal{E}_M(\mathbf{R}^n)$ and $P_\varepsilon(D)$ is a differential operator with $\mathcal{E}_M(\mathbf{R}^n)$-coefficients is solved in $\mathcal{E}_M(\mathbf{R}^n)$. This enabled us to study in [15] the local properties of solutions in $\mathcal{G}$, for example the hypoellipticity of (1.1), similarly as it is done in $\mathcal{D}'$. If $P_\varepsilon(D)$ is hypoelliptic for every fixed ε, then the solution to (1.1), G_ε, is in $C^\infty(\Omega)$ but if we suppose that for every $K \subset\subset \Omega$, $\sup_{x\in K\subset\subset\Omega} |D^\alpha F_\varepsilon(x)|$ satisfies the power growth condition $\mathcal{O}(\varepsilon^{-N_K})$ for every α, whether it is true that the derivatives of G_ε satisfy similar estimates. This property defines the hypoellipticity of a family $P_\varepsilon(D)$; it implies the hypoellipticity of every equation with fixed ε but conversely it does not hold in general.

Second, we follow Hörmander's approach to microlocal analysis and find some essential properties of Colombeau generalized functions which appear as a consequence of a very rich structure of $\mathcal{G}$ (cf. [19], [24], [10], [6], [14], [17], [13], [18]). The dot wave front set $\dot{\mathrm{WF}}_g$, introduced in [6], intrinsically characterizes the singularities of elements in $\mathcal{G}$ and we compare this notion with the generalized wave front set WF_g which is in fact a straightforward generalization of the distributional wave front set. This notion is used in the analysis of propagation of singularities for differential operators with coefficients in $\dot{\mathcal{G}}^\infty$. Our investigations of microlocal properties include the analysis of suppleness and non–flabbiness of the sheaf of Colombeau functions.

Third, we introduce the quasiasymptotics in the space of Colombeau generalized functions. This notion generalizes the corresponding notion of distribution theory but here it is used for the analysis of generalized solutions to a class of non-linear equations ([21], [22]). As an example, we give the generalized asymptotic expansion of delta square.

2 Notations

We will use the so-called simplified version of Colombeau's theory (cf. [3], [1], [16], [24]) and for the definition of spaces $\mathcal{E}$, $\mathcal{N}$, $\mathcal{E}_M$, $\mathcal{E}'$, $\mathcal{N}'$ we refer to these monographs. Recall, Colombeau spaces of generalized functions and generalized (complex) numbers are defined by

$$\mathcal{G}(\mathbf{R}^n) = \mathcal{E}_M(\mathbf{R}^n)/\mathcal{N}(\mathbf{R}^n), \quad \overline{\mathbf{C}} = \mathcal{E}^0/\mathcal{N}^0.$$

The presheaf $\Omega \to \mathcal{G}(\Omega)$ (Ω is open in $\mathbf{R}^n$) is a sheaf and the support of a generalized function R is defined as the complement of the largest open subset $\Omega' \subset \Omega$ such that $R|_{\Omega'} = 0$.

The algebra $\mathcal{G}^\infty(\Omega)$ is defined by Oberguggenberger in [16]. With this space the local analysis in $\mathcal{G}$ has started. $\mathcal{E}_M^\infty(\Omega)$ is defined to be the space of $U_\varepsilon \in \mathcal{E}_M$ such that for every $K \subset\subset \Omega$ there exists $N \in \mathbf{N}$ such that for every $\alpha \in \mathbf{N}_0^n$, $\sup_{x\in K} |U_\varepsilon^{(\alpha)}(x)| = \mathcal{O}(\varepsilon^{-N})$. $\mathcal{G}^\infty(\Omega)$ is a subalgebra of $\mathcal{G}(\Omega)$ defined by $\mathcal{E}_M^\infty(\Omega)/\mathcal{N}(\Omega)$; $\mathcal{G}^\infty(\Omega)$, $\Omega \subset \mathbf{R}^n$ is a sheaf.

It is said that $U \in \mathcal{G}(\Omega)$ is $\mathcal{G}^\infty$ at $x_0 \in \Omega$ if there exists an open set Ω', $x_0 \in \Omega' \subset \Omega$, such that $U|_{\Omega'} \in \mathcal{G}^\infty(\Omega')$. It is said that $G \in \mathcal{G}(\Omega)$ is $\mathcal{G}^\infty$ in $\Omega_1 \subset \Omega$ if $G|_{\Omega_1} \in \mathcal{G}^\infty(\Omega_1)$. Let $G \in \mathcal{G}(\Omega)$. The complement of the largest open set of Ω in which G is $\mathcal{G}^\infty$ is called the singular support of G and denoted by $\operatorname{ssupp}_g G$.
If $T \in \mathcal{D}'(\Omega)$, then $\operatorname{ssupp} T = \operatorname{ssupp}_g[T]$.
Let $\phi \in \mathcal{S}(\mathbf{R}^n)$ such that $\mathcal{F}(\phi) = \hat{\phi} \in \mathcal{D}(\mathbf{R}^n)$ and $\hat{\phi} \equiv 1$ on a neighbourhood of zero. Put $\phi_\varepsilon(x) = \frac{1}{\varepsilon^n}\phi(\frac{x}{\varepsilon})$, $x \in \mathbf{R}^n, \varepsilon \in (0,1)$. We fix once for all such a function ϕ and call it the vision function.
If $T \in \mathcal{D}'_c(\Omega) = \mathcal{E}'(\Omega) \subset \mathcal{D}'_c(\mathbf{R}^n)$ we put $I_\phi(T) = [T_\varepsilon]$, where $T_\varepsilon = T * \phi_\varepsilon$, $\varepsilon \in (0,1)$. To simplify notation, we will use $[T]$ for $I_\phi(T)$. We will also use the following embedding: $\mathcal{E}'(\Omega) \to \mathcal{G}(\Omega) : g \mapsto I_{2,\phi}(g) = g * \phi_{\varepsilon^2}$.
The s-association ($\overset{s}{\sim}$), $s \geq 0$ of generalized complex numbers is defined as follows. For $G \in \overline{\mathbf{C}}$, $G \overset{s}{\sim} 0$ means that there exists its representative G_ε such that $G_\varepsilon = o(\varepsilon^s)$ as $\varepsilon \to 0$. If G_1, G_2 are in $\mathcal{G}(\Omega)$, then $G_1 \overset{s,\mathcal{D}}{\sim} G_2$ if $\langle G_1 - G_2, \varphi\rangle \overset{s}{\sim} 0$ for every $\varphi \in \mathcal{D}$. If $s = 0$, then instead of 0-association we use the notation association.
Let $G, F \in \mathcal{G}(\Omega)$. It is said that they are strongly associated $G \approx H$, if there exists representatives G_ε and H_ε of G and H, respectively, such that for every open set $\Omega' \subset\subset \Omega$ there exists $\beta > 0$ such that $\langle G_\varepsilon(x) - F_\varepsilon(x), \varphi(x)\rangle = \mathcal{O}(\varepsilon^\beta)$, for any $\varphi \in C_c^\infty(\Omega')$. Clearly, this definition does not depend on the choice of representatives. Note that the association of a generalized function in $\mathcal{G}$ and a distribution in $\mathcal{D}'$ is of a special interest for the theory.

3 Generalized differential operators

A generalized polynomial is a generalized function represented by

$$\sum_{|\alpha| \leq m,} a_{\alpha,\varepsilon} x^\alpha, \quad a_{\alpha,\varepsilon} \in \mathcal{E}^0 \ (\alpha \in \mathbf{N}).$$

We say that such a generalized polynomial is of degree m if there exists β, $|\beta| = m$, such that $a_{\beta,\varepsilon} \notin \mathcal{N}^0$.
If $H^i_\varepsilon(x) = \sum_{|\alpha|\leq m} a^i_{\alpha,\varepsilon} x^\alpha$, $i = 1, 2$ are representatives of a generalized polynomial, then $a^1_{\alpha,\varepsilon} - a^2_{\alpha,\varepsilon} \in \mathcal{N}^0$, $|\alpha| \leq m$.
The corresponding differential operator is of the form $P(D) = \sum_{|\alpha|\leq m} a_\alpha D^\alpha$, $a_\alpha \in$

$\overline{\mathbf{C}}, D^\alpha = (-i\partial)^\alpha$, is represented by

$$P_\varepsilon(D) = \sum_{|\alpha|\le m} a_{\alpha,\varepsilon} D^\alpha, \quad a_{\alpha,\varepsilon} \in \mathcal{E}^0. \tag{3.1}$$

A differential operator of the form $P(x,D) = \sum_{|\alpha|\le m} a_\alpha(x) D^\alpha$, where $a_\alpha \in \mathcal{G}(\mathbf{R}^n)$, is called a generalized differential operator. Thus, a representative of $P(x,D)$ is given by $P_\varepsilon(x,D) = \sum_{|\alpha|\le m} a_{\alpha,\varepsilon} D^\alpha$, where $a_{\alpha,\varepsilon}(x)$ is a representative of $a_\alpha(x)$, $|\alpha| \le m$. Note that if $\tilde{a}_{\alpha,\varepsilon}(x)$ is another representative of $a_\alpha(x)$, $|\alpha| \le m$, then $\sum_{|\alpha|\le m}(a_{\alpha,\varepsilon}(x) - \tilde{a}_{\alpha,\varepsilon}(x)) D^\alpha G_\varepsilon \in \mathcal{N}(\mathbf{R}^n)$, for every $G_\varepsilon \in \mathcal{E}_M(\mathbf{R}^n)$. Also, if $G_\varepsilon \in \mathcal{N}(\mathbf{R}^n)$, then $P_\varepsilon(x,D) G_\varepsilon \in \mathcal{N}(\mathbf{R}^n)$.

4 Hypoellipticity

Let P_ε be a representative of a generalized differential operator with coefficients in $\mathcal{E}^0$ of the form (3.1) such that

$$|\sum_{|\alpha|=m} a_{\alpha,\varepsilon} c^\alpha| \ge C\varepsilon^r, \quad \varepsilon \in (0,1), \tag{4.1}$$

for some $c \in \mathbf{R}^n$, $C > 0$ and $r \in \mathbf{R}$.
First we recall a theorem from [20].

Theorem 4.1 *For every $H_\varepsilon \in \mathcal{E}_M(\mathbf{R}^n)$ there exists a solution $E_\varepsilon \in \mathcal{E}_M(\mathbf{R}^n)$ to $P_\varepsilon(D)U_\varepsilon = H_\varepsilon$. In particular, $[E_\varepsilon]$ is the solution to $P(D)U = H$.*

The hypoellipticity in $\mathcal{G}$ is studied in [15].

Definition 4.2 *An operator determined by (3.1) is called hypoelliptic in $\mathcal{G}$ if for every open $\Omega \subset \mathbf{R}^n$ and every solution $G \in \mathcal{G}(\Omega)$ to $P(D)G = 0$, the generalized function $G * [\delta_\varepsilon]$ is in $\mathcal{G}^\infty(\Omega)$, where $\delta_\varepsilon = \varepsilon^{-n}\psi(x/\varepsilon)$ and ψ is an arbitrary function in $\mathcal{D}(\mathbf{R}^n)$ with integral equal to 1.*

Proposition 4.3
a)If for every $[\delta_\varepsilon]$ there exists $E = [E_\varepsilon]$ which satisfy $P_\varepsilon(D)E_\varepsilon = \delta_\varepsilon$ and $\operatorname{ssupp}_g E = \{0\}$, *then $P(D)$ is hypoelliptic in $\mathcal{G}$.*
b) Let $[P_\varepsilon(D)]$ be hypoelliptic in $\mathcal{G}$. Then, for every open set Ω and $G \in \mathcal{G}(\Omega)$, $P(D)G \in \mathcal{G}^\infty(\Omega)$ implies $G * [\delta_\varepsilon] \in \mathcal{G}^\infty(\Omega)$ for every $[\delta_\varepsilon]$.

Assertion b) is usually used as the definition of hypoellipticity. This means that (3.1) is hypoelliptic if $P_\varepsilon(D)G_\varepsilon \in \mathcal{E}_M^\infty(\mathbf{R}^n)$ implies $G_\varepsilon * \delta_\varepsilon \in \mathcal{E}_M^\infty(\mathbf{R}^n)$ for every δ_ε. The following result is the extension of the corresponding well-known result of distribution theory ([8]).

We will use the notation $V(P_\varepsilon) = \{\sigma_\varepsilon + i\tau_\varepsilon;\ P_\varepsilon(\sigma_\varepsilon + i\tau_\varepsilon) = 0\}$; it is the variety of P_ε.

Theorem 4.4 *The operator $P(D)$ is hypoelliptic in $\mathcal{G}$ if and only if there exist $N > 0$ such that for every $A > 0$ there exist $\eta > 0$ and $B > 0$ such that*

$$|\tau| \geq A(\log|\sigma| + N\log\varepsilon) - B,\ \sigma + i\tau \in V(P_\varepsilon),\ \varepsilon \in (0,\eta). \tag{4.2}$$

Example 4.5 *The operator ∂_x^2 is not hypoelliptic while $[\partial_x^2 + \varepsilon\partial_y^2]$ is hypoelliptic in $\mathcal{G}$. A perturbation of the non-hypoelliptic operator ∂_x leads to a family of operators $\partial_x + a_\varepsilon\partial_y^2$ which does not satisfy the assumption (4.2) although for every fixed ε every perturbed equation is hypoelliptic. Let $\varepsilon_{n,k}$, $n, k \in \mathbf{N}$ be a positive matrix such that $\varepsilon_{n,k}$ monotonically decreases to zero as $n \to \infty$, for every fixed k, and $\varepsilon_{n,k}$ monotonically decreases to zero as $k \to \infty$, for every fixed n. Let*

$$a_\varepsilon = \begin{cases} \varepsilon_{n,k}^n,\ \varepsilon = \varepsilon_{n,k},\ n,k \in \mathbf{N} \\ 0,\ \textit{otherwise} \end{cases}$$

Then $a_\varepsilon \in \mathcal{E}^0$ and one can prove that (4.2) does not hold. This means that the corresponding differential operator is not hypoelliptic in $\mathcal{G}$.

A convenient relation describing the hypoellipticity of a differential operator is given by the next theorem.

Theorem 4.6 *An operator $[P_\varepsilon(D)]$ is hypoelliptic if there exist $N > 0$, $h > 0$, $\eta > 0$ and $b \in \mathbf{R}$ such that: $\sigma + i\tau \in V(P_\varepsilon) \Rightarrow |\tau| \geq \varepsilon^{Nh}|\sigma|^h - b$, $\varepsilon \in (0,\eta)$.*

5 Microlocalization

As a motivation we give the following remark.

Remark 5.1 *Let us say that a function $f : \Omega \to \mathbf{C}$ is C^∞ at $x_0 \in \Omega$ if for every $k \in \mathbf{N}_0^n$ there exists $U_k \in \mathcal{V}(x_0)$ (the set of neighbourhoods of x_0) such that $f \in C^k(U_k)$.*

Then it is easy to prove that if f is C^∞ at every point in an open set then it is C^∞ in this open set.
Note that one can simply construct a function f on Ω such that f is C^∞ at $x_0 \in \Omega$ but there does not exist $\omega \in \mathcal{V}(x_0)$ such that $f \in C^\infty(\omega)$.
Clearly, $f \in \mathcal{G}^\infty(\Omega)$ if and only if f is $\mathcal{G}^\infty$ at every point of Ω.

Recall,
(i) $\mathcal{G}^\infty(\Omega) \cap \mathcal{D}'(\Omega) = C^\infty(\Omega)$ ([16]).
(ii) If $G \in \mathcal{G}^\infty(\Omega)$, $f \in \mathcal{D}'(\Omega)$ and $G \approx [f]$, then $f \in C^\infty(\Omega)$ ([24]).
If we replace in (ii) the strong association with the association, then the assertion in (ii) does not hold.
Now we will introduce a new concept of microlocal regularity ([6]).

Definition 5.2 *Let $x_0 \in \Omega$. Then*

$$\dot{\mathcal{G}}^\infty_{x_0}(\Omega) = \{[G_\varepsilon(x)] \in \mathcal{G}(\Omega) \mid \exists r \in \mathbf{R}\, \forall \alpha \in \mathbf{N}_0^n\, \exists V_\alpha \in \mathcal{V}_{x_0}(\sup_{x \in V_\alpha} |G_\varepsilon^\alpha(x)| = \mathcal{O}(\varepsilon^r))\};$$

$$\dot{\mathcal{G}}^\infty(\Omega) = \bigcap_{x \in \Omega} \dot{\mathcal{G}}^\infty_x(\Omega).$$

It is said that $G \in \mathcal{G}(\Omega)$ is $\dot{\mathcal{G}}^\infty$-regular at x_0 if $G \in \dot{\mathcal{G}}^\infty_{x_0}$; G is $\dot{\mathcal{G}}^\infty$-regular in Ω if it is $\dot{\mathcal{G}}^\infty$-regular at every $x_0 \in \Omega$.

Definition 5.3 *The dot singular support of $G \in \mathcal{G}(\Omega)$, $\mathrm{s\dot{s}upp}_g G$ is defined as the complement of the union of all open sets where G is $\dot{\mathcal{G}}^\infty$-regular.*

Clearly, we have

$$\mathrm{s\dot{s}upp}_g G \subset \mathrm{ssupp}_g G, \quad G \in \mathcal{G}(\Omega).$$

Definition 5.4 *It is said that $G \in \mathcal{G}(\Omega)$ is $\dot{\mathcal{G}}^\infty$-microlocally regular in an open conic set $\gamma \subset \Omega \times (\mathbf{R}^n \setminus \{0\})$ if for any $(x_0, \xi_0) \in \gamma$ there exist families of open neighbourhoods $\{(\Omega_\alpha)\}_{\alpha \in \mathbf{N}_0^n}$ of x_0, of open cone neighbourhoods $\{(\Gamma_\alpha)\}_{\alpha \in \mathbf{N}_0^n}$ of ξ_0, $\{\varphi_\alpha \in C_0^\infty(\Omega)\}_{\alpha \in \mathbf{N}_0^n}$ and $\{\psi_\alpha \in C^\infty(\mathbf{R}^n)\}_{\alpha \in \mathbf{N}_0^n}$ such that*

(a) $\mathrm{supp}\,\varphi_\alpha \subset \Omega_\alpha$, *$\varphi \equiv 1$ on $\Omega'_\alpha \subset\subset \Omega_\alpha$, $\alpha \in \mathbf{N}_0^n$;*

(b) $\mathrm{supp}\,\psi_\alpha \subset \Gamma_\alpha$, *$\psi_\alpha$ is positively homogeneous of degree 0 in Γ_α, $\psi \equiv 1$ in a cone $\Gamma'_\alpha \subset\subset \Gamma_\alpha$, $\alpha \in \mathbf{N}_0^n$, such that*

$$\exists b \in \mathbf{R}\, \forall \alpha \in \mathbf{N}_0^n\, \forall \beta \in \mathbf{N}_0^n \sup_{\xi \in \mathbf{R}^n} \langle \xi \rangle^\alpha |D^\beta(\psi_\alpha(\xi)\mathcal{F}(\varphi_\alpha G_\varepsilon)(\xi))| = \mathcal{O}(\varepsilon^b).$$

The dot wave front set of $G \in \mathcal{G}$, denoted by $\dot{\mathrm{W}}\mathrm{F}_g G$, is the complement of the union of all conic open sets γ where G is $\dot{\mathcal{G}}^\infty$-microlocally regular.

If $\Omega_\alpha = \Omega, \Gamma_\alpha = \Gamma, \alpha \in \mathbf{N}_0^n$, then the above definition reduces to the definition of $\mathrm{WF}_g G$ ([19]).

Let us recall the properties of the wave front set for a generalized function on $\mathbf{R}^n$ (cf. [19]):

(i) If $h \in C_0^\infty(\mathbf{R}^n)$ and $G \in \mathcal{G}(\mathbf{R}^n)$, then $\mathrm{WF}_g(hG) \subset \mathrm{WF}_g(G)$.

(ii) $\pi\, \mathrm{WF}_g G = \mathrm{ssupp}_g G$ (π is the first projection).

(iii) Let $f \in \mathcal{D}'(\Omega)$. Then, $\mathrm{WF}\, f = \mathrm{WF}_g[f]$.

Assertions which are to follow are taken from [6].

Theorem 5.5

(i) $\Omega \mapsto \dot{\mathcal{G}}^\infty(\Omega)$ is a sheaf (denoted by $\dot{\mathcal{G}}^\infty$).

(ii) $\mathcal{G}^\infty$ is a proper subsheaf of $\dot{\mathcal{G}}^\infty$.

Theorem 5.6 *Let $G \in \mathcal{G}(\Omega)$ and $T \in \mathcal{D}'(\Omega)$ such that $[T] \approx G$. If $G \in \dot{\mathcal{G}}_{x_0}^\infty$, $(x \in \Omega)$ then $T \in C^\infty$ at x_0.*

Note that this assertion does not hold if the strong association is replaced by association.

Corollary 5.7 *If $T \in \mathcal{D}'(\mathbf{R}^n)$ and $T \approx G$, then* $\mathrm{ssupp}\, T = \mathrm{ss\dot{u}pp}_g G = \mathrm{ssupp}_g G$.

Theorem 5.8 *Let $T \in \mathcal{D}'(\Omega)$. If $T \approx G$ and G is $\dot{\mathcal{G}}^\infty$-microlocally regular in a conic set $\gamma \subset \Omega \times (\mathbf{R}^n \backslash \{0\})$. Then T is C^∞-microlocally regular in γ. ($\mathrm{WF}\, T \subset$ compl. γ).*

Corollary 5.9 *If $T \in \mathcal{D}'(\mathbf{R})^n$ and $T \approx G$, then* $\mathrm{WF}\, T = \dot{\mathrm{W}}\mathrm{F}_g G = \mathrm{WF}_g G$.

Example 5.10 *We recall an example from [6] which is illustrative in our analysis. Let*

$$(a_n)_{n \in \mathbf{N}_0} \downarrow 0, a_0 > 2, a_1 > 1,$$

$$(\varepsilon_{\alpha,n})_{n \in \mathbf{N}} \downarrow 0 \text{ and } (\varepsilon_{\alpha,n})_{n \in \mathbf{N}} \cap (\varepsilon_{\beta,n})_{n \in \mathbf{N}} = \emptyset \text{ if } \alpha \neq \beta, \alpha, \beta \in \mathbf{N}_0,$$

$(n_i)_{i\in\mathbf{N}_0} \uparrow \infty, \lim_{\nu\to\infty} \varepsilon_\nu = 0,$ *where* $\varepsilon_\nu = \varepsilon_{j\nu},\ \nu \in [n_j, n_{j+1}),$

$$a_{\alpha,n} = \varepsilon_{\alpha,n}^{\alpha+1}, \alpha \in \mathbf{N}_0, n \in \mathbf{N}.$$

Put

$F_\varepsilon(x) = 0, x \in \mathbf{R}, \varepsilon \notin \bigcup_{\alpha\in\mathbf{N}_0}\{\varepsilon_{\alpha,n}; n \in \mathbf{N}, n \in [n_\alpha, n_{\alpha+1})\},$

$$F_{\varepsilon_{0,n}}(x) = \frac{1}{a_{0,n}}\phi(x - a_0), x \in \mathbf{R}, n \in \mathbf{N}, n \in [n_0, n_1), \dots$$

$$F_{\varepsilon_{\alpha,n}}(x) = \frac{1}{a_{\alpha,n}}\int_0^t \frac{(x-t)^{\alpha-1}}{(\alpha-1)!}\phi\frac{(t-a_\alpha)}{a_{\alpha,n}}dt, x \in \mathbf{R}, n \in \mathbf{N}, n \in [n_\alpha, n_{\alpha+1}), \alpha \in \mathbf{N}.$$

Then $F = [F_\varepsilon] \in \mathcal{G}(\mathbf{R})$ *has the following properties:*

$$F \text{ is } \mathcal{G}^\infty - \text{regular at every } x \neq 0 \text{ and } F \text{ is not } \dot{\mathcal{G}}^\infty - \text{regular at } 0. \tag{5.1}$$

6 Differential operators with $\dot{\mathcal{G}}^\infty$-coefficients

Let

$$P_\varepsilon(x, D) = \sum_{|\alpha|=m} a_{\alpha\varepsilon}(x)D^\alpha + \sum_{|\alpha|<m} a_{\alpha\varepsilon}(x)D^\alpha = P_{m\varepsilon}(x, D) + Q_\varepsilon(x, D) \tag{6.1}$$

be a family of differential operators such that $a_{\alpha\varepsilon} \in \mathcal{G}_M^\infty$, $|\alpha| < m$.
Then we denote by $P(x, D)$ the corresponding differential operator with coefficients in $\mathcal{G}^\infty(\Omega)$, $P(x, D) = [P_\varepsilon(x, D)]$. The propagation of singularities in $\mathcal{G}$ related to this operator is analysed in [6].

Definition 6.1 *A point* $(x_0, \xi_0) \in \Omega \times (\mathbf{R}^n \setminus \{0\})$ *is not in the characteristic set of* $P_{m\varepsilon}(x, D)$, $(x_0, \xi_0) \notin \operatorname{Char} P_{m\varepsilon}(x, D)$, *if there exist neighbourhoods* V_{x_0} *and* W_{ξ_0} *of* x_0 *and* ξ_0*, respectively, and constants* $r \in \mathbf{R}$, $\eta > 0$ *and* $C > 0$ *such that*

$$|P_{m\varepsilon}(x, \xi)| \geq C\varepsilon^r|\xi|^m, \quad x \in V_{x_0}, \xi \in V_{\xi_0}, \varepsilon \in (0, \eta). \tag{6.2}$$

The complement of the set of such points is called the characteristic set of P_ε, $\operatorname{Char} P_\varepsilon$.

Homogeneity of $P_{m\varepsilon}$ implies that we can take $\Gamma_{\xi_0} \in \mathcal{W}(\xi_0)$ (the set of conic neighbourhood of ξ_0) instead of V_{ξ_0} such that (6.2) holds for $x \in V_{x_0}$, $\xi \in \Gamma_{\xi_0}$ and $\varepsilon \in (0, \eta)$.

Clearly, if $P_\varepsilon(x,D)$ does not depend on ε ($a_\alpha \in C^\infty(\Omega)$, $|\alpha| \leq m$), then the above definition reduces to the definition of the well-known characteristics set of a standard differential operator.

We note that there exists $r_0 \in \mathbf{R}$ such that for every $\alpha, \beta \in \mathbf{N}^n$, there exists $C_{\alpha,\beta} > 0$ such that

$$|\partial_x^\alpha \partial_\xi^\beta \left(\frac{1}{P_{m\varepsilon}(x,\xi)} \right)| \leq C_{\alpha,\beta} \varepsilon^{r_0} \langle \xi \rangle^{-m}, \quad x \in V_{x_0}, \xi \in V_{\xi_0}, \varepsilon \in (0,\eta). \tag{6.3}$$

If $\tilde{P}_\varepsilon(x,D)$ is another representative of $P(x,D) = [P_\varepsilon(x,D)]$, then

$$(x_0,\xi_0) \in \operatorname{Char} P_\varepsilon \iff (x_0,\xi_0) \in \operatorname{Char} \tilde{P}_\varepsilon.$$

Thus the above definition naturally implies the definition of $\operatorname{Char} P(x,D)$ in the sense of generalized function: $(x_0,\xi_0) \in \operatorname{Char} P$ if $(x_0,\xi_0) \in \operatorname{Char} P_\varepsilon$.

Theorem 6.2 *Let $P(x,D)$ be a differential operator with coefficients in $\mathcal{G}^\infty(\Omega)$. Then*

$$\dot{\mathrm{W}}\mathrm{F}U \subset \operatorname{Char} P \cup \dot{\mathrm{W}}\mathrm{F}P(D)U \quad \textit{for every} \quad U \in \mathcal{G}(\Omega).$$

Theorem 6.3 *Let $P(x,D)$ be a differential operator with coefficients in $\dot{\mathcal{G}}^\infty(\Omega)$. Then for every $U \in \mathcal{G}(\Omega)$,*

$$\dot{\mathrm{W}}\mathrm{F}U \subset \operatorname{Char} P \cup \dot{\mathrm{W}}\mathrm{F}P(x,D)U.$$

Remark 6.4 *Since $\dot{\mathrm{W}}\mathrm{F}U \subset \mathrm{WF}\,U$, if $P(x,D)$ has coefficients in $\dot{\mathcal{G}}^\infty(\Omega)$, then we could not estimate $\mathrm{WF}\,U$ as above.*

Definition 6.5 *A differential operator in $\mathcal{G}$ determined by (6.1) is elliptic if* $\operatorname{Char} P_m = \emptyset$.

Now we have

Theorem 6.6 *If $P(x,D)$ with coefficients in $\dot{\mathcal{G}}^\infty$ is elliptic, then*

$$\dot{\mathrm{W}}\mathrm{F}U = \dot{\mathrm{W}}\mathrm{F}P(x,D)U.$$

7 Sheaf properties of Colombeau generalized functions

The sheaf properties of $\mathcal{G}$ are studied in [18]. It is well known that $\Omega \to \mathcal{G}(\Omega)$ is a fine sheaf.

Let us recall that a sheaf is supple if for every $f \in \mathcal{G}(\Omega)$ the following holds. If $\operatorname{supp} f = Z = Z_1 \cup Z_2$ where Z_1 and Z_2 are arbitrary closed sets, then there are $f_1, f_2 \in \mathcal{G}(\Omega)$ with the properties $\operatorname{supp} f_1 \subset Z_1$, $\operatorname{supp} f_2 \subset Z_2$ and $f = f_1 - f_2$.

A sheaf is flabby if for every Ω the restriction mapping $\mathcal{G}(\mathbf{R}^n) \to \mathcal{G}(\Omega)$ is surjective.

Remark 7.1 *Clearly, if a sheaf is flabby, then it is supple. Note, sheaves of Sato's hyperfunctions $\mathcal{B}$, Fourier hyperfunctions $\mathcal{D}$ and of Rosinger's generalized functions $\hat{\mathcal{M}}^m_T(\Omega)^*$ are flabby sheaves.*

Theorem 7.2

(i) The sheaf $\Omega \to \mathcal{G}(\Omega)$ is supple but not flabby.

(ii) Let $f \in \mathcal{G}(\Omega)$. Then there exists $F \in \mathcal{G}(\mathbf{R}^n)$ such that $F|_\Omega = f$ if and only if $\forall \alpha \in \mathbf{N}_0$, $\forall K \subset\subset \mathbf{R}^n$ $\exists N > 0$ $\forall r \in (0,1)$ $\exists d > 1$

$$\sup\{|f_\varepsilon^{(i)}(x)|; |i| \le \alpha,\ x \in (\Omega \cap K)_{-r}\} \le \frac{1}{\varepsilon^N r^N}, \varepsilon \in (0, r^d). \tag{7.1}$$

8 Quasiasymptotics

The quasiasymptotics in $\mathcal{G}$ is introduced and analyzed in [21]. We recall the definition.

Let $\mathcal{K}$ be a set of positive measurable functions defined on $(0,1)$ with the property $A^{-1}\varepsilon^p \le c(\varepsilon) \le A\varepsilon^{-p}$, $\varepsilon \in (0,1)$ for some $A > 0$ and $p > 0$.

Definition 8.1 *Let $F \in \mathcal{G}(\omega)$, where ω is an open set containing zero. It is said that F has the quasiasymptotics at zero with respect to $c(\varepsilon) \in \mathcal{K}$ (c–quasiasymptotics) if there is F_ε, a representative of F, and $g \in \mathcal{D}'(\omega)$, $g \neq 0$ such that for every $\theta \in \mathcal{D}(\omega)$*

$$\lim_{\varepsilon \to 0} \langle \frac{F_\varepsilon(\varepsilon x)}{c(\varepsilon)}, \theta(x) \rangle = \langle g, \theta \rangle. \tag{8.1}$$

It is said that F has the c–quasiasymptotics at zero as g.

Let $R_\varepsilon \in \mathcal{N}(\omega)$. Then, for every $c \in \mathcal{K}$ and every $\theta \in \mathcal{D}(\omega)$,

$$\lim_{\varepsilon\to 0}\langle\frac{R_\varepsilon(\varepsilon x_1,...,\varepsilon x_n)}{c(\varepsilon)},\theta(x_1,...,x_n)\rangle = 0.$$

Let $f \in \mathcal{E}'(\mathbf{R}^n)$. Recall, $I_{2,\varepsilon}(f) = [f * \phi_{\varepsilon^2}]$.

Proposition 8.2 *Let $f \in \mathcal{E}'(\mathbf{R}^n)$. Then f has the quasiasymptotics as g at zero with respect to $c(\varepsilon)$ in the sense of distributions if and only if*

$$I_{2,\phi}(f) \text{ has the c-quasiasymptotics as } g \text{ at zero with respect to } c(\varepsilon). \tag{8.2}$$

We will give an application of this notion in the analysis of a system of nonlinear Volterra integral equations of the second kind ([22])

$$f^i(x) = g^i(x) + \int_0^x K^i(x,y,f(y))dy,\ x \in I \subset \mathbf{R},\ i = 1,...,n, \tag{8.3}$$

where $f(y) = (f^1(y),...,f^n(y))$, is a n-tuple of generalized functions to be found, g^i are distributions and K^i are smooth functions of Lipschitz class $i = 1,...,n$. We regularize $g = (g^1,...,g^n)$ in order to avoid singularities of g and find a generalized solution to (8.2) Thus we consider

$$F^i_\varepsilon(x) = g^i_\varepsilon(x) + \int_0^x K^i(x,y,F_\varepsilon(y))dy + d^i_\varepsilon(x),\ x \in I,\ \varepsilon \in (0,\varepsilon_0), i = 1,...,n, \tag{8.4}$$

where $g^i_\varepsilon = I_{2,\varepsilon}(g^i)$, $F_\varepsilon = (F^1_\varepsilon,...,F^n_\varepsilon)$ and $d^i_\varepsilon \in \mathcal{N}(I)$, $i = 1,...,n$.

Theorem 8.3 *Let $K^i \in C^\infty(x,y,u_1,...,u_n,v_1,...,v_n)$, $x,y \in I$ (I is open and $I \ni 0$)), $(u,v) \in \mathbf{R}^{2n}$ satisfy the Lipschitz condition with respect to (u,v) on the compact interval of I^2 , $i = 1,...,n$, $g \in (\mathcal{D}'(I))^n$ and $c(\varepsilon) \in \mathcal{K}$ such that*

$$\lim_{\varepsilon\to 0}\frac{\varepsilon log\varepsilon}{c(\varepsilon)} = 0,\ \lim_{\varepsilon\to 0}\frac{g^i(\varepsilon x)}{c(\varepsilon)} \text{ exist in } \mathcal{D}'(I),\ i = 1,2,...,n.$$

Then, the solution $f_\varepsilon(x) = [(f^1_\varepsilon(x),...,f^n_\varepsilon(x))]$ to (8.3), has the quasiasymptotics at zero with respect to $c(\varepsilon)$, i.e.,

$$\lim_{\varepsilon\to 0} = \langle\frac{f^i_\varepsilon(\varepsilon x)}{c(\varepsilon)},\psi(x)\rangle = C^i_\psi,\ C^i_\psi \in \mathbf{C},\ \psi \in \mathcal{D}(I),\ i = 1,...,n.$$

In order to give the quasiasymptotic expansion of $\delta^2 \in \mathcal{G}\backslash\mathcal{D}'$, we introduce the following notation. Denote by Λ the set $\mathbf{N}$ or a finite set of the form $\{1, 2, ..., N\}$, $N \in \mathbf{N}$. Let $c_k \in \mathcal{K}$, $k \in \Lambda$, such that

$$\lim_{\varepsilon\to 0} \frac{c_{i+1}(\varepsilon)}{c_i(\varepsilon)} \to 0,\ i = 1, ..., N-1 \text{ (or } i \in \mathbf{N},\ \text{if } \Lambda = \mathbf{N}) \tag{8.5}$$

and

$$P_i = [P_{i\varepsilon}] \in \mathcal{G}(\Omega),\ i \in \Lambda.$$

Then $G = [G_\varepsilon] \in \mathcal{G}(\Omega)$ has the quasiasymptotic expansion at zero as $\sum_{k\in\Lambda} P_k$ with respect to $\{c_k(\varepsilon);\ k \in \Lambda\}$ if

$$\frac{(G_\varepsilon - \sum_{k=1}^m P_{k\varepsilon})(\varepsilon x)}{c_m(\varepsilon)} \to 0,\ \varepsilon \to 0^+,\ \text{in } \mathcal{D}'(\Omega) \text{ for every } m \in \Lambda. \tag{8.6}$$

In this case we write

$$G(\varepsilon x) \overset{q.e.c.}{\sim} \sum_{k\in\Lambda} P_k(\varepsilon x) \text{ with respect to } \{c_k(\varepsilon),\ k \in \Lambda\} \tag{8.7}$$

and say that G has the quasiasymptotic expansion in Colombeau sense.
One can simply prove that this definition does not depend on representatives. Similarly as in Theorem 8.3 we can apply the concept of quasiasymptotic expansion for the analysis of solutions. Here we give just an example of an expansion.

Example 8.4 *Let $\delta^2 = [\frac{1}{\varepsilon^2}\phi^2(\frac{t}{\varepsilon^2})]$, where $\phi \in C_0^\infty$, $\int \phi = 1$, $\int t^m \phi = 0$, $i = 1, ..., N$, $N \in \mathbf{N}$. Then*

$$\delta^2(\varepsilon x) \overset{q.e.c.}{\sim} \sum_{k=0}^{m} (-1)^k \frac{\mu_k}{k!} [\frac{1}{\varepsilon^4} \phi^{(k)}(\frac{t}{\varepsilon^2})](\varepsilon x)$$

with respect to the scale $\{\varepsilon^{-3+m}, m \le N\}$, where $\mu_k = \langle x^k, \phi^2 \rangle$.

References

[1] Biagioni, H.A.: A Nonlinear Theory of Generalized Functions, Springer-Verlag, Berlin-Heildelberg-New York, 1990.

[2] Biagioni, H.A., Colombeau, J.F.: Whitney's extension theorem for generalized functions, J. Math. Anal. Appl. 114, 2 (1986), 574–583.

[3] COLOMBEAU, J.F.: Elementary Introduction in New Generalized Functions, North Holland, Amsterdam, 1985.

[4] COLOMBEAU, J.F.: A Multiplication of Distributions, Bull. Am. Math. Soc. 23 (1990), 251–268.

[5] DJAPIĆ, N., PILIPOVIĆ, S.: Colombeau's generalized functions on a manifold—the microlocal properties Indag. Math. 7(3) 1996, 293–309.

[6] DJAPIĆ, N., PILIPOVIĆ, S., SCARPALÉZOS, D.: Intrinsic Microlocal Characterization of Colombeau's Generalized Functions, J. Anal. Math. (1998), to appear.

[7] EGOROV, YU.V.: A contribution to the theory of generalized functions, Russian Math. Surveys 45:5 (1990), 1–49. translated from: Uspehi Mat. Nauk 45;5 (1990), 3–40.

[8] FRIEDMAN, A.: Generalized Functions and Partial Differential Equations, Prentice-Hall, Inc., Englewood Cliffs, New York, 1963.

[9] GRAMCHEV, T.: Nonlinear maps in spaces of distributions, Math. Zeitschr. 209 (1992), 101–114.

[10] GRAMCHEV, T., OBERGUGGENBERGER, M.: Regularity theory and pseudodifferential operators in algebras of generalized functions, in preparation.

[11] HÖRMANDER, L.: The Analysis of Linear Partial Differential Operators III, Springer, Berlin, Heidelberg, New York, Tokyo, 1985.

[12] IVANOV, V.K.: An associative algebra of the simplest generalized functions, Siberian Math. J. 20 (1980), 509–516. Translated from: Sib. Mat. Z. 20 (1979), 731–740.

[13] KUNZINGER, M., OBERGUGGENBERGER, M.: Characterization of Colombeau generalized functions by their pointvalues, in Math. Nachr. (1999), to appear.

[14] MARTI J.-A., NUIRO S.P., VALMORIN V.: Algèbres différentielles et problème de Goursat non linéaire à données irrégulières, J. Math. Pure Appl., Preprint.

[15] NEDELJKOV, M., PILIPOVIĆ, S.: Hypoelliptic differential operators with generalized constant coefficients, Proc. Edinburgh Math. Soc. (1998), to appear.

[16] OBERGUGGENBERGER, M.: Multiplications of Distributions and Applications to Partial Differential Equations, Longman, 1992.

[17] OBERGUGGENBERGER, M.: Lecture Notes on Pseudo-differential Calculus, Preprint.

[18] OBERGUGGENBERGER, M., PILIPOVIĆ, S.: Local properties of Colombeau generalized functions, Preprint.

[19] PILIPOVIĆ, S.: Pseudo-differential operators and the microlocalization in the space of Colombeau's generalized functions, Bull. Serbian Acad. Sci. 20 (1995), 13–27.

[20] PILIPOVIĆ, S., SCARPALÉZOS, D.: Differential operators with generalized constant coefficients Portugaliae Math. 53 (1996), 305–324.

[21] PILIPOVIĆ, S., STOJANOVIĆ, M.: On The Behaviour of Colombeau's Generalized Functions at a Point, Publ. Math. Debrecen 51, 1–2 (1997), 111–126.

[22] PILIPOVIĆ, S., STOJANOVIĆ, M.: Generalized Solutions to Nonlinear Volterra Integral Equation with a non-Lipshitz Nonlinearity, Nonlinear Anal. Theory and Application (1998), to appear.

[23] ROSINGER, E.E.: Nonlinear Partial Differential Equations – An Algebraic View of Generalized Solutions, North Holland, Amsterdam, 1990.

[24] SCARPALÉZOS, D.: Colombeau's generalized functions: topological structures, microlocal properties. A simplified point of view, Preprint, Paris, 1993.

S. PILIPOVIĆ
University of Novi Sad, Faculty of Science, Institute of Mathematics,
Trg D. Obradovića 4, 21000 Novi Sad, Yugoslavia, E-mail: pilipovic@unsim.ns.ac.yu

Basics of a General Spectral Theory of Banach Modules

NIKOLAI YA. RADYNO

1 Introduction

When solving the problem of the multiplication of generalized functions, physicists and mathematicians regard a generalized function as an equivalence class of sequences of C^∞-functions. The equivalence relation is defined by approximation. This implies that these classes form an algebra which is called the algebra of mnemofunctions. A general method of constructing such algebras was obtained by A. B. Antonevich and Ya. V. Radyno in [1]. It has been found that this method can be applied not only to generalized functions but also to objects of different character, for example, to unbounded operators. In this paper we try to do this. We expect that the proposed ideas and approaches will permit one to solve nonlinear problems and in particular, certain problems of Quantum field theory.

2 Generalized normed spaces and pseudo-norms

Let X be a normed complex space. For a given $m \in \mathbf{Z}$ denote by $X^{[m]}$ the space of sequences $x = (x_k)$ (where $x_k \in X$) such that

$$\sup_{k \geq 1} \left(\|x_k\| k^{-m} \right) < +\infty,$$

or, alternatively, $X^{[m]} = l_\infty(X; k^{-m})$.

The following is an easy exercise:

Proposition 2.1. The set $X^{[m]}$ (with the coordinate-wise addition and the scalar multiplication) is a normed space. By the norm $\|x\|_m$ of an element $X^{[m]}$ we mean the quantity

$$\|x\| = \sup_{k \geq 1} \left(\|x_k\| k^{-m} \right).$$

Furthermore, if X is a Banach space, then $X^{[m]}$ is a Banach space.
Let us introduce the following spaces:

$$X^{[+\infty]} = \bigcup_{m\in Z} X^{[m]}, \qquad X^{[-\infty]} = \bigcap_{m\in Z} X^{[m]}, \qquad X_* = X^{[+\infty]}/X^{[-\infty]}.$$

Proposition 2.2. If X is a normed algebra (in particular, X is a Banach algebra), then $X^{[+\infty]}$ and $X^{[-\infty]}$ are algebras where the multiplication is defined coordinatewise, and $X^{[-\infty]}$ is an ideal in $X^{[+\infty]}$. Hence X_* is an algebra. It should be observed that $X^{[m]}$ (except $X^{[0]}$) is not an algebra.
If $X = \mathbf{C}$, then we obtain the ring $\mathbf{C}_*$ which we shall call the ring of generalized numbers.
Let $x = (x_k)$ be in $X^{[+\infty]}$. The symbol $[x]$ will denote the class of equivalent sequences in X_* which contain the element x.
The field $\mathbf{C}$ is embedded in $\mathbf{C}_*$ by the natural injection :

$$\mathbf{C} \ni \lambda \longrightarrow [(\lambda, \lambda, ...\lambda, ...)] \in \mathbf{C}_*$$

Let $[\lambda] = [(\lambda_k)]$ be in C_* and $[x] = [(x_k)] \in X_*$. We define the "generalized scalar multiplication" by the equality

$$[\lambda] \cdot [x] = [(\lambda_k x_k)].$$

Then X_* will be called a $\mathbf{C}_*$-module and a $\mathbf{C}_*$-algebra, if X is an algebra.
Remark 2.1. The element $\tilde{x} = (x_1, x_2, ...x_{n_0}, 0, 0, ...)$ belongs to $X^{[-\infty]}$. Thus, for $\tilde{x}$ and $\tilde{y}$ in X_*, we have $\tilde{x} = \tilde{y}$ if the inequality $\|x_n - y_n\| \le c \cdot n^{-m}$ holds for every $n \ge n_0$,$n \in N$.
Proposition 2.3. The ring C_* is not a field.
Since the ring of generalized complex numbers $\mathbf{C}_*$ is not a field, we shall describe which elements from $\mathbf{C}_*$ are invertible and which elements are not. The following proposition explains this.
Proposition 2.4. Let X be a Banach algebra. If an element $\tilde{x} = [(x_k)] \in X_*$ is invertible in X_*, then for any representative (x_k) there exist an integer $n_0 > 0$ and two constants $c > 0$ and m such that x_k is invertible in X for $k \ge n_0$ and

$$\|x_k^{-1}\| \le ck^m, \quad k \ge n_0.$$

Conversely, if there is some representative of $\tilde{x}$ for which the above property holds, the element $\tilde{x} = [(x_k)] \in X_*$ is invertible in X_*.

Corollary 2.1. If an element $\tilde{\lambda} \in \mathbf{C}_*$ is invertible in $\mathbf{C}_*$, then for any representative (λ_k) there exist an integer n_0 and two constants $c > 0$ and m such that λ_k is invertible in $\mathbf{C}$ for all $k \geq n_0$ and

$$|\lambda_k^{-1}| \leq ck^m, \quad for \ all \ k \geq n_0.$$

Conversely, if there is some representative (λ_k) for which the above property holds, then the element $\tilde{\lambda} \in \mathbf{C}_*$ is invertible in $\mathbf{C}_*$.

Definition 2.1. We call an element $\tilde{\lambda} = [(\lambda_k)] \in \mathbf{C}_*$ non-negative and write $\tilde{\lambda} \geq 0$ if there is a representative (λ'_k) of $\tilde{\lambda}$ such that $\lambda'_k \geq 0$ for all k. We call an element $\tilde{\lambda} \in \mathbf{C}_*$ positive and write $\tilde{\lambda} > 0$ if $\tilde{\lambda} \geq 0$ and $\tilde{\lambda}$ is invertible in $\mathbf{C}_*$.

If $\tilde{\lambda} > 0$, then for any of its representatives (λ_k) there exists an integer $n_0 > 0$ and two constants $c > 0$ and m such that λ_k is invertible in $\mathbf{C}$ for $k \geq n_0$ and

$$\lambda_k^{-1} > \frac{c}{k^m}, \quad for \ any \ k \geq n_0.$$

It is possible that $\tilde{\lambda} \geq 0$, but $\tilde{\lambda}$ not positive. For instance take as a representative the sequence

$$1, e^{-1}, 1, e^{-2}, 1, \ldots, e^{-k}, 1, \ldots.$$

The positive generalized numbers will play an important role in our calculations. Positive numbers in $\mathbf{C}_*$ play the same role as the positive real numbers in $\mathbf{C}$.

Since X_* is a $\mathbf{C}_*$- module, the functions on X_* are assumed to be $\mathbf{C}_*$-valued.

We define a function by the formula

$$\|\cdot\|_* : X_* \ni \tilde{x} = [(x_k)] \longrightarrow \|\tilde{x}\|_* = [(\|x_k\|)] \in \mathbf{C}_*.$$

We will use this function for defining a topology on X_*. In what follows we give some more details about $\|\cdot\|_*$.

Proposition 2.5. The function $\|\cdot\|_*$ has the following properties:

for any $\tilde{x}, \tilde{y} \in X_*, \tilde{\lambda} \in \mathbf{C}_*$ we have

1) $\|\tilde{x}\|_* \geq 0$,

2) If $\|\tilde{x}\|_* = 0$, then $\tilde{x} = \tilde{0}$,

3) $\|\tilde{x} + \tilde{y}\|_* \leq \|\tilde{x}\|_* + \|\tilde{y}\|_*$,

4) $\|\tilde{\lambda}\tilde{x}\|_* = |\tilde{\lambda}|_* \|\tilde{x}\|_*$,

5) $\|\tilde{x}\tilde{y}\|_* \leq \|\tilde{x}\|_* \|\tilde{y}\|_*$.

Since these properties are completely similar to those of a norm function, we call $\|\cdot\|_*$ a pseudo-norm.

Proposition 2.6. Let X be a normed algebra and $\tilde{x} \in X_*$ an invertible element. Then $\|\tilde{x}\|_*$ is a positive element of $\mathbf{C}_*$.

3 The topology on X_* generated by the pseudo-norm

We define a topology on X_* which we believe is natural for our objects.

Definition 3.1. The open ball with the center $\tilde{x}_0 \in X_*$ and the radius $0 < \tilde{r}$ in $(X_*, \|\cdot\|_*)$ is the following set

$$B(\tilde{x}_0, \tilde{r}) = \{\tilde{x} \in X_* : \|\tilde{x} - \tilde{x}_0\|_* < \tilde{r}, \quad \tilde{r} > 0\}.$$

The set $U \subset X_*$ is open if for any $\tilde{x}_0 \in U$ there exists $\tilde{r} > 0$ such that $B(\tilde{x}_0, \tilde{r}) \subset U$.

Proposition 3.1. This family of open sets defines a Hausdorff topology on X_*.

In this topology X_* is a topological group with respect to the addition. If X is a normed algebra, then X_* is a topological ring (for more details see [2], [3], [9], [5]). As soon as X_* becomes a topological group, we can consider the concept of a Cauchy sequence.

Definition 3.2. A sequence $(\tilde{x}_n) \in X_*$ is called a Cauchy sequence if, for each $\tilde{\epsilon} > 0$, there is N such that $\|\tilde{x}_n - \tilde{x}_m\|_* < \tilde{\epsilon}$ for all $n, m \geq N$.

Proposition 3.2. If X is a Banach space, then X_* is a complete group and the topological ring $\mathbf{C}_*$ is complete as well.

Definition 3.3. Let $\tilde{a} \in X_*$ and $\tilde{e}$ be the unit of X_*.The resolvent set of $\tilde{a}$ is

$$\rho(\tilde{a}) = \left\{\tilde{\lambda} \in \mathbf{C}_* : \tilde{\lambda}\tilde{e} - \tilde{a} \ \textit{is invertible in} \ X_*\right\}.$$

The spectrum of $\tilde{a}$ is the set $\sigma(\tilde{a}) = \mathbf{C}_* \setminus \rho(\tilde{a})$.

In the next section we classify the points of a resolvent set. Now we consider an analogue of one important theorem on Banach algebras [4], [8], [6].

Theorem 3.1. Let X be a Banach algebra, $\tilde{x} \in X_*$, and $\|\tilde{x}\|_* < \tilde{1}$. Then the element $\tilde{y} = \tilde{e} - \tilde{x}$ is invertible in X_*. Furthermore, the set of invertible elements X_*^{-1} of the algebra X_* is open in X_* and the mapping $\tilde{x} \longmapsto \tilde{x}^{-1}$ is continuous.

Corollary 3.1. Let $\tilde{a} \in X_*$. The resolvent set $\rho(\tilde{a})$, is open in $\mathbf{C}_*$.

4 Non-Archimedean topology on X_*

In this section we reintroduce the topology on the space X_* in a different way [4], [7]. Let us denote

$$X_m = X^{[m]}/X^{-\infty}.$$

We have

$$0 = X_{-\infty} \subset \ldots \subset X_m \subset X_{m+1} \subset \ldots \subset X_*.$$

Definition 4.1. The order of an element $x \in X_*$ is

$$\nu(x) = \min\{m \in \mathbf{Z} : x \in X_m\}.$$

The order $\nu(x)$ has the following properties:

1) $\nu(x) = \nu(-x)$,
2) $\nu(x+y) \leq \max\{\nu(x), \nu(y)\}$,
3) $\nu(xy) \leq \nu(x) + \nu(y)$,
4) $\nu(x) = -\infty$ is equivalent to $x = 0$,
5) $\nu(\|x\|_*) = \nu(x)$,
6) $\nu(\lambda x) \leq \nu(\lambda) + \nu(x)$,

where $x, y \in X_*$ and $\lambda \in \mathbf{C}_*$.

Remark 4.1. There exist x, y such that $\nu(xy) \neq \nu(x) + \nu(y)$.
For instance, take $x = (1, 1/2, 3, 1/4, 5, \ldots)$ and $y = x^{-1} = (1, 2, 1/3, 4, 1/5, \ldots)$. It is clear that $\nu(x) = \nu(x^{-1}) = 1$, but $xx^{-1} = 1$ and

$$\nu(xx^{-1}) = 0 < 2 = \nu(x) + \nu(x^{-1}).$$

The following definition classifies the points of the resolvent set. An example is given in the next section.

Definition 4.2. Let $\tilde{x} \in X_*$ be invertible and $\tilde{y} = [(y_k)]$ be its inverse. The order $\nu(\tilde{y})$ of the element $\tilde{y}$ is called the order of reversibility of $\tilde{x}$.
We denote by $\rho^m(\tilde{a})$ the set of all $\tilde{\lambda} \in \mathbf{C}_*$ such that the element $\tilde{\lambda}\tilde{e} - \tilde{a}$ has order of reversibility m. If $\tilde{\lambda} \in \rho^m(\tilde{a})$, we say that $\tilde{\lambda}$ has order of regularity m. It is easy to see that

$$\rho(\tilde{a}) = \bigcup_{m \in \mathbf{Z}} \rho^m(\tilde{a}).$$

Definition 4.3. Let $p > 1$ and $x \in X_*$. Define the function $\|\cdot\|_p : X_* \longmapsto R_+$ by

$$\|x\|_p = p^{\nu(x)}.$$

Proposition 4.1. The function $\|\cdot\|_p$ is a norm on X_* which defines a metric on X_*, so that X_* is a non-Archimedean normed group (an NN-group) (or an NN-ring if X is a normed algebra). Norms corresponding to different p's are equivalent.

Proposition 4.2. The topology on X_* determined by $\|\cdot\|_p$ is the same as that determined by the pseudo-norm $\|\cdot\|_*$. If X is a Banach space, then X_* is a non-Archimedean Banach $\mathbf{C}_*$-module.

Theorem 4.1. Let X be a Banach algebra. Then X_* is a non-Archimedean Banach $\mathbf{C}_*$-algebra and the spectrum of any element $\tilde{a} \in X_*$ is nonempty and closed in $\mathbf{C}_*$. By $\mathbf{C}_*$-algebra we mean a $\mathbf{C}_*$-module with an operation of multiplication.

5 Example of a generalized operator

In this section we give some examples that illustrate some aspects of the previous theory.

First we present an algebra X_* which, in some sense, contains unbounded operators.

Example. Consider the sequence $\{A_n\}$ of bounded operators on the space of continuous functions $C[0,1]$ defined by

$$A_n x(t) = nx(t) - n^2 \int_0^t x(s) e^{n(s-t)} ds.$$

The norms of such operators $\|A_n\|$ grow not faster than some power of n. Indeed,

$$A_n = nI - nT_n, \quad where \quad T_n x(t) = n \int_0^t x(s) e^{n(s-t)} ds.$$

Hence

$$\|A_n\| = \|nI - nT_n\| \leq n + n\|T_n\|$$

and

$$\|T_n\| = \max_{t \in [0,1]} \left| n \int_0^t e^{n(s-t)} ds \right| = \max_{t \in [0,1]} (1 - e^{-nt}) = 1.$$

Therefore

$$\|A_n\| \leq 2n.$$

Thus, the sequence of the operators A_n defines some element $\tilde{A}$ of X_* if X is the algebra of the bounded operators on $C[0,1]$. Let us take the operator $A = d/dt$ with the domain

$$D(A) = \{x(t) \in C[0,1] : x'(t) \in C[0,1], \quad x(0) = 0\}.$$

The sequence $\{A_n\}$ strongly converges to the operator A.

The spectrum of the operator A is empty in $\mathbf{C}$. The existence of the inverse operator $(A - \lambda I)^{-1}$ is equivalent to the solvability of the initial value problem

$$\begin{cases} u'(t) - \lambda u(t) = f(t); \\ u(0) = 0 \end{cases}$$

and this problem is solvable for all $\lambda \in \mathbf{C}$.

However, according to Theorem 4.1, the spectrum of the generalized operator $\tilde{A}$ is not empty in $\mathbf{C}_*$. For example, the generalized complex number generated by the sequence $\lambda_n = n$ belongs to it since the spectrum of the operator A_n is $\{n\}$.

Theorem 5.1. The spectrum of $\tilde{A}$ is contained in the set of the generalized complex numbers μ having the representatives μ_n with one of the following properties:

(i) The sequence $n-\mu_n$ represents a generalized complex number which is irreversible in $\mathbf{C}_*$;

(ii) The sequence μ_n satisfies the inequality

$$\left|\mu_n - \left(n - \frac{n^2}{2(n + c_1 + c_2 \ln n)}\right)\right| \leq \frac{n^2}{2(n + c_1 + c_2 \ln n)},$$

where c_1 and c_2 are constants.

Remark. We should note that such generalized operators can be applied in models of quantum mechanics. It is possible to consider generalized operators instead of unbounded operators and use generalized numbers instead of real numbers.

Acknowledgment. The author wants to thank the Erwin Schrödinger International Institute for the financial support of participation in Project "Nonlinear Theory of Generalized Functions".

References

[1] ANTONEVICH A. AND RADYNO YA.: On a general method for constructing algebras of generalized functions, Docl. Academy Nauk SSSR, 1991, v.318, no 2 (5), pp. 267-270. (Russian).

[2] BOURBAKI N.: Algebra, Hermann & C, Paris, 1962, 183 pp.

[3] GELFAND, I., RAIKOV, D., AND SHILOV G.: Commutative normed rings, New York, 1964, 306 pp.

[4] GULETSKAYA, O. AND RADYNO, N.: On a general spectral theory of Banach modules, Docl. RAN, 1995, v.343, no.1, pp.3-9 (Russian).

[5] NAIMARK, M.: Normed rings, Nauka, Moscow, 1968, 664 pp. (Russian).

[6] RUDIN, W.: Functional analysis, New York, 1973, 397 pp.

[7] VLADIMIROV, V., VOLOVICH, I., AND ZELENOV E.: p -adic analysis and mathematical physics, Nauka, Moscow, 1994, 352 pp.

[8] YOSIDA, K.: Functional analysis, Springer-Verlag, Berlin, 1965, 624 pp.

[9] ZARISKI, O. AND SAMUEL, P.: Commutative algebra, Volume 1, Princeton, New Jersey, 1958, 329 pp.

NIKOLAI YA. RADYNO
Institute of Mathematics National Academy of Sciences ul. Surganova 11, 220072 Minsk, BELARUS, E-mail: kolya@im.bas-net.by

Extensions of Algebras, Mnemofunctions, and their Applications

Yakov V. Radyno

1 Introduction

One of the basic problems of mathematical analysis is the problem of extending the operation of differentiation to a probably wider class of functions. One of the possible ways of solving this problem consists, for instance, in the following. Let $C(\mathbb{R})$ be the space of continuous functions and denote by $d = d/dx : C(\mathbb{R}) \to C(\mathbb{R})$ a differentiation with domain $D(d) = C^1(\mathbb{R})$. Let us consider the operation ∂ which is the extension of the differentiation d. The differentiation ∂ has a domain $D(\partial)$, namely the set of all functions which have a derivative almost everywhere. Such an extension has two disadvantages. First, it is not "wide", meaning that not only a locally integrable function is almost everywhere differentiable. Second, it is not "good" enough in the sense that the trivial differential equation $du = 0$ has as solutions only constants but the differential equation $\partial u = 0$ has different step functions and even Cantor's function as solutions [1].

The solution of the above problem is given by the theory of generalized functions with generalized differentiation. This theory removes both disadvantages. However, this theory has one significant defect. Namely, "it is known" that Laurent Schwartz [2] proved that "on the space of distributions $\mathcal{D}'(\mathbb{R})$ it is impossible to define an associative multiplication "! Actually Schwartz did not prove anything like this!

First S.L.Sobolev and then L.Schwartz introduced the multiplication (becoming the classical multiplication) between an infinitely differentiable function $a \in C^\infty(\mathbb{R})$ and a distribution $u \in \mathcal{D}'(\mathbb{R})$. According to this multiplication in the space $\mathcal{D}'(\mathbb{R})$ there is an object $\delta \neq 0$ such that $x \cdot \delta = 0$. This object is the classical δ - function or Dirac function! Actually in the mentioned paper [2] Schwartz proved that *on the space of distributions* $\mathcal{D}'(\mathbb{R})$ *it is impossible to introduce any associative* *multiplication $\odot$ which is compatible with the above product.* In general, $a \odot u \neq a \cdot u$ for $a \in C^\infty(\mathbb{R})$, $u \in \mathcal{D}'(\mathbb{R})$.

The successes of the theory of generalized functions in various areas of mathematics have apparently accustomed scientists to the idea that the original product $a \cdot u$, $a \in C^\infty(\mathbb{R})$, $u \in \mathcal{D}'(\mathbb{R})$ is the "unique correct" one. After all one question comes up : Is there an extension of the differentiation without all the mentioned deficiencies?

2 Extensions of algebras with differentiation

Definition 1. An algebra is a vector space A with multiplication.

Let us denote this algebra as $A(+,\cdot)$. We assume that all our algebras are associative.

Definition 2. A differentiation d on an algebra $A(+,\cdot)$ is a linear operator $d : A \to A$ with domain $D(d)$, for which the following equality holds

$$d(x \cdot y) = dx \cdot y + x \cdot dy, \qquad \forall x, y \in D(d);$$

$D(d)$ is a subalgebra of A.

An algebra A with differentiation d is denoted by $A(+,\cdot,d)$.

Problem 1. Let $A(+,\cdot,d)$ be an algebra with differentiation, $D(d) \subset A$. Is there an algebra $G(\oplus,\odot,\partial)$ with **everywhere defined** differentiation ∂, which is a nontrivial extension of the algebra $A(+,\cdot,d)$ and such that $\ker d = \ker \partial$?

In other words, is there an algebra G satisfying the following conditions:

$$A \subset G,\ A \neq G,\ \oplus|_A = +,\ \odot|_A = \cdot,\ \partial\,\big|_{D(d)} = d,\ D(\partial) = G,\ \ker d = \ker \partial\,?$$

Theorem 1. *For any commutative algebra A with everywhere defined differentiation d, i.e., $D(d) = A$ problem 1 has a solution.*

Proof. Let's assume $G = A[x]$ is the algebra of polynomials with coefficients from A with the natural operations of addition and multiplication. As differentiation we take $\partial x = -x^2$ and extend to the polynomials using the Leibniz rule. The algebra G satisfies all conditions of problem 1. We shall verify just the last one. Thus let us check that $\ker d = \ker \partial$. As ∂ is the continuation of d, it follows that $\ker d \subset \ker \partial$. Let

$$g = a_0 + a_1 x + \ldots + a_n x^n \in G = A[x], \quad a_k \in A.$$

then

$$\partial g = da_0 + da_1 x + (da_2 - a_1)x^2 + \ldots + (da_n - (n-1)a_{n-1})x^n - na_n x^{n+1}.$$

If $\partial g = 0$ then

$$da_0 = 0,\ da_1 = 0,\ da_2 - a_1 = 0, \ldots, da_n - (n-1)a_{n-1} = 0,\ na_n = 0.$$

We get that $a_n = a_{n-1} = \ldots = a_1 = 0$, $da_0 = 0$ i.e., $g = a_0$ and $da_0 = 0$. Thus $\ker \partial = \ker d$.

Remark 1. Condition $D(d) = A$ in Theorem 1 is essential. Indeed, small modifications in Schwartz' proofs [2] allow to formulate the following statement.

Theorem 2. *There is a commutative algebra A with differentiation d and $D(d) \neq A$ for which Problem 1 has no solution.*

Proof. Indeed, let's assume $A = C(\mathbb{R})$, $d = d/dx$, $D(d) = C^1(\mathbb{R}) \neq C(\mathbb{R})$ and that there exists an algebra $G(\oplus, \odot, \partial)$ being an extension of the algebra $A = C(\mathbb{R})$ as required in Problem 1. Let us choose three functions $f(x)$, $g(x)$, $h(x)$:

$$f(x) = \begin{cases} x[\ln|x| - 1], & x \neq 0 \\ 0, & x = 0 \end{cases}$$

$$g(x) = x \cdot f(x) = \begin{cases} x^2[\ln|x| - 1], & x \neq 0 \\ 0, & x = 0 \end{cases}$$

$$h(x) = x \cdot |x|.$$

It is obvious that $f \in C(\mathbb{R}) \setminus C^1(\mathbb{R})$, $g \in C^1(\mathbb{R}) \setminus C^2(\mathbb{R})$, $h \in C^1(\mathbb{R}) \setminus C^2(\mathbb{R})$ and

$$g'(x) = dg = \begin{cases} 2x[\ln|x| - 1] + x, & x \neq 0 \\ 0, & x = 0 \end{cases} = 2f + x, \qquad h'(x) = dh = 2|x| \qquad (1)$$

Applying the operator ∂ to equality (1) we obtain

$$\partial^2 g = 2\partial f + 1, \qquad \partial^2 h = 2\partial|x| \qquad (2)$$

On the other hand, taking the derivative of these equalities we get

$$\partial^2 g = \partial^2(x \odot f) = \partial[\partial x \odot f + x \odot \partial f] = 2\partial f + x \odot \partial^2 f, \qquad (3)$$

$$\partial^2 h = \partial^2(x \odot |x|) = \partial[\partial x \odot |x| + x \odot \partial|x|] = 2\partial|x| + x \odot \partial^2|x|. \qquad (4)$$

Comparing (2), (3) and (4) we have

$$x \odot \partial^2 f = 1 \quad \text{and} \quad x \odot \partial^2 |x| = 0. \tag{5}$$

Moreover, $\partial^2|x| \neq 0$. For if $\partial^2|x| = 0$, $\partial|x| \in \ker\partial = \ker d = \text{const} = d(\alpha x + \beta) = \partial(\alpha x + \beta)$, i.e., $|x| - \alpha x - \beta \in \ker\partial = \ker d$. Hence, $|x|$ is a linear function. Thus the algebra G contains elements 1, x, $x^{-1} = \partial^2 f$ and $2\delta = \partial^2|x| \neq 0$ which is impossible as is well known.

Remark 2. A basic theorem proved by Schwartz in [2] says that the algebra $C(\mathbb{R})$ with differentiation $d = d/dx$ cannot be embedded into any G with everywhere defined differentiation and containing the elements $\delta \neq 0$, $x \odot \delta = 0$.

Thus, any commutative algebra with everywhere defined differentiation can be extended to a similar nontrivial algebra.

This raises the following question: how big can we make such an extension?

Problem 2. Consider a commutative algebra with differentiation $A(+, \cdot, d)$; $D(d) = A$ and let E be any vector space containing A as a subspace. Is there an algebra with differentiation $G(\oplus, \odot, \partial)$ such that

1) $A \subset E \subset G$; 2) A is a subalgebra of G; 3) E is a subspace in G; 4) $D(\partial) = G$; 5) $\ker\partial = \ker d$?

Theorem 3. *For each pair $A \subset E$ where A is a commutative algebra with unit* **1** *and E is a vector space, problem 2 has a solution.*

Proof. Choose some algebraic basis $(a_i)_{i\in I}$ in the vector space A and add to it elements $(e_j)_{j\in J} \in E$ such that the system $\{(a_i)_{i\in I}, (e_j)_{j\in J}\}$ forms an algebraic basis in the space E. Let's assume as definition that $G = A[(e_j)_{j\in J}]$ is an algebra of polynomials of the infinitely many variables $(e_j)_{j\in J}$ with coefficients from the algebra A. To extend a differentiation to G we should employ the following procedure. Let's assume $A_1 = A[e_1]$ and ∂_1 on A_1 according to Theorem 1 such that $\ker\partial_1 = \ker d$. Then let $A_2 = A_1[e_2]$ and choose ∂_2 on A_2 according to Theorem 1 such that $\ker\partial_2 = \ker\partial_1 = \ker d$, etc.. Using transfinite induction prove that there is a differentiation ∂ on $G = A[(e_j)_{j\in J}]$ such that $\ker\partial = \ker d$. To finish our proof let us show how to include E in G. Let $x \in E$. Then $x = \sum_{i\in I}\lambda_i a_i + \sum_{j\in J}\mu_j e_j = a_0 + \sum \mu_j \cdot \mathbf{1} \cdot e_j \in G$.

Remark 3. According to Theorem 3, given an algebra $A(+, \cdot, d)$ which satisfies the conditions of the Theorem it is possible to embed A into an algebra $G(\oplus, \odot, \partial)$ so that between them we can include a vector space E. Multiplication $\odot$ and differentiation ∂ is automatically defined on the space E. In particular, if $A =$

$C^\infty(\mathbb{R})$, $d = d/dx$, $E = \mathcal{D}'(\mathbb{R})$, Theorem 3 solves the problem of multiplication of distributions.

Remark 4. Theorem 3 is an analogue of the Hahn-Banach theorem in the following sense. As the Hahn-Banach theorem is the basis of duality theory for locally convex spaces so Theorem 3 is a basis for constructing algebras of generalized functions. Moreover, the proofs of these theorems are similar and both are nonconstructive. Therefore

Problem 3. To give a constructive description of the algebra G.

First, a constructive description of an algebra G for the algebra $A = C^\infty(\mathbb{R})$, $d = d/dx$, $E = \mathcal{D}'(\mathbb{R})$ was given by J.-F.Colombeau. (see, for example [3]). A general construction for any Hausdorff locally convex algebra A and Hausdorff locally convex vector space E such that $\overline{A} = E$ is given in the paper [5].

3 The statement of the problem and the general scheme of the solution

Let E' be some space of distributions and let A be an algebra densely embedded in E' consisting of infinitely differentiable functions. We want to construct a new algebra G and a linear embedding $j : E' \to G$ such that A is embedded as an algebra, i.e.,

$$j(ab) = j(a)\, j(b), \quad a, b \in A. \tag{6}$$

There is a certain delicacy in the statement of the problem. The Schwartz example shows that there are no embeddings of the space $D'(\mathbf{R})$ in an algebra satisfying the condition: $j(au) = j(a)\, j(u)$, $a \in A, u \in E'$ in the case when $E' = D'(\mathbf{R}), A = C^\infty(\mathbf{R})$. Therefore we have to require the preservation of multiplication only for the elements of the algebra of smooth functions. The construction presented below does not use the fact that the space E' consists of distributions and has a more general character.

The initial objects in our construction are:

a) a Hausdorff topological vector space E' (for whose elements we want to define a product);

b) a locally convex algebra A, everywhere densely embedded in E';

c) a family of continuous linear operators $R_{\phi,\epsilon} : E' \to A$, where $\phi \in \Phi, \epsilon \in I$; Φ is a given set, I is a given set with a given filter F, and for each fixed ϕ $R_{\phi,\epsilon}(u) \to u$ with respect to F in the topology of E' for any $u \in E'$. For a fixed ϕ the family $R_{\phi,\epsilon}, \epsilon \in I$, is called a method of regularization of elements of E' by the elements of A and condition c) is in fact a specification of some set of regularization methods.

Let $\widetilde{G}$ be the set of all the mappings from $\Phi \times I$ into A. This set is an algebra and the space E' is linearly embedded into this algebra by means of the mapping $j_0 : u \to R_{\phi,\epsilon}(u)$. In some situations even such a "trivial" embedding enables us to get valuable results. However, this construction does not solve the problem posed above, since the equality (6) is not satisfied. Moreover, the algebra $\widetilde{G}$ is too big: in concrete problems different elements of this algebra describe one and the same object and should therefore be identified.

The construction goes as follows. In the algebra $\widetilde{G}$ we specify a certain subalgebra G_M and a certain ideal N in it. The quotient algebra G_M/N is the desired object. Thus the construction is reduced to the choice of a suitable pair (G_M, N).

Two elements f and g of $\widetilde{G}$ are said to be weakly equivalent if $f(\phi,\epsilon) - g(\phi,\epsilon) \to 0$ in E' with respect to F for any $\phi \in \Phi$. The set N_0 of all the elements weakly equivalent to zero is not even a subalgebra. Therefore the weak equivalence relation being natural in other problems does not solve this one.

Theorem 4. *Suppose that the subalgebra G_M and the ideal N satisfy the following conditions:*

i) for any $u \in E'$ the element $R_{\phi,\epsilon}(u) \in G_M$;

ii) $N \subset N_0$;

iii) the elements of the form $R_{\phi,\epsilon}(ab) - R_{\phi\epsilon}(a)R_{\phi\epsilon}(b)$ belong to N for any $a, b \in A$.

Then the space E' is embedded in the quotient algebra $G = G_M/N$ and A is embedded as a subalgebra. If G_M and N are invariant with respect to a certain continuous linear operator $L : A \to A$, then this operator L can be extended in a natural way to the algebra G.

Condition (i) requires the ideal N to be rather small while condition (ii) requires the same ideal to be rather large. Therefore the existence of such ideals is not obvious. Its explicit description depends on the choice of the family of regularization methods.

A way of choosing the subalgebra G_M and the ideal N is suggested by the topology of A. Let the topology on A be defined by a system of seminorms $(p_\alpha), \alpha \in \Lambda$.

The scalar function $p_\alpha(R_{\phi,\epsilon}(u)), u \in E'$, gives the typical growth of the elements of G_M with respect to ϵ and the functions $p_\alpha(R_{\phi,\epsilon}(ab) - R_{\phi,\epsilon}(a)R_{\phi,\epsilon}(b)), a, b \in A$, give a rate of descent of elements of N.

Let H_α be some algebra of scalar functions defined on $\Phi \times I$, generated by the functions of the form $p_\alpha(R_{\phi,\epsilon}(u)), u \in E'$ and let N_α be an algebra of scalar functions on $\Phi \times I$, generated by the functions of the form

$$p_\alpha(R_{\phi,\epsilon}(ab) - R_{\phi,\epsilon}(a)R_{\phi,\epsilon}(b)), a, b \in A.$$

The following theorem enables us to reduce the analysis to a rather simple case of scalar functions.

Theorem 5. *For every α let the algebra N_α be an ideal in H_α. Then the sets*

$$G_N(A) = \{f \in \widetilde{G} : \forall\alpha\ \exists h \in H_\alpha, p_\alpha(f(\phi,\epsilon)) \leq h(\phi,\epsilon)\},$$

$$N(A) = \{f \in \widetilde{G} : \forall\alpha\ \exists h \in N_\alpha, p_\alpha(f(\phi,\epsilon)) \leq h(\phi,\epsilon)\}$$

satisfy the condition of Theorem 4 and the quotient algebra $G = G_M(A)/N(A)$ gives the solution of the problem.

If $L : A \to A$ is a continuous linear operator and $B : A \times A \to A$ is a continuous bilinear operator, then they extend to operators $L : G \to G$ and $B : G \times G \to G$.

Remark 5. In particular, if $A = \mathcal{S}(\mathbb{R})$, L is differentiation or Fourier transformation, and B is multiplication or convolution then the algebra G containing all these operations was constructed in the paper [6].

4 Differential equations in the space of mnemofunctions

Let us consider as a model the differential equation

$$u' = \delta u + \delta,\ u(-1) = 1. \tag{7}$$

This equation has no solutions in the space of functions continuous at zero. If the function u is discontinuous , then the product δu is not defined. This means that there is no solution concept for this equation in the classical theory.

To consider the equation (7) in the space of mnemofunctions it is necessary to choose some mnemofunctions among those associated with the δ-function, for example δ_Φ and δ_Ψ, and to analyze the equation

$$u' = \delta_\phi u + \delta_\psi. \tag{8}$$

The solution of this equation is associated with an ordinary function of the form

$$u_0(x) = \begin{cases} 1, & x \leq 0 \\ \mu, & x > 0 \end{cases},$$

where the number μ depends on the choice we have made. We emphasize that the information about the number μ has been obtained as the result of the choice of mnemofunction, i.e., as the result of the refinement of the statement of the problem.

5 Spectral theory of operators

Let A be a Banach algebra with unit e, in particular, we can take the algebra $\mathcal{L}(X)$ of bounded linear operators on a Banach space X. Let $G_M(A)$ be the algebra of slowly (not faster than some degree) growing sequences from A and N the ideal in G_M consisting of rapidly (faster than any degree) decreasing sequences from A. Then we denote the quotient algebra $G = G_M/N$ as A_* and the corresponding ring of generalized numbers as $\mathbb{C}_*$.

The algebra A_* is a $\mathbb{C}_*$ - algebra that is a module over the ring $\mathbb{C}_*$ and its natural topology is non-Archimedean [7].

If $\widetilde{a} \in A_*$ then the generalized spectrum of an element $\widetilde{a}$ is the subset $\widetilde{\sigma}(\widetilde{a})$ in $\mathbb{C}_*$

$$\widetilde{\sigma}(\widetilde{a}) = \left\{ \widetilde{\lambda} \in \mathbb{C}_* : \widetilde{\lambda}\widetilde{e} - \widetilde{a} \textit{ is not invertible in } A_* \right\}.$$

It was found out in the paper [8] that A_* is a Banach non-Archimedean normed $\mathbb{C}_*$ - algebra and the generalized spectrum of any element from A_* is closed and not empty in $\mathbb{C}_*$. This allows to develop the spectral theory of elements of the algebra A_*.

The elements of the algebra $\mathcal{L}(X)_*$ are called generalized operators. This approach allows to take a new look on many questions of the spectral theory and functional calculus for unbounded operators because we can interpret some unbounded operators as generalized operators.

6 Algebra of generalized stochastic processes

Parallel to the development of the theory of generalized functions the development of the theory of generalized stochastic processes took place. The theory of generalized stochastic processes was created by I.M. Gelfand on the basis of the theory of distributions. K. Urbanik proposed to interpret generalized stochastic processes using Mikusinskii's sequential approach. Despite the importance of these theories in the modern theory of stochastic processes they have not found wide use in stochastic differential equations because they cannot be applied to nonlinear problems.

In the papers [9], [10] the construction of the algebra of generalized stochastic problems based on the theory of mnemofunctions $G(\widetilde{T}, \Omega)$[4] was proposed.

In [10] a uniform approach which applied to all known classes of differential equations with stochastic functions was proposed. This approach is based on the apparatus of algebras of stochastic mnemoprocesses and on the concept of generalized stochastic differentials in $G(\widetilde{T}; \Omega)$.

It is known that differential equations containing on the right-hand side a generalized stochastic process like "white noise" cannot be investigated by classical methods of the theory of the ordinary differential equations. For this purpose Ito developed the theory of stochastic differential equations based on concepts of stochastic integrals by Ito and Stratanovich. It was shown in [11] that the solutions of Ito and Stratanovich stochastic differential equations can be approximated by the solutions of the equations in differentials of the algebra $G(\widetilde{T}; \Omega)$. The method of proof of these statements is based on concepts of generalized stochastic differentials and Ito and Stratanovich sets.

The paper [12] is devoted to these methods which can be applied successfully to solutions of new classes of nonlinear stochastic differential equations.

The author thanks the Erwin Schrödinger International Institute for the financial support of his participation in the Project "Nonlinear Theory of Generalized Functions".

References

[1] STONE M. H.: Amer. Math. Soc. Coll. Pull., New York, 1932, v.15.

[2] SCHWARTZ L.: Sur l'impossibilite' de la multiplication des distributions. C.R. Acad. Sci. Paris. 1954, v.239, 874 - 848.

[3] COLOMBEAU J.-F.: New generalized functions and multiplication of distributions. North Holland, 1989.

[4] Egorov Yu.Y.: On the theory of generalized functions. Uspekhi Math. Nauk. 1990, v.45, No 5(275), 3 - 40.

[5] Antonevich A.B., Radyno Ya.V.: Dokl. Ac. Nauk SSSR 1991, 318, No 2, pp.267-270.

[6] Radyno Ya.V., Ngo Fu Thchan, Sabra R.: The Fourier transform in the algebra of new generalized functions. Dokl. RAN. 1992, v.327, n.1, 20 - 24.

[7] Radyno N.Ya.: Banach modules on the ring of the generalized numbers and mnemofunctions. Ph.D. dissertation, Minsk, 1996.

[8] Guletskaya O.I., Radyno N.Ya.: On general spectral theory of Banach modules. Dokl RAN, 1995, v.343, No.1, pp. 7 - 9.

[9] Lazakovich N.V., Radyno Ya.V.: Algebra generalized stochastic processes. Conf. "Theoretical and applied problems of mathematics", Abstracts, Tartu, 1990, pp. 281 - 283.

[10] Lazakovich N.V.: Stochastic differentials in algebra of generalized stochastic processes. Dokl. AN Belarusi, 1994, v.38, No 5, pp. 21 - 27.

[11] Lazakovich N.V.: Approximation of stochastic differential equations by finite difference equations. Dokl. AN Belarusi, 1995, v.39, No 3, pp. 20 - 22.

[12] Lazakovich N.V., Stashulyonok S.P.: Approximation of stochastic differential equations and integrals in the algebra of generalized stochastic processes. Dokl. AN Belarusi, 1994, v.39, No 6, pp. 34 - 38.

Yakov V. Radyno
Belorussian State University, Mechanics and Mathematics Faculty, Department of Functional Analysis, Belarus, Minsk, 220050, pr. F. Skoriny 4, E-mail: yakov@mmf.bsu.unibel.by

On the Multiplication of Periodic Hyperfunctions of one Variable

VINCENT VALMORIN

1 Introduction

In his book ('[5]), M. Oberguggenberger posed the problem of the construction of algebras of generalized functions containing the space of hyperfunctions. He also asked to consider distributional products arising from analytic regularization in this framework. In the papers ('[6], '[7], '[8]), the author constructed algebras of periodic generalized functions containing the space of distributions, in the n-dimensional case. But, it may be seen that none of these algebras contains the space of periodic hyperfunctions.

The purpose of this note is to give a first answer to these questions in the one-dimensional periodic case. We construct an algebra of simplified Colombeau type ('[2], '[3]) by taking as basic algebra the one of 2π-periodic analytic functions on the real line. Our construction is based on Fourier representation of periodic analytic functions and periodic hyperfunctions. As for distributions, we shall be able to deal with nonlinear differential problems in the space of periodic hyperfunctions. This note is only concerned with the construction and some basic properties of this algebra. Other topics will be presented in a forthcoming paper.

2 Notations

All the functions are supposed to be complex valued. $\mathcal{D}_{2\pi}$ is the space of C^∞ and 2π-periodic functions on $\mathbb{R}$ and $\mathcal{D}'_{2\pi}$ denotes the corresponding space of periodic distributions. $L^r_{2\pi}$, $1 \le r \le \infty$, are the Lebesgue spaces with norms defined, with usual modification if $r = \infty$, by $\|f\|_r = \left[\frac{1}{2\pi}\int_{-\pi}^{\pi} |f(x)|^r dx\right]^{\frac{1}{r}}$. For $p \in \mathbb{Z}$ let m_p denote the function defined on $\mathbb{R}$ by $m_p(x) = e^{ipx}$. If $f \in L^1_{2\pi}$ (resp. $T \in \mathcal{D}'_{2\pi}$)

and $p \in \mathbb{Z}$, the pth Fourier coefficient of f (resp. T) is $\widehat{f}_p = \frac{1}{2\pi}\int_{-\pi}^{\pi} f(x)e^{-ipx}dx$ (resp. $T(m_{-p})$). Let Ω be an open subset of $\mathbb{C}$. Then $\mathcal{O}(\Omega)$ denotes the space of holomorphic functions on Ω.

3 The space of 2π-periodic hyperfunctions on $\mathbb{R}$

In this section we refer to ('[1], chap. 1) and ('[4], chap. 4).

3.1 The space $C^{\omega}_{2\pi}$

The space $C^{\omega}_{2\pi}$ of 2π-periodic analytic functions on $\mathbb{R}$ is identified with the topological inductive limit of the Banach spaces $\mathcal{O}(U)$ where U runs over the open neighborhoods of the unit circle $\mathbb{T}$. If $T \in \mathcal{D}'_{2\pi}$, then it is well known that $T \in C^{\omega}_{2\pi}$ if and only if its Fourier coefficients verify:

$$\exists \alpha > 0\,,\, \exists c_\alpha > 0\,,\, \forall p \in \mathbb{Z}:\ |\widehat{T}_p| \le c_\alpha e^{-\alpha|p|}. \tag{1}$$

Consequently, via the Fourier transform, we may identify $C^{\omega}_{2\pi}$ with the topological inductive limit $\mathcal{A}$ of the Banach spaces $\mathcal{A}_\alpha$, $(\alpha > 0)$ where $\mathcal{A}_\alpha$ is the space of sequences $(c_p)_p$ such that $\|(c_p)_p\|_\alpha = \sup_{p\in\mathbb{Z}} |c_p|e^{\alpha|p|} < \infty$. Hence, if $f \in C^{\omega}_{2\pi}$, its Fourier series $\sum_{-\infty}^{+\infty} \widehat{f}_p e^{ipx}$ converges to it in $C^{\omega}_{2\pi}$.

3.2 The space $\mathcal{B}_{2\pi}$

The space $\mathcal{B}_{2\pi}$ of 2π-periodic hyperfunctions on $\mathbb{R}$ is the topological dual $[C^{\omega}_{2\pi}]'$ of $C^{\omega}_{2\pi}$. Hence, via the transpose of the inverse Fourier transform, one identifies $\mathcal{B}_{2\pi}$ with the space $\mathcal{A}'$ of sequences $(b_p)_p$ such that for every $\beta > 0$ the series $\sum_{-\infty}^{+\infty} |b_p|e^{-\beta|p|}$ converges; that is, the sequences $(b_p)_p$ such that:

$$\forall r > 0\,,\, \exists c_r > 0\,,\, \forall p \in \mathbb{Z}:\ |b_p| \le c_r e^{r|p|}. \tag{2}$$

The pth Fourier coefficient of $H \in \mathcal{B}_{2\pi}$ is the complex number $\widehat{H}_p = H(m_{-p})$. It may be seen that the pth Fourier coefficient of its derivative H' is $ip\widehat{H}_p$. Moreover, if $f \in C^{\omega}_{2\pi}$, one has

$$H(f) = \sum_{-\infty}^{+\infty} \widehat{H}_{-p}\widehat{f}_p. \tag{3}$$

On the other hand, H has the Fourier expansion $H = \sum_{-\infty}^{+\infty} \widehat{H}_p e^{ipx}$, the series being convergent in $\mathcal{B}_{2\pi}$. Note that the definition of $\mathcal{A}'$ gives a characterization of the Fourier coefficients of elements of $\mathcal{B}_{2\pi}$.

4 The algebra $\mathcal{X}_{e,M}$ and the ideal $\mathcal{N}_e$

4.1 Definitions and first properties

Let $E =]0,1]$ and $\mathcal{X} = [C^{\omega}_{2\pi}]^E$. Let $\mathcal{X}_{i,M}$ be the set of elements $(f_\varepsilon)_\varepsilon \in \mathcal{X}$ such that for all $\alpha > 0$, there exist nonnegative numbers a_α and b_α such that for all $m \in \mathbb{N}$ and all $\varepsilon \in E$, we have :

$$\left\| f_\varepsilon^{(m)} \right\|_\infty \leq a_\alpha b_\alpha^m m! e^{\frac{\alpha}{\varepsilon}}. \tag{4}$$

Let $\mathcal{N}_e$ denote the set of elements $(f_\varepsilon)_\varepsilon \in \mathcal{X}$ such that there exists $\beta > 0$ and nonnegative numbers a and b such that for all $m \in \mathbb{N}$ and all $\varepsilon \in E$, the following inequality holds :

$$\left\| f_\varepsilon^{(m)} \right\|_\infty \leq a b^m m! e^{-\frac{\beta}{\varepsilon}}. \tag{5}$$

Elements of $\mathcal{X}_{i,M}$ are said to have *infra-exponential growth*, and those of $\mathcal{N}_e$ are said to have *exponential decrease*.
Note that if $(f_\varepsilon)_\varepsilon \in \mathcal{X}_{i,M}$, then $(f_\varepsilon)_\varepsilon \in \mathcal{N}_e$ if and only if :

$$\exists \beta > 0, \exists b > 0, \exists \eta > 0 : \forall m \in \mathbb{N}, \forall \varepsilon \in]0, \eta[, \ \left\| f_\varepsilon^{(m)} \right\|_\infty \leq b^m m! e^{-\frac{\beta}{\varepsilon}}.$$

It follows straightforwardly from the definition of $\mathcal{X}_{i,M}$ and $\mathcal{N}_e$ that if $(f_\varepsilon)_\varepsilon \in \mathcal{X}_{i,M}$ (resp. $(f_\varepsilon)_\varepsilon \in \mathcal{N}_e$), then for every $\varepsilon \in E$, f_ε can be extended to a holomorphic function in the strip $\{z \in \mathbb{C} : |\Im z| < 1/b_\alpha\}$ (resp. $\{z \in \mathbb{C} : |\Im z| < 1/b\}$).

Proposition 4.1 . *$\mathcal{X}_{i,M}$ is a subalgebra of $\mathcal{X}$ and $\mathcal{N}_e$ is an ideal of $\mathcal{X}_{i,M}$.*

Proof. It is a direct application of Leibniz's formula. □

Later we use the following lemma which is a consequence of Stirling's formula :

Lemma 4.2 . *Let $\varphi(t) = t^m e^{-\lambda t}$ with $m \in \mathbb{N}$, $\lambda > 0$ and $t > 0$. Then there exists a constant $C > 0$ such that* $\sup_{t \in \mathbb{R}^+} \varphi(t) \leq C \lambda^{-m} m!$.

4.2 Fourier characterization of $\mathcal{X}_{i,M}$ and $\mathcal{N}_e$

It is important and even indispensable, due to the techniques used for solving our problem, to know whether an element is in $\mathcal{X}_{i,M}$ or in $\mathcal{N}_e$ or not, by a simple analysis of its Fourier coefficients. We have the following

Proposition 4.3. *Let $(f_\varepsilon)_\varepsilon \in \mathcal{X}$. Then, we have*
(i) $(f_\varepsilon)_\varepsilon \in \mathcal{X}_{i,M}$ *if and only if for any $\alpha > 0$, there exist positive numbers c_α and d_α such that for all $\varepsilon \in E$:*

$$\left|\widehat{f}_{\varepsilon,p}\right| \leq c_\alpha e^{\frac{\alpha}{\varepsilon}-|p|d_\alpha}, \qquad p \in \mathbb{Z}. \tag{6}$$

(ii) $(f_\varepsilon)_\varepsilon \in \mathcal{N}_e$ *if and only if there exist positive numbers β, c and d such that for all $\varepsilon \in E$:*

$$\left|\widehat{f}_{\varepsilon,p}\right| \leq ce^{-\left(\frac{\beta}{\varepsilon}+|p|d\right)}, \qquad p \in \mathbb{Z}. \tag{7}$$

Proof. **(i)** Suppose $(f_\varepsilon)_\varepsilon \in \mathcal{X}_{i,M}$. Let $\alpha > 0$. There exist nonnegative constants a_α, b_α such that for all $m \in \mathbb{N}$, $\left\|f_\varepsilon^{(m)}\right\|_\infty \leq a_\alpha b_\alpha^m m! e^{\frac{\alpha}{\varepsilon}}$. We have $\widehat{f_{\varepsilon,p}^{(m)}} = (ip)^m \widehat{f}_{\varepsilon,p}$. Then it follows from the above inequality that $|p|^m \left|\widehat{f}_{\varepsilon,p}\right| \leq a_\alpha b_\alpha^m m! e^{\frac{\alpha}{\varepsilon}}$, for all $m \in \mathbb{N}$ and $p \in \mathbb{Z}$. Writing $b_\alpha = rs_\alpha$ with $0 < r < 1$, we have $\frac{1}{m!}\left(\frac{|p|}{s_\alpha}\right)^m \left|\widehat{f}_{\varepsilon,p}\right| \leq a_\alpha r^m e^{\frac{\alpha}{\varepsilon}}$. By summation on m from 0 to ∞, we obtain $e^{\frac{|p|}{s_\alpha}} \left|\widehat{f}_{\varepsilon,p}\right| \leq \frac{a_\alpha}{1-r} e^{\frac{\alpha}{\varepsilon}}$. Put $c_\alpha = \dfrac{a_\alpha}{1-r}$ and $d_\alpha = \dfrac{1}{s_\alpha}$; then we have for any $p \in \mathbb{Z}$ and $\varepsilon \in E$, $\left|\widehat{f}_{\varepsilon,p}\right| \leq c_\alpha e^{\frac{\alpha}{\varepsilon}-|p|d_\alpha}$. Conversely, if (6) is fulfilled, then we have $\left\|f_\varepsilon^{(m)}\right\|_\infty \leq c_\alpha e^{\frac{\alpha}{\varepsilon}} \sum_{p=-\infty}^{+\infty} |p|^m e^{-|p|d_\alpha}$. From Lemma 4.2, there exists a constant C such that

$$|p|^m e^{-|p|\frac{d_\alpha}{2}} \leq C\left(\frac{2}{d_\alpha}\right)^m m!.$$

Set $a_\alpha = Cc_\alpha \sum\limits_{p=-\infty}^{+\infty} e^{-|p|\frac{d_\alpha}{2}}$ and $b_\alpha = \dfrac{2}{d_\alpha}$. We obtain $\left\|f_\varepsilon^{(m)}\right\|_\infty \leq a_\alpha b_\alpha^m m! e^{\frac{\alpha}{\varepsilon}}$, for all $m \in \mathbb{N}$ and $\varepsilon \in E$. Hence $(f_\varepsilon)_\varepsilon \in \mathcal{X}_{i,M}$.
(ii) The proof of this second part is a direct repetition of the first one. □

Now, we have the necessary material to build a differential algebra of generalized functions which contains the space of (periodic) hyperfunctions as a subspace. This will enable one to define a class of nonlinear operations on hyperfunctions.

5 Generalized functions of infra-exponential type

5.1 Definitions and notations

Definition 5.1 *The algebra of periodic generalized functions of infra-exponential type, is the quotient algebra* $\mathcal{G}_i = \mathcal{X}_{i,M}/\mathcal{N}_e$.

If $F \in \mathcal{G}_i$ has $(f_\varepsilon)_\varepsilon$ as representative in $\mathcal{X}_{i,M}$, we write $F = [f_\varepsilon]$. With the above notations, putting $F' = [f'_\varepsilon]$, we then define a differential structure on $\mathcal{G}_i$. In the sequel, we shall refer to an element of $\mathcal{G}_i$ as a generalized function.

5.2 Embedding of $\mathcal{B}_{2\pi}$ and $C^\omega_{2\pi}$ in $\mathcal{G}_i$

Consider the maps $\mathbf{i} : \mathcal{B}_{2\pi} \to \mathcal{X},\ H \mapsto (H_\varepsilon)_\varepsilon$ and $\mathbf{i}_0 : C^\omega_{2\pi} \to \mathcal{X},\ f \mapsto (f_\varepsilon)_\varepsilon$, where $f_\varepsilon = f$ and $H_\varepsilon = \sum_{|p| \le 1/\varepsilon} \widehat{H}_p e^{ipx}$. Obviously, we have $\lim_{\varepsilon \to 0} H_\varepsilon = H$ in $\mathcal{B}_{2\pi}$.

Proposition 5.1 $\mathbf{i}$ *is a linear injective map and* $\mathbf{i}_0$ *is an injective morphism of algebras. Moreover* $\mathbf{i}(\mathcal{B}_{2\pi})$ *and* $\mathbf{i}_0(\mathcal{B}_{2\pi})$ *are subsets of* $\mathcal{X}_{i,M}$.

Proof. We just prove the inclusion $\mathbf{i}(\mathcal{B}_{2\pi}) \subset \mathcal{X}_{i,M}$. For this, let $\alpha > 0$. There exists $c_\alpha > 0$ such that $|\widehat{H}_p| \le c_\alpha e^{\alpha|p|}$ for all $\forall p \in \mathbb{Z}$. It follows that for any $m \in \mathbb{N}$,

$$\left\|H_\varepsilon^{(m)}\right\|_\infty \le \sum_{|p| \le 1/\varepsilon} c_\alpha |p|^m e^{\alpha|p|} \le c_\alpha \left(\frac{1}{\varepsilon}\right)^m \sum_{|p| \le 1/\varepsilon} (e^\alpha)^{|p|} \le c_\alpha \left(\frac{1}{\varepsilon}\right)^m \left[\frac{2e^{\alpha(1+1/\varepsilon)}}{e^\alpha - 1}\right].$$

Consequently, for any positive α, there exists a positive constant d_α such that

$$\forall \varepsilon \in E,\ \left\|H_\varepsilon^{(m)}\right\|_\infty \le d_\alpha \left(\frac{1}{\varepsilon}\right)^m e^{\frac{\alpha}{\varepsilon}}.$$

According to Lemma 4.2, we have $\left(\frac{1}{\varepsilon}\right)^m e^{-\frac{\alpha}{2\varepsilon}} \le C \left(\frac{2}{\alpha}\right)^m m!$; hence

$$\forall \alpha > 0,\ \exists a_\alpha > 0,\ b_\alpha > 0,\ \forall \varepsilon \in E:\ \left\|H_\varepsilon^{(m)}\right\|_\infty \le a_\alpha b_\alpha^m m! e^{\frac{3\alpha}{2\varepsilon}}.$$

This above inequality shows that $(H_\varepsilon)_\varepsilon \in \mathcal{X}_{i,M}$. □

As $\mathbf{i}\left(C^\omega_{2\pi}\right) \subset \mathcal{B}_{2\pi}$, then we may consider $\mathbf{i}(f)$. We have

Proposition 5.2 *Let* $\bar{\mathbf{i}} : \mathcal{B}_{2\pi} \to \mathcal{G}_i,\ H \mapsto [H_\varepsilon]$ *and* $\bar{\mathbf{i}}_0 : C^\omega_{2\pi} \to \mathcal{G}_i,\ f \mapsto [f]$ *. Then,* $\bar{\mathbf{i}}$ *is a linear embedding and* $\bar{\mathbf{i}}_0$ *is an injective morphism of algebras. Moreover,* $\bar{\mathbf{i}}|_{C^\omega_{2\pi}} = \bar{\mathbf{i}}_0$ *and* $\bar{\mathbf{i}}(H') = \left(\bar{\mathbf{i}}(H)\right)'$ *for* $H \in \mathcal{B}_{2\pi}$.

Proof. From Proposition 5.1, it is obvious that $\bar{\mathbf{i}}$ is a linear embedding and that $\bar{\mathbf{i}}_0$ is an injective morphism of algebras. We just have to prove the second part of the statement, that is $(\mathbf{i}_0 - \mathbf{i})(C^{\omega}_{2\pi}) \subset \mathcal{N}_e$. Let $(\mathbf{i}_0 - \mathbf{i})(f) = (g_\varepsilon)_\varepsilon$. We have $g_\varepsilon(x) = \sum_{|p|>1/\varepsilon} \widehat{f}_p e^{ipx}$, and for $m \in \mathbb{N}$, $g_\varepsilon^{(m)}(x) = \sum_{|p|>1/\varepsilon} (ip)^m \widehat{f}_p e^{ipx}$. As $f \in C^{\omega}_{2\pi}$, it follows : $\exists r > 0, c_0 > 0, \forall p \in \mathbb{Z}, \left|\widehat{f}_p\right| \leq c_0 e^{-3r|p|}$. Therefore, $\left\|g_\varepsilon^{(m)}\right\|_\infty \leq \sum_{|p|>1/\varepsilon} |p|^m \left|\widehat{f}_p\right| \leq c_0 \sum_{|p|>1/\varepsilon} |p|^m e^{-3r|p|}$. If $|p| > 1/\varepsilon$, then we have $e^{-r|p|} \leq e^{-\frac{r}{\varepsilon}}$. Further, from Lemma 4.2, there exists $C > 0$ such that $|p|^m e^{-r|p|} \leq C \left(\frac{1}{r}\right)^m m!$. Hence we have for all $\varepsilon \in E$:

$$\left\|g_\varepsilon^{(m)}\right\|_\infty \leq c_0 \sum_{|p|>1/\varepsilon} C \left(\frac{1}{r}\right)^m m! e^{-\frac{r}{\varepsilon}} e^{-r|p|}.$$

Consequently, we finally obtain

$$\left\|g_\varepsilon^{(m)}\right\|_\infty \leq c_0 C \left[\sum_{p=-\infty}^{+\infty} e^{-r|p|}\right] \left(\frac{1}{r}\right)^m m! e^{-\frac{r}{\varepsilon}}.$$

This last inequality shows that $(g_\varepsilon)_\varepsilon \in \mathcal{N}_e$, which proves the expected inclusion. The last equality is obvious. □

In the sequel we shall identify H and $\bar{\mathbf{i}}(H)$ for any $H \in \mathcal{B}_{2\pi}$.

5.3 Generalized numbers of infra-exponential type

Let $\mathcal{C}$ be the set of complex valued sequences $(z_\varepsilon)_{\varepsilon \in E}$. As previously, such an element will be simply denoted $(z_\varepsilon)_\varepsilon$. Let $\mathcal{C}_{i,M}$ be set of $(z_\varepsilon)_\varepsilon \in \mathcal{C}$ such that for all positive α, there exists a positive number a_α such that: $\forall \varepsilon \in E$, $|z_\varepsilon| \leq a_\alpha e^{\frac{\alpha}{\varepsilon}}$. Elements of $\mathcal{C}_{i,M}$ are said to be of infra-exponential growth. In the same way, we define $\mathcal{I}_e$ as the set of elements $(z_\varepsilon)_\varepsilon \in \mathcal{C}_{i,M}$ for which there exist nonnegative constants β, b such that: $\forall \varepsilon \in E$, $|z_\varepsilon| \leq b e^{-\frac{\beta}{\varepsilon}}$. The elements of $\mathcal{I}_e$ are said to be of exponential decrease. It may be seen that $\mathcal{C}_{i,M}$ is a subalgebra of $\mathcal{C}$ and that $\mathcal{I}_e$ is an ideal of $\mathcal{C}_{i,M}$.

Definition 5.2 *The algebra of complex generalized constants of* infra-exponential type, *is the quotient algebra* $\mathcal{C}_i = \mathcal{C}_{i,M}/\mathcal{I}_e$.

We shall refer to an element of $\mathcal{A}_i$ as a generalized number or a generalized constant. Every complex number z is identified with a generalized number in a natural way.

Let $\mathcal{R} = \mathbb{R}^E$, $\mathcal{R}_{i,M} = \mathcal{C}_{i,M} \bigcap \mathcal{R}$, $\mathcal{J}_e = \mathcal{I}_e \bigcap \mathcal{R}_{i,M}$. It may be seen that $\mathcal{R}_{i,M}$ is a subalgebra of $\mathcal{R}$ and that $\mathcal{J}_e$ is an ideal of $\mathcal{R}_{i,M}$. We define the algebra of real generalized numbers of *infra-exponential type* as the factor algebra $\mathcal{R}_i = \frac{\mathcal{R}_{i,M}}{\mathcal{J}_e}$. Following ('[8]), we obtain the identification $\mathcal{C}_i = \mathcal{R}_i \oplus i\mathcal{R}_i$.
Let $f \in \mathcal{G}_i$ and $x \in \mathcal{R}_i$. Let $(f_\varepsilon)_\varepsilon$ and $(x_\varepsilon)_\varepsilon$ be representatives of f and x respectively. Consider $g_\varepsilon = f_\varepsilon + h_\varepsilon$ and $y_\varepsilon = x_\varepsilon + z_\varepsilon$ with $(h_\varepsilon)_\varepsilon \in \mathcal{N}_e$ and $(z_\varepsilon)_\varepsilon \in \mathcal{J}_e$. One easily obtains $[g_\varepsilon(y_\varepsilon)] = [f_\varepsilon(x_\varepsilon)]$, justifying the following

Definition 5.3 *Let $f \in \mathcal{G}_i$ and $x \in \mathcal{R}_i$. The value of f at x, denoted $f(x)$, is the generalized number in $\mathcal{C}_i$ defined by $f(x) = [(f_\varepsilon(x_\varepsilon))_\varepsilon]$ where $(f_\varepsilon)_\varepsilon$ and $(x_\varepsilon)_\varepsilon$ are arbitrary representatives of f and x, respectively.*

Finally, we define association between generalized numbers in the usual way :

Definition 5.4 *Two generalized constants a and b are associated, denoted by $a \approx b$, if they admit representatives $(a_\varepsilon)_\varepsilon$ and $(b_\varepsilon)_\varepsilon$ such that* $\lim_{\varepsilon \to 0}(a_\varepsilon - b_\varepsilon) = 0$.

5.4 Vanishing criteria

It may be seen that the following definition does not depend on the representatives.

Definition 5.5 *A generalized function f is said to* vanish analytically *at a point $x_0 \in \mathcal{C}_i$, if f and x_0 admit representatives $(f_\varepsilon)_\varepsilon$ and $(x_{0,\varepsilon})_\varepsilon$ for which there exist positive constants α, a and b such that: $\forall \varepsilon \in E$, $\forall m \in \mathbb{N}$, $\left|f_\varepsilon^{(m)}(x_{0,\varepsilon})\right| \leq ab^m m! e^{-\frac{\alpha}{\varepsilon}}$.*

It is clear that the zero generalized function vanishes analytically at any point. In the following, we examine some cases where the converse is true.

Theorem 5.3 *Suppose that a generalized function f vanishes analytically at a classical point x_0 with $b \leq (\pi e)^{-1}$. Then f is the zero function.*

Proof. Let $f = [f_\varepsilon]$ and $k \in \mathbb{N}$. Choose a, b, α such that for every $(m, \varepsilon) \in \mathbb{N} \times E$, $\left|f_\varepsilon^{(m)}(x_0)\right| \leq ab^m m! e^{-\frac{\alpha}{\varepsilon}}$. From Taylor's formula, there exists $\rho > 0$ such that

$$\left|f_\varepsilon^{(k)}(x)\right| \leq \sum_{m=k}^{\infty} \left|\frac{f_\varepsilon^{(m)}(x_0)}{(m-k)!}\right| \rho^{m-k} \quad (|x - x_0| < \rho).$$

Hence, we have $\left|f_\varepsilon^{(k)}(x)\right| \leq ab^k k! e^{-\frac{\alpha}{\varepsilon}} \sum_{m=0}^{\infty} \frac{(m+k)!}{m!k!}(b\rho)^m \quad (|x - x_0| < \rho)$. From Stirling's formula, there is a constant C such that for $m \neq 0$ and $k \neq 0$: $\frac{(m+k)!}{m!k!} \leq$

$C\frac{(m+k)^{m+k+1/2}}{m^{m+1/2}k^{k+1/2}} = C\frac{e^{\phi(m)}}{k^{k+1/2}}$, where $\phi(x) = \left(x+k+\frac{1}{2}\right)\ln(x+k) - \left(x+\frac{1}{2}\right)\ln x \ (x \geq 1)$. Due to $\ln(x+k) - \ln x \leq \dfrac{k}{x}$, we find $\phi(x) \leq k\left[\frac{2x+1}{2x} + \ln(x+k)\right] \leq k[2+\ln(x+k)]$. Hence, we have $e^{\phi(x)} \leq e^{2k}(x+k)^k$ and then $\frac{e^{\phi(m)}}{k^{k+1/2}} \leq \frac{e^{2k}}{\sqrt{k}}\left(1+\frac{m}{k}\right)^k \leq e^{m+2k}$. Finally, if we take $\rho < (eb)^{-1}$, then we obtain

$$\left|f_\varepsilon^{(k)}(x)\right| \leq Ca\left[\sum_{m=0}^{\infty}(eb\rho)^m\right](e^2b)^k k! e^{-\frac{\alpha}{\varepsilon}}.$$

According to the continuity of the $f_\varepsilon^{(k)}$'s, it follows that the above inequality holds in $[x_0, x_0+\rho]$ and consequently in $[x_0-(eb)^{-1}, x_0+(eb)^{-1}]$. In the same way as for x_0, we obtain the same type of inequality in the closed interval of center x_0 and radius $(eb)^{-1}+(e^3b)^{-1}$. By induction we obtain this type of inequality in the interval $[x_0-r, x_0+r]$ where $r = \sum_{m=0}^{\infty}(e^{2m+1}b)^{-1}$. We have $r \geq \pi$ if $b \leq (\pi e)^{-1}$. Consequently the same type of inequality holds on the interval $[x_0-\pi, x_0+\pi]$, hence $(f_\varepsilon)_\varepsilon \in \mathcal{N}_e$, that is $f = 0$. □

Corollary 5.4 *Let $f \in \mathcal{G}_i$. We have $f = 0$ if and only if f vanishes analytically at some classical point with $b \leq (\pi e)^{-1}$.*

We now consider the case where a generalized function vanishes analytically at a generalized point. According to the periodicity, we may suppose that this point admits a bounded representative.

Theorem 5.5 *Let $f \in \mathcal{G}_i$ vanish analytically at $x_0 \in \mathcal{R}_i$. Suppose that there exist $\xi \in \mathbb{R}$, $\rho > 0$ and representatives $(x_{0,\varepsilon})_\varepsilon$ and $(f_\varepsilon)_\varepsilon$ of x_0 and f such that:*
(i) $\forall \varepsilon \in E$, $|x_{0,\varepsilon} - \xi| < \rho$,
(ii) $\exists a > 0$, $\exists b \in]0, (e\rho)^{-1}[\cap[0, (\pi e)^{-1}]$, $\exists \alpha > 0 : \quad \forall \varepsilon \in E, \forall m \in \mathbb{N}, \left|f_\varepsilon^{(m)}(x_{0,\varepsilon})\right| \leq ab^m m! e^{-\frac{\alpha}{\varepsilon}}$.
Then, we have $f \equiv 0$.

Proof. Let $\varepsilon \in E$ and $k \in \mathbb{N}$ be fixed. We have :

$$f_\varepsilon^{(k)}(x) = \sum_{m=k}^{\infty}\frac{f_\varepsilon^{(m)}(x_{0,\varepsilon})}{(m-k)!}(x-x_{0,\varepsilon})^{m-k} \quad (|x-x_{0,\varepsilon}| < \rho).$$

Due to (ii), the above series converges absolutely. It follows that

$$\left|f_\varepsilon^{(k)}(x)\right| \leq \sum_{m=0}^{\infty}\frac{ab^{m+k}(m+k)!}{m!}e^{-\frac{\alpha}{\varepsilon}}\rho^m \quad (|x-x_{0,\varepsilon}| < \rho).$$

Now, as $|\xi - x_{0,\varepsilon}| < \rho$ for every $\varepsilon \in E$, we may substitute ξ for x in the above inequality, getting

$$\left|f_\varepsilon^{(k)}(\xi)\right| \le ab^k k! e^{-\frac{\alpha}{\varepsilon}} \sum_{m=0}^{\infty} \frac{(m+k)!}{m!k!}(b\rho)^m.$$

From the proof of Theorem 5.3, it follows that f vanishes analytically at the classical point ξ with $b \le (\pi e)^{-1}$, consequently $f = 0$. □

Remark 5.1 *Note that one might replace hypothesis (i) by :*
(i') $\exists \eta > 0 : \forall \varepsilon \in]0, \eta[,\ |x_{0,\varepsilon} - \xi| < \rho$.

Corollary 5.6 *Suppose that a generalized function f vanishes analytically, with $b \le (\pi e)^{-1}$, at a generalized point which is associated with a classical number. Then f is the zero function.*

Proof. Let f vanish analytically at x_0. One has representatives $(x_{0,\varepsilon})_\varepsilon$ and $(f_\varepsilon)_\varepsilon$ fulfilling a relation of the type :

$$\forall m \in \mathbb{N},\ \forall \varepsilon \in E,\ \left|f_\varepsilon^{(m)}(x_{0,\varepsilon})\right| \le ab^m m! e^{-\frac{\alpha}{\varepsilon}}.$$

Choose ρ such that $eb\rho < 1$; we can therefore take η such that (i') holds. Now the assertion follows easily. □

Acknowledgment. The author would like to thank Professor M. Oberguggenberger and the Erwin-Schrödinger-Institut for their kind invitation and support.

References

[1] Berenstein C. A. and R. Gay R.: Complex Analysis and Special Topics in Harmonic Analysis, Springer, 1995.

[2] Colombeau J-F.: New Generalized Functions and Multiplication of Distributions, North-Holland Math. Stud. 84, 1984.

[3] Colombeau J-F.: Elementary Introduction to New Generalized Functions, North-Holland Math. Stud. 113, 1985.

[4] KANEKO A.: Introduction to Hyperfunctions. Kluwer Academic Publishers, 1988.

[5] OBERGUGGENBERGER M.: Multiplication of Distributions and Applications to Partial Differential Equations. Pitman Res. Notes Math. Ser. 259, Longman, Harlow, 1992.

[6] VALMORIN V.: Fonctions généralisées périodiques et problème de Goursat, C. R. Acad. Sci. Paris Série I 320 (1995), 537-540.

[7] VALMORIN V.: A new algebra of periodic generalized functions, Journal for Analysis and its Applications 15 (1996), N^0 1, 57-74.

[8] VALMORIN V.: Fonctions généralisées périodiques et applications. Dissertationes Mathematicae, N^0 CCCLXI (1997).

V. VALMORIN
Université des Antilles et de la Guyane, Département Math-Info, Campus de Fouillole: 97159 Pointe à Pitre Cedex, France., E-mail : vincent.valmorin@univ-ag.fr

GEOMETRY,
GENERAL RELATIVITY

Distributional Aspects of General Relativity: The Example of the Energy-Momentum Tensor of the Extended Kerr-geometry

HERBERT BALASIN

Introduction

General relativity represents the physical framework to describe gravitational phenomena. One of its exciting and at the same time most disturbing features is the prediction of spacetime singularities, regions where the geometry becomes singular. Although it was originally believed that these singularities are artifacts of the high degree of symmetry present in the explicitly known solutions like the Schwarzschild black hole or the Friedmann-Robertson-Walker (FRW) cosmological models, Penrose and Hawking [1, 2] were able to show that singularities arise under physically reasonable assumptions as a generic feature of the theory. In the standard approach to general relativity the singular regions are excluded from the physical spacetime manifold, since concepts such as curvature lose their (classical) meaning. However, mathematically, distributional techniques have precisely been developed to deal with singular situations, like point-charges in electromagnetism, which seems to make them the natural tool for the analysis of space-time singularities. Unfortunately, general relativity is an inherently non-linear theory, which is in sharp contrast to the linear distributional framework, and therefore puts an immediate stopgap to the applicability of distributional concepts.

On the other hand there have been various investigations of distributional sources, representing idealized matter-distributions. Most prominently, Israel's framework [3] of thin-shells or surface-layers provides an unambiguous treatment of geometries, whose induced 3-metric remains continuous across a three-dimensional hypersurface representing the singular shell of matter, but whose extrinsic curvature suffers a discontinuous jump. Another important example of an even more singular geometry is provided by the Aichelburg Sexl (AS)-metric [4], which describes a spacetime that is flat apart from a null-hyperplane, where the gravitational field is concentrated. This geometry arises as the ultrarelativistic limit of the Schwarzschild black hole and belongs to a class of geometries, which may be described by Penroses cut-and-paste construction [5], that glues flat spacetime along a null-hypersurface. From a physical point of view the AS-geometry is of particular interest, since it allows an exact treatment of (quantum) particle-scattering at Planckian energies as was first observed by 't Hooft [6]. This fact has in turn increased the interest in a generalization of the AS-results to other (stationary) black hole geometries such as Reissner Nordström and Kerr [7, 8, 9, 10, 11]. An important feature of the AS-geometry is its energy-momentum tensor whose support is on a null-line. This is a crucial aspect for the physical interpretation of the AS-metric as the gravitational field of a massless particle, since the metric-limit contains an ambiguity, which has to be fixed by a physical condition [4, 10]. Actually, the existence of the distributional energy-momentum tensor emphasizes a certain imbalance between the original Schwarzschild geometry, which is considered to be a vacuum solution, and its ultrarelativistic limit, the AS-geometry. Precisely this imbalance led to investigations, which allowed the assigning of a distributional energy-momentum tensor to the Schwarzschild geometry (and, in fact, the whole Kerr-Newman family [9]), which is supported on the singular region. Moreover, its (unambiguous) limit coincides with the AS-result.

1 Distributions on arbitrary (paracompact) C^∞-manifolds

Distributions on $\mathbb{R}^n$ are usually defined as topological dual $\mathcal{D}'$ of test-function space $\mathcal{D} = C_0^\infty(\mathbb{R}^n)$. Ordinary (locally integrable) functions are embedded via integration

$$f \in L^1_{loc}(\mathbb{R}^n), \qquad (f,\varphi) := \int f\varphi \, d^n x \qquad \forall \varphi \in \mathcal{D}$$

Linear operations, multiplication by C^∞-functions and the derivative of a distribution make use of the corresponding actions on test-function space.

$$(gf,\varphi) := (f, g\varphi); \quad (\partial_j f, \varphi) := -(f, \partial_j \varphi); \qquad \varphi \in \mathcal{D}, f \in \mathcal{D}', g \in C^\infty(\mathbb{R}^n),$$

Due to the rich structure of $\mathbb{R}^n$ test-function spaces with different boundary conditions, such as test-functions of rapid decrease, may be introduced thus giving rise to different distribution spaces (and multiplier concepts). However, since our main concern will be the generalization to arbitrary manifolds, which in general will come without any preferred boundary structure, we will stick to C_0^∞ in the following. Since distributions are a "super-concept" of C^∞-functions a generalization to arbitrary manifolds should only depend on the differentiable structure. The space of test-functions has an obvious extension to an arbitrary manifold M as $C_0^\infty(M)$. However, in order to embed locally integrable functions it seems that orientability is required to define

$$(f,\varphi) = \int f\varphi\,\omega \qquad f \in L^1_{loc}(M), \varphi \in C_0^\infty(M), \omega \in \Omega^\infty(M),$$

where ω denotes a volume form on M. A closer look reveals that this requirement is actually unnecessary, if we consider test-n-forms instead of test-functions. That is, we replace test-function space by Ω_0^∞, the space of C^∞ n-forms. This has in addition the important consequence that distributions generalize functions rather than n-forms. Linear operations as well as multiplication by C^∞-functions have obvious extensions. Finally, the Lie-derivative along an arbitrary C^∞ vector-field X

$$(L_X f, \tilde{\varphi}) := -(f, L_X\tilde{\varphi}) \qquad \varphi \in \Omega_0^\infty, X \in \Gamma^\infty(TM)$$

which is motivated by the classical identity

$$\int L_X f\tilde{\varphi} = \int \underbrace{L_X(f\tilde{\varphi})}_{d(X\rfloor f\tilde{\varphi})} - \int f L_X\tilde{\varphi} = -\int f L_X\tilde{\varphi} \qquad f \in C^1(M)$$

extends the ordinary derivative operation. If the manifold is orientable, there exists a volume form ω and the above test-form space $\Omega_0^\infty(M)$ is isomorphic to $C_0^\infty(M)$ via

$$\tilde{\varphi} = \varphi\omega \qquad \tilde{\varphi} \in \Omega_0^\infty, \varphi \in C_0^\infty,$$

which is especially true for $\mathbb{R}^n$ with the preferred volume form $d^n x$. Intuitively it is also more natural to consider n-forms as the argument of a distribution rather than a function, since the n-form, being thought of as an infinitesimal volume, represents a "smeared-out" version of a point. This construction allows an immediate

generalization by considering np-forms[1] [12] as test-forms, which in turn give rise to generalized nq-forms as distribution space with $p + q = n$.
The above construction is specifically important for the applicability of distributional concepts to general relativity, since the metric, that will be allowed to have a more singular behavior, may not be used in the definition of test-function space.

2 Colombeau theory on arbitrary manifolds

As has already been mentioned in the introduction the applicability of distributional concepts to general realtivity necessitates a well-defined framework for nonlinear operations on generalized functions. Mathematically there have been several investigations on the extension of distribution theory which admit nonlinear operations [13, 14, 15]. In the following we will make use of the specific construction given by Colombeau [16], whose simplified version is particularly well suited for practical calculations.
The Colombeau algebra on $\mathbb{R}^n$ is usually defined in terms of equivalence classes of one-parameter families $(f_\epsilon)_{0<\epsilon<1}$ of $C^\infty(\mathbb{R}^n)$ functions, subject to a so-called moderate growth condition in ϵ, i.e.,

$$K \subset \mathbb{R}^n \textit{ compact},\ j_1, \dots, j_p \in \{1, \dots, n\} \quad \exists\, c > 0,\ \eta > 0,\ N \in \mathbb{N}$$
$$s.t. \quad \sup_{x \in K} |\partial_{j_1} \dots \partial_{j_p} f_\epsilon(x)| \le c\epsilon^{-N} \quad \forall\, 0 < \epsilon < \eta.$$

The equivalence relation is defined by employing the ideal consisting of those families $(f_\epsilon)_{0<\epsilon<1}$ satisfying

$$K \subset \mathbb{R}^n \textit{ compact},\ j_1, \dots, j_p \in \{1, \dots, n\},\ q \in \mathbb{N} \quad \exists\, c > 0,\ \eta > 0$$
$$s.t. \quad \sup_{x \in K} |\partial_{j_1} \dots \partial_{j_p} f_\epsilon(x)| \le c\epsilon^{q} \quad \forall\, 0 < \epsilon < \eta.$$

It is easily shown that the moderate functions form a subalgebra of the algebra of one-parameter families of C^∞-functions with the usual pointwise definition of the algebraic operations.
Two additional ingredients allow the contact with the usual distribution theory: the embedding of continuous functions f (and distributions) via convolution with a

[1] n-forms of weight $p \in \mathbb{R}$, which have the coordinate representation $\tilde{\varphi} = \varphi(x)(d^n x)^p$

smoothing kernel ϕ

$$f_\epsilon(x) := \int f(y)\frac{1}{\epsilon^n}\phi(\frac{y-x}{\epsilon})d^n y \quad \text{where } \phi \in C^\infty(\mathbb{R}^n) \text{ satisfies}$$

$$\int \phi(x)d^n x = 1, \int \phi(x)x^\alpha d^n x = 0 \quad \forall \alpha \in \mathbb{N}^n |\alpha| \geq 1$$

and the association relation, which coarse grains the algebra, by identifying those objects which have the same "distributional limit":

$$(f_\epsilon) \approx (g_\epsilon) \quad \textit{iff} \quad \lim_{\epsilon \to 0} \int (f_\epsilon(x) - g_\epsilon(x))\varphi(x)d^n x = 0 \quad \forall \varphi \in C_0^\infty(\mathbb{R}^n).$$

Association respects the linear operations and differentiation, but is in general incompatible with multiplication.

The above definitions have been given in a manner tailored to allow a pretty straightforward extension to an arbitrary manifold M with the notable exception of the embedding via the smoothing kernel[2]. This is due to the fact that scaling and shifting operations and the condition of vanishing momenta, draw heavily upon the vector space structure of $\mathbb{R}^n$, which is lost in the passage to arbitrary manifolds.

The moderate functions are defined by

$$K \subset M \textit{ compact}, X_1, \ldots, X_p \in \Gamma^\infty(TM) \quad \exists c > 0, \eta > 0, N \in \mathbb{N}$$
$$s.t. \quad \sup_{x \in K} |L_{X_1} \ldots L_{X_p} f_\epsilon(x)| \leq c\epsilon^{-N} \quad \forall 0 < \epsilon < \eta.$$

and the elements of the ideal have to obey

$$K \subset M \textit{ compact}, X_1, \ldots, X_p \in \Gamma^\infty(TM), q \in \mathbb{N} \quad \exists c > 0, \eta > 0$$
$$s.t. \quad \sup_{x \in K} |L_{X_1} \ldots L_{X_p} f_\epsilon(x)| \leq c\epsilon^q \quad \forall 0 < \epsilon < \eta.$$

$$(f_\epsilon) \text{ and } (g_\epsilon) \text{ are associated if} \quad \lim_{\epsilon \to 0} \int (f_\epsilon - g_\epsilon)\tilde{\varphi} = 0 \quad \forall \tilde{\varphi} \in \Omega_0^\infty.$$

As was the case for distribution space the Colombeau algebra is a concept that depends only on the differentiable structure of the manifold M, which is important for its applicability to general relativity.

[2]Proposals to generalize these concepts may be found in [17, 18].

3 Distributional energy-momentum tensor of the extended Kerr-geometry

The Kerr-geometry represents the spacetime that is generated by a rotating (stationary) black hole. It belongs to a specific class of geometries, the so-called Kerr-Schild geometries [19], which may be characterized by the decomposition of the metric tensor

$$g_{ab} = \eta_{ab} + f\, k_a k_b; \qquad k_a := \eta_{ab} k^b,\ k^a k^b \eta_{ab} = 0; \quad (k\nabla)k^a = (k\partial)k^a = 0,$$

where ∇_a and ∂_a denote the (torsion-free) covariant derivatives associated with g_{ab} and η_{ab}, respectively. k^a is a geodetic null vector-field, f denotes a scalar functions and η_{ab} is the flat part of the decomposition. For the specific case of the Kerr-geometry they are given with respect to coordinates associated with the flat part of the decomposition by

$$\eta = -dt^2 + dx^2 + dy^2 + dz^2, \quad r^4 - (x^2 + y^2 + z^2 - a^2)r^2 - a^2 z^2 = 0$$
$$k = \partial_t + \frac{r}{r^2 + a^2}(x\partial_x + y\partial_y) - \frac{a}{r^2 + a^2}(x\partial_y - y\partial_x) + \frac{z}{r}\partial_z, \quad f = \frac{2mr^3}{r^4 + a^2 z^2},$$

where m and a respectively denote the mass and the rotation-parameter of the black hole. An important feature of the Kerr-Schild class is the fact that the mixed Ricci-tensor is expressable as

$$R^a{}_b = \frac{1}{2}(\partial^a \partial_c (f k^c k_b) + \partial_b \partial_c (f k^a k^c) - \partial^2 (f k^a k_b))$$

The linearity in f allows an immediate distributional evaluation of this expression, whenever $f k^a k_b$ is locally integrable. Moreover, the distributional existence of the mixed form of the Ricci-tensor implies the existence of the mixed Einstein-tensor

$$G^a{}_b = R^a{}_b - \frac{1}{2}\delta^a{}_b R = 8\pi T^a{}_b$$

which in turn defines the energy-momentum tensor $T^a{}_b$. This is precisely the reason that allows calculation of $T^a{}_b$ for the Schwarzschild, Reissner-Nordström, Kerr and Kerr-Newman black holes [9]. Although it seems that these considerations are of purely distributional nature, the intermediate steps necessarily involve expressions nonlinear in f. However, the final result is independent of these nonlinearities and therefore represents a well-defined distributional object. The explicit calculation for Kerr-geometry yields [9]

$$T^a{}_b = \frac{m\delta(z)}{8\pi}\left\{\frac{2}{a}\left(\left[\frac{a^2\theta(a-\rho)}{\sqrt{a^2-\rho^2}^3}\right] - \frac{\theta(a-\rho)}{\sqrt{a^2-\rho^2}} - \delta(\rho-a)\right)\partial_t^a dt_b\right.$$
$$+\left(2\left[\frac{\rho\theta(a-\rho)}{\sqrt{a^2-\rho^2}^3}\right] - \frac{\pi}{a}\delta(\rho-a)\right)(\partial_t^a e_b^\phi - e_\phi^a dt_b) - \frac{2}{a}\frac{\theta(a-\rho)}{\sqrt{a^2-\rho^2}}e_\rho^a e_b^\rho$$
$$\left.-\frac{2}{a}\left(\left[\frac{\rho^2\theta(a-\rho)}{\sqrt{a^2-\rho^2}^3}\right] + \frac{\theta(a-\rho)}{\sqrt{a^2-\rho^2}} - 2\delta(\rho-a)\right)e_\phi^a e_b^\phi\right\}$$

where ρ and ϕ refer to circular polar-coordinates in the x, y-plane and e_ρ, e_ϕ denote the corresponding unit vectors. The energy-momentum tensor is supported on a disk-like region of radius a given by the rotation parameter. There exists however an extension of the Kerr-manifold through the disk-like region to negative values of r, which is a double-covering of the original $\mathbb{R}^4$ manifold-structure. In this extension the singular region becomes a branching line which coincides with ring boundary of the singular disk in the original (positive r) sheet. An immediate consequence of this extension is a loss of regularity in the flat part of the Kerr-Schild decomposition, which becomes manifest in a branch-singularity along the ring [20]. In this regard the extended Kerr-geometry does no longer belong to the Kerr-Schild class. There exists, however, a remarkable generalization of the Kerr-Schild class, which allows for a non-flat part in the metric decomposition, but retains most of the properties of the ordinary Kerr-Schild class

$$\tilde{g}_{ab} = g_{ab} + f\,k_a k_b;\quad k_a := g_{ab}k^b,\ k^a k^b g_{ab} = 0,\ (k\tilde{\nabla})k^a = (k\nabla)k^a = 0$$
$$\tilde{R}^a{}_b = R^a{}_b - fk^a k^c R_{cb} + \frac{1}{2}(\nabla_c\nabla^a(fk^c k_b) + \nabla_c\nabla_b(fk^c k^a) - \nabla^2(fk^a k_b)).$$

The expression for the full Ricci-tensor involves now products of the non-flat part of the decomposition with the scalar function f, both of which are singular expressions on the ring-like region. It is however, possible to find a regularized version of the extended Kerr-geometry [20] that stays within the generalized Kerr-Schild class and is in the Colombeau-sense associated to extended Kerr-geometry. The explicit calculation yields a mixed Ricci-tensor (density) that is supported on the singular ring

$$\omega_{\tilde{g}}\tilde{R}^a{}_b = -2\pi\delta(\cos\theta)\omega_0\left(a\delta(r)(\partial_r^a dr_b + \partial_\theta^a d\theta_b) - ma\delta'(r)(\partial_t^a - \frac{1}{a}\partial_\phi^a)(dt_b + ad\phi_b)\right),$$

where θ and ϕ refer to standard coordinates on S^2 and ω_0 denotes the volume form on $\mathbb{R}^2 \times S^2$. Physically, the negative r-region corresponds to a region of negative mass, which is reflected in the appearance of the $\delta'(r)$-term, whereas the part proportional to $\delta(r)$ represents the curvature of the non-flat part of the metric-decomposition, which again emphasizes the fact that the extended Kerr-geometry belongs to the generalized rather than the ordinary Kerr-Schild class. The above example provides a non-trivial application of the Colombeau-formalism adapted to general manifolds within general relativity, which allows some light to be shed on the singularity structure of the extended Kerr-spacetime.

Conclusion

Although the results are encouraging for the applicability of (generalized) distributional techniques to general relativity, the key ingredient for singling out a Colombeau-object associated with the extended Kerr-metric was provided by the (generalized) Kerr-Schild class. In order to deal with more general geometric situations it would be preferable to have a construction for embedding geometries into the algebra that plays the role of the convolution embedding in $\mathbb{R}^n$. On the other hand even in $\mathbb{R}^n$ the convolution embedding depends on the smoothing kernel and may thus be thought of as a particular way of modellizing the microscopic structure of a singular object within the algebra, which, from a physical point of view requires some additional justification.

Acknowledgment: The author would like to thank M. Oberguggenberger and M. Kunzinger for introducing him to the subject of Colombeau theory during his stay in Innsbruck in spring 1996.

References

[1] PENROSE, R.:*Phys. Rev. Lett.* **14**, 57, 1965.

[2] HAWKING, S. AND PENROSE, R.:*Proc. Roy. Soc. Lond.* **A314**, 529, 1970.

[3] ISRAEL, W.:*Nuovo Cimento* **44B**, 1, 1966.

[4] AICHELBURG, P. AND SEXL, R.:*Gen. Rel. Grav.* **2**, 303, 1970.

[5] PENROSE, R.:in *Studies in relativity* Oxford Univ. Press, London 1972.

[6] 'T HOOFT, G.:*Phys. Lett.* **198B**, 61 1987.

[7] LOUSTO, C. AND SANCHEZ N.:*Nucl. Phys.* **383**, 377, 1992.

[8] BALASIN, H. AND NACHBAGAUER, H.: *Class. Quantum Grav.* **10**, 2271, 1993.

[9] BALASIN, H. AND NACHBAGAUER, H.: *Class. Quantum Grav.* **11**, 1453, 1994.

[10] BALASIN, H. AND NACHBAGAUER, H.: *Class. Quantum Grav.* **12**, 707, 1995.

[11] STEINBAUER, R.:*J. Math. Phys.* **38**, 1614, 1997.

[12] BALASIN, H. AND KLÖSCH T.: *Beyond the Schwarzschild singularity* in prep.

[13] ROSINGER, E.:*Math. Studies* **146**, North Holland, 1987.

[14] COLOMBEAU, J.:*Math. Studies* **84**, North Holland, 1984.

[15] OBERGUGGENBERGER, M.:*Research Notes in Math.* **259**, Pitman, 1992.

[16] COLOMBEAU, J.:*Lect. Notes in Math.* **1532**, Springer, 1992.

[17] COLOMBEAU, J. AND MERIL, A.: *J. Math. Anal. and Apps.* **186**, 357, 1994.

[18] BALASIN, H.:Alberta-Thy-35-96, gr-qc9610017, 1996.

[19] DEBNEY, G., KERR, R. AND SCHILD A.: *J.Math. Phys.* **10**, 183, 1969.

[20] BALASIN, H.:*Class. Quantum Grav.* **14**, 3353, 1997.

H. BALASIN
Technische Universität Wien, Institut für Theoretische Physik, Wiedner Hauptstraße 8-10, 1040 WIEN, AUSTRIA, E-mail: hbalasin@tph.tuwien.ac.at

Lie Symmetries of Differential Equations in a Generalized Functions Setting

MICHAEL KUNZINGER

1 Introduction

Group analysis of differential equations provides a vast variety of methods for analyzing linear and nonlinear PDEs, with applications ranging from the construction of explicit or invariant solutions to particular equations, Lie Bäcklund symmetries, variational symmetries and conservation laws to the study of geometric properties of PDEs, to name only a few. For an introduction to these topics we refer to [17], [18], [9].

Starting from the 1950s research has also been directed towards utilizing symmetry methods for the study of differential equations within the framework of generalized functions. Here we mention the investigations of Methée, Tengstrand, Szmydt and Ziemian ([13], [24], [22, 23], [25]), whose main goal lay in the construction of group invariant fundamental solutions to linear PDE.

The first systematic approach towards a generalization of Lie group analysis of linear differential equations into a distributional framework is due to Berest and Ibragimov ([1, 2, 3, 4]). Finally, using the theory of algebras of generalized functions ([5], [14], [19]) it is possible to extend the range of applicability of symmetry group analysis even beyond the distributional setting to nonlinear equations and to symmetries that are themselves generalized functions ([16], [20], [21], [10], [11], [12]). It is the aim of this article to introduce some of the main ideas of these approaches.

To begin with, let us fix some notations (following [17] and [14]). We shall consider regular actions of a Lie group G on some open subset M of $\mathcal{X} \times \mathcal{U}$ (with $\dim(\mathcal{X}) = p$ and $\dim(\mathcal{U}) = q$) with $g \cdot (x, u) = (\Xi_g(x, u), \Phi_g(x, u))$. The group action is called projectable if $\Xi_g = \Xi_g(x)$. Elements of the Lie algebra of G are identified with vector fields on M and denoted by $\mathbf{v}$. $M^{(n)}$ denotes the n-jet space of M; the n-th prolongation of a group action g or vector field $\mathbf{v}$ is written as $\mathrm{pr}^{(n)}g$ or $\mathrm{pr}^{(n)}\mathbf{v}$, respectively. For an n-th order system S of differential equations in p dependent and q independent variables we write

$$\Delta_\nu(x, u^{(n)}) = 0, \quad 1 \leq \nu \leq l. \tag{1}$$

Thus S is identified with the subvariety $S_\Delta = \{(x, u^{(n)}) : \Delta(x, u^{(n)}) = 0\}$ of $\mathcal{X} \times \mathcal{U}^{(n)}$. For any $f : \Omega \subseteq \mathcal{X} \to \mathcal{U}$, Γ_f is the graph of f and $\Gamma_f^{(n)} := \{(x, \mathrm{pr}^{(n)} f(x)) : x \in \Omega\}$ is the graph of the n-jet of f.

Colombeau generalized functions will be considered in the so called 'simplified' version $\mathcal{G}(\Omega) = \mathcal{E}_M(\Omega)/\mathcal{N}(\Omega)$, see [14], p. 110. Also, we will make use of the (simplified) algebra $\mathcal{G}_\tau(\Omega) = \mathcal{E}_\tau(\Omega)/\mathcal{N}_\tau(\Omega)$ of tempered generalized functions as well as a 'mixed' variant $\widetilde{\mathcal{G}}_\tau(\Omega \times \Omega')$ whose elements satisfy $\mathcal{G}$-bounds in the Ω-variables and $\mathcal{G}_\tau$-bounds in the Ω'-variables (cf. [7] or [10]). Equivalence classes of elements of any of the above algebras will be denoted by $U = \mathrm{cl}[(u_\varepsilon)_\varepsilon]$. $\mathcal{O}_M(\Omega)$ denotes the space of smooth functions all whose derivatives are polynomially bounded towards the boundary of Ω. Its elements are called slowly increasing.

2 Permanence properties of classical symmetry groups

If we consider generalized solutions to some classical PDE (1), then in order to apply symmetry methods it is important to have criteria at hand that allow to decide if classical symmetry groups will transform generalized solutions into generalized solutions as well. In order to be able to apply some projectable transformation group G (whose action we write as $g \cdot (x, u) = (\Xi_g(x), \Phi_g(x, u))$) to arbitrary elements $U \in \mathcal{G}(\Omega)$ we want to set

$$gU := \mathrm{cl}[((\Phi_g \circ (id \times u_\varepsilon)) \circ \Xi_g^{-1})_\varepsilon] \tag{2}$$

To make this a well-defined element of $\mathcal{G}(\Xi_g(\Omega))$ we have to put some growth restriction on the group action, namely, that $u \to \Phi_g(x, u)$ is slowly increasing, uniformly for x in compact sets. Such group actions will be called slowly increasing. A slowly increasing group G will be called a $\mathcal{G}$-symmetry group of (1) if it transforms generalized solutions to generalized solutions.

The basic problem in detecting such permanence properties of classical symmetry groups stems from the quotient structure of the Colombeau algebra $\mathcal{G}$ itself: in fact, if $U \in \mathcal{G}$ is a solution to (1), then for any representative $(u_\varepsilon)_\varepsilon$ of U there exists some $(n_\varepsilon)_\varepsilon \in \mathcal{N}(\Omega)$ such that for all x and all ε we have $\Delta(x, \mathrm{pr}^{(n)} u_\varepsilon(x)) = n_\varepsilon(x)$. On the other hand, the fact that G is a classical symmetry group only yields that from $\Delta_\nu(x, \mathrm{pr}^{(n)} f(x)) = 0$ for all x we can conclude $\Delta_\nu(x, \mathrm{pr}^{(n)}(g \cdot f)(x)) = 0$. Since by the above the original differential equations need not be satisfied for any component of any given representative of a $\mathcal{G}$-solution, a direct translation of classical results seems quite difficult to do. There are, however, alternative properties of classical symmetry groups that lend themselves better to a study of permanence properties. First, we mention the following factorization property (cf. [12]):

2.1 Theorem *Suppose that (1) is nondegenerate in the sense of [17], Def. 2.70. Let $G = \{g_\eta : \eta \in (-\eta_o, \eta_o)\}$ be a one parameter symmetry group of (1) and set $g_\eta \cdot (x, u) = (\Xi_\eta(x, u), \Phi_\eta(x, u))$. Then there exist C^∞-functions $Q_{\mu\nu} : (-\eta_o, \eta_o) \times \mathcal{V} \to \mathbb{R}$ $(1 \le \mu, \nu \le l$, $\mathcal{V}$ an open subset of $\mathcal{M}^{(n)})$ such that if $u : \Omega \subseteq \mathbb{R}^p \to \mathbb{R}^q$ is smooth and $g_\eta u$ exists we have*

$$
\begin{aligned}
&\Delta_\nu(\Xi_\eta(x, u(x)), \mathrm{pr}^{(n)}(g_\eta u)(\Xi_\eta(x, u(x)))) = \\
&= \sum_{\mu=1}^{l} Q_{\mu\nu}(\eta, x, \mathrm{pr}^{(n)} u(x)) \Delta_\mu(x, \mathrm{pr}^{(n)} u(x))
\end{aligned} \tag{3}
$$

on the domain of $g_\eta u$ for $1 \le \nu \le l$.

2.2 Example Let us illustrate this result by considering the inviscid Burgers equation $u_t + uu_x = 0$, so that $\Delta(x, t, u, u_x, u_t) = u_t + uu_x$. The vector field $X = x^2 \partial_x + xt\partial_t + (ux - u^2 t)\partial_u$ generates the local one parameter symmetry group

$$
g_\eta : (x, t, u) \to \left(\frac{x}{1 - \eta x}, \frac{t}{1 - \eta x}, \frac{ux}{ut + (1 - \eta x)(x - ut)} \right)
$$

with $\Xi_\eta(x, t) = \left(\frac{x}{1-\eta x}, \frac{t}{1-\eta x} \right) =: (\widetilde{x}, \widetilde{t})$. For any smooth $u : \mathbb{R}^2 \to \mathbb{R}$ we get

$$
g_\eta u =: \widetilde{u} = (x, t) \to \frac{u\left(\frac{x}{1+\eta x}, \frac{t}{1+\eta x}\right) x}{u\left(\frac{x}{1+\eta x}, \frac{t}{1+\eta x}\right) t - u\left(\frac{x}{1+\eta x}, \frac{t}{1+\eta x}\right) \frac{x}{1+\eta x} + \frac{x}{1+\eta x}}
$$

In this case we have:

$$
\Delta(\Xi_\eta(x, t), \mathrm{pr}^{(1)}(g_\eta u)(\Xi_\eta(x, t))) = (\partial_2 \widetilde{u} + \widetilde{u} \partial_1 \widetilde{u})(\widetilde{x}, \widetilde{t}) =
$$

$$\left(\partial_2 u\left(\tfrac{\tilde{x}}{1+\eta\tilde{x}}, \tfrac{\tilde{t}}{1+\eta\tilde{x}}\right) + u\left(\tfrac{\tilde{x}}{1+\eta\tilde{x}}, \tfrac{\tilde{t}}{1+\eta\tilde{x}}\right)\partial_1 u\left(\tfrac{\tilde{x}}{1+\eta\tilde{x}}, \tfrac{\tilde{t}}{1+\eta\tilde{x}}\right)\right)\frac{1}{\left(u\left(\frac{\tilde{x}}{1+\eta\tilde{x}}, \frac{\tilde{t}}{1+\eta\tilde{x}}\right)t\eta+1\right)^3} =$$
$$= (\partial_2 u + u\partial_1 u)(x,t)\frac{1}{(u(x,t)t\eta+1)^3} = \frac{1}{(u(x,t)t\eta+1)^3}\Delta(x,t,u^{(1)}),$$

so $Q(\eta, x, t, u^{(n)}) = \frac{1}{(ut\eta+1)^3}$.

A direct application of 2.1 yields our first criterion for a transfer of classical symmetry groups to the Colombeau setting:

2.3 Proposition *Suppose that $\eta \to g_\eta$ is a slowly increasing one parameter symmetry group of (1) such that*

$$(x, u^{(n)}) \to Q_{\mu\nu}(\eta, x, u^{(n)})$$

is slowly increasing in the $u^{(n)}$-variables, uniformly in x on compact sets for $1 \leq \mu, \nu \leq l$ and every η. Then $\eta \to g_\eta$ is a $\mathcal{G}$-symmetry group of (1) as well.

Thus any classical symmetry group whose factorization in the sense of 2.1 behaves 'nicely' will remain a symmetry group in the $\mathcal{G}$-setting. Moreover, for many PDEs there are a priori statements on the factorization of their symmetry groups ([12]):

2.4 Theorem *Let $\eta \to g_\eta$ be a slowly increasing symmetry group of the single equation*

$$\Delta(x, u^{(n)}) = 0. \tag{4}$$

Set $N = dim(\mathcal{M}^{(n)})$ and let $\frac{\partial\Delta}{\partial z_k} \equiv c \neq 0$ for some $p+1 \leq k \leq N$. Then $\eta \to g_\eta$ is a $\mathcal{G}$-symmetry group of (4).

A generalization of this result to systems of PDEs can be found in [6]. Finally, we mention the following 'topological' criterion for permanence properties of symmetry groups of linear PDEs ([12]):

2.5 Theorem *If (1) is a system of linear PDEs with slowly increasing symmetry group $\eta \to g_\eta$ such that the operator defined by its left hand side is surjective $(\mathcal{C}^\infty(\Omega))^l \to (\mathcal{C}^\infty(\Omega))^l$ then $\eta \to g_\eta$ is also a $\mathcal{G}$-symmetry group for (1).*

3 Symmetries of PDEs in generalized functions

In this section we will address the question of group analysis of systems of differential equations of the form (1), where the Δ_ν themselves will be allowed to be generalized functions.

3.1 Distributional symmetries of linear equations

Following Berest and Ibragimov ([1, 2, 3, 4]), let us first investigate symmetry groups of linear PDEs in $\mathcal{D}'$ exclusively by means of classical distribution theory. Since in such an approach only linear operations may be applied to the dependent variables, we have to restrict the admissible group actions to the following class of projectable (one parameter) transformations:

$$g_\eta(x,u) = (\Xi(\eta,x), \Phi(\eta,x)\cdot u + \Psi(\eta,x)) \,. \tag{5}$$

We are looking for symmetries of linear systems of the form

$$\mathbf{L}(x,\partial)u = \sum_{|\alpha|\le m} a_\alpha(x)\partial^\alpha u = F \tag{6}$$

where the a_α are smooth and $F = (F_1,\dots,F_q) \in (\mathcal{D}'(\mathbb{R}^n))^q$. Hence $\Delta(x,u^{(n)}) = \mathbf{L}u - F$.

A direct translation of results from classical group analysis to the distributional setting is again impossible, basically due to the fact that distributions don't possess graphs and can therefore not be treated 'geometrically'. This difficulty is circumvented in [3] by implementing a factorization property into the *definition* of distributional symmetry groups. Thus (5) is called a symmetry group of (6) if there exists a matrix $Q(\eta,x)$ of smooth functions such that

$$\Delta_\nu(\Xi_\eta(x), \mathrm{pr}^{(n)}(g_\eta u)(\Xi_\eta(x))) = \sum_{\mu=1}^{q} Q_{\mu\nu}(\eta,x)\Delta_\mu(x,\mathrm{pr}^{(n)}u(x)), \tag{7}$$

or, for short: $\overline{\mathbf{L}}\overline{u} - \overline{F} = Q(\mathbf{L}u - F)$. Writing the infinitesimal generator of (5) in the form

$$\mathbf{v} = \sum_{i=1}^{p} \xi^i(x)\partial_{x_i} - \sum_{k=1}^{q}(\sum_{l=1}^{q}\varphi_l^k(x)u^l + \psi^k(x))\partial_{u_k}$$

(with $\xi(x) = \partial_\eta\Xi(0,x)$, $\psi(x) = -\partial_\eta\Psi(0,x)$, $\varphi(x) = -\partial_\eta\Phi(0,x)$) and setting

$$\mathbf{B} = \sum_{i=1}^{p}\xi^i(x)\partial_{x_i} + \varphi(x)$$

it follows that (7) is equivalent to

$$\begin{aligned} [\mathbf{L},\mathbf{B}] &= \lambda\mathbf{L} \\ (\mathbf{B}+\lambda)F &= -\mathbf{L}\psi \end{aligned} \tag{8}$$

for some smooth function λ. Moreover, a distribution u is invariant under the group action (5) iff $\mathbf{B}u = -\psi$. Applying this criterion together with (8) to (6) (with $F = \delta$) one can compute group invariant fundamental solutions for linear differential equations (for an overview of such applications, cf. [8]).

To formulate purely distributional symmetry group analysis of PDEs in the Colombeau framework one has to resort to the concept of association. Investigations on this matter can be found in [10] and [6].

3.2 Generalized transformation groups

In this final section we show that by using Colombeau algebras of generalized functions it is possible to adapt classical geometric methods of group analysis to (linear and nonlinear) differential equations involving generalized functions. We shall see that in this case the corresponding symmetry group actions will usually be generalized functions themselves.

A distinguishing feature of Colombeau functions (compared to distributions) rests on the pointvalue description obtained in [15] which allows their characterization by their graphs. Denoting by $\mathcal{R}$ resp. $\mathcal{R}_c$ the generalized numbers (resp. the compactly supported generalized numbers) any element U of $\mathcal{G}_\tau(\mathbb{R}^p)^q$ is uniquely determined by its graph $\Gamma_U = \{(\tilde{x}, U(\tilde{x})) : \tilde{x} \in \mathcal{R}^p\}$, and analogously for $U \in \mathcal{G}(\mathbb{R}^p)^q$ with $\mathcal{R}$ replaced by $\mathcal{R}_c$.

In order to implement the geometric approach to Lie point symmetries in the Colombeau framework, we first introduce the following terminology:

3.1 Definition $\Phi \in (\widetilde{\mathcal{G}}_\tau(\mathbb{R}^{1+n}))^n$ *is called generalized group action on* $\mathbb{R}^n$ *if*

(i) $\Phi(0, \,.\,) = \mathrm{id}$ *in* $(\mathcal{G}_\tau(\mathbb{R}^n))^n$.

(ii) $\Phi(\eta_1 + \eta_2, \,.\,) = \Phi(\eta_1, \Phi(\eta_2, \,.\,))$ *in* $(\mathcal{G}_\tau(\mathbb{R}^n))^n$, $\forall \eta_1, \eta_2 \in \mathbb{R}$.

3.2 Definition *A generalized vector field* $X = \sum_{i=1}^{n} \xi_i(x)\partial_{x_i}$ *(with* $\xi_i \in \mathcal{G}_\tau(\mathbb{R}^n)$*) is called* $\mathcal{G}$*-complete if the initial value problem*

$$\begin{aligned} \dot{x}(t) &= \xi(x(t)) \\ x(t_o) &= \tilde{x}_o \end{aligned} \tag{9}$$

is uniquely solvable in $\mathcal{G}(\mathbb{R})^n$ *for any* $\tilde{x}_o \in \mathcal{R}^n$ *and any* $t_o \in \mathbb{R}$.

Note that (9) actually requires composition of generalized functions (cf. [7], [10]). A generalized group action is called $\mathcal{G}$-complete if the corresponding vector field $\xi = \partial_\eta \Phi(0, \,.\,)$ is $\mathcal{G}$-complete. From the pointvalue characterization in [15] it follows that any $\mathcal{G}$-complete group action is uniquely determined by its infinitesimal generator. Having introduced the necessary geometric tools, let us now turn to the determination of symmetries of algebraic equations in generalized functions in the following sense: A generalized group action Φ is called a symmetry group of the equation $F(x) = 0$ (with $F \in \mathcal{G}_\tau(\mathbb{R}^n)$) if for any $\widetilde{x} \in \mathcal{R}^n$ with $F(\widetilde{x}) = 0 \in \mathcal{R}$ it follows that $\eta \to F(\Phi(\eta, \widetilde{x})) = 0$ in $\mathcal{G}(\mathbb{R})$. Then we obtain the following infinitesimal criterion ([12]):

3.3 Theorem *Suppose that $F \in \mathcal{G}_\tau(\mathbb{R}^n)$ is of the form*

$$F(x_1, \ldots, x_n) = x_i - f(x_1, \ldots, x_{i-1}, x_{i+1}, \ldots, x_n)$$

for some $1 \leq i \leq n$ and $f \in \mathcal{G}_\tau(\mathbb{R}^{n-1})$. Let Φ be a $\mathcal{G}$-complete generalized group action with infinitesimal generator $X = \sum_{i=1}^n \xi_i(x)\partial_{x_i}$ such that $x' \to \xi(x', f(x'))$ defines a generalized vector field on $\mathbb{R}^{n-1}$ whose corresponding system of ODEs possesses a flow (cf. [7]) in $(\widetilde{\mathcal{G}}_\tau(\mathbb{R}^{1+(n-1)}))^{n-1}$. The following conditions are equivalent:

(i) Φ is a symmetry group of $F(x) = 0$.

(ii) If $\widetilde{x} \in \mathcal{R}^n$ with $F(\widetilde{x}) = 0 \in \mathcal{R}$ it follows that $X(F)(\widetilde{x}) = 0$ in $\mathcal{R}$.

Now turning to symmetry group analysis of differential equations we consider generalized group actions $\Phi \in (\widetilde{\mathcal{G}}_\tau(\mathbb{R} \times \mathbb{R}^{p+q}))^{p+q}$ which are projectable in the sense that $\Phi(\eta, (x, u)) = (\Xi_\eta(x), \Psi_\eta(x, u))$, where $\Xi \in (\widetilde{\mathcal{G}}_\tau(\mathbb{R} \times \mathbb{R}^p))^p$ and $\Psi \in (\widetilde{\mathcal{G}}_\tau(\mathbb{R} \times \mathbb{R}^{p+q}))^q$. The action of Φ on some $U \in (\mathcal{G}_\tau(\mathbb{R}^p))^q$ is defined as

$$\Phi_\eta(U) : x \to \Psi_\eta(\Xi_{-\eta}(x), U \circ \Xi_{-\eta}(x)) \in (\mathcal{G}_\tau(\mathbb{R}^p))^q .$$

One can show that $\Phi_\eta(\Gamma_U) = \Gamma_{\Phi_\eta(U)}$, so the geometric meaning of group actions is preserved. Furthermore, the solution sets of systems of PDEs in $\mathcal{G}$ also have a geometric characterization analogous to the smooth case: Suppose that Δ in (1) is in $(\mathcal{G}_\tau((\mathbb{R}^p \times \mathbb{R}^q)^{(n)}))^l$. The solution set is then defined as $\mathcal{S}_\Delta := \{\widetilde{z} \in \mathcal{R}^{(n)} : \Delta_\nu(\widetilde{z}) = 0 \ (1 \leq \nu \leq l)\}$ and by the pointvalue characterization it follows that $U \in (\mathcal{G}_\tau(\mathbb{R}^p))^q$ is a solution of the system iff $\Gamma_{\mathrm{pr}^{(n)}U} \subseteq \mathcal{S}_\Delta$.

Similar to the classical case, for any generalized group action Φ on $\mathbb{R}^p \times \mathbb{R}^q$ and any $n \in \mathbb{N}$ one can define its n-th prolongation $\mathrm{pr}^{(n)}\Phi$ (which is a generalized group action on $(\mathbb{R}^p \times \mathbb{R}^q)^{(n)}$) such that for any $U \in \mathcal{G}_\tau(\mathbb{R}^p)$ we have $\Gamma_{\mathrm{pr}^{(n)}(\Phi_\eta U)} = \mathrm{pr}^{(n)}\Phi_\eta(\Gamma_{\mathrm{pr}^{(n)}U})$. The above construction shows:

3.4 Proposition *If Φ is a projectable generalized group action on $\mathbb{R}^p \times \mathbb{R}^q$ such that $\mathrm{pr}^{(n)}\Phi$ is a symmetry group of the algebraic equation $\Delta(z) = 0$ (with $\Delta \in (\mathcal{G}_\tau((\mathbb{R}^p \times \mathbb{R}^q)^{(n)}))^l$) then Φ is a symmetry group of the corresponding system of PDEs.*

If X is a $\mathcal{G}$-complete generalized vector field, then its n-th prolongation is defined as

$$\mathrm{pr}^{(n)}X|_z = \frac{d}{d\eta}\Big|_0 \mathrm{pr}^{(n)}\Phi_\eta(z),$$

provided that $\mathrm{pr}^{(n)}\Phi$ is $\mathcal{G}$-complete as well. X and Φ are then called $\mathcal{G}$-n-complete. From 3.3 and 3.4 we obtain the following main result:

3.5 Theorem *Let Δ be as in 3.4 and let Φ be a $\mathcal{G}$-n-complete generalized group action on $\mathbb{R}^p \times \mathbb{R}^q$ with infinitesimal generator X satisfying the conditions of 3.3. If*

$$\mathrm{pr}^{(n)}X(\Delta)(\widetilde{z}) = 0 \quad \forall \widetilde{z} \in (\mathcal{R}^p \times \mathcal{R}^q)^{(n)} \text{ with } \Delta(\widetilde{z}) = 0,$$

then Φ is a symmetry group of (1). □

Thus the same algorithm as in classical group analysis can be applied for determining symmetries of PDEs in generalized functions: Make an ansatz for the infinitesimal generator, calculate its n-th prolongation and then derive the defining equations (in this case: ODEs in $\mathcal{G}_\tau$) from 3.5. For applications and examples see [10], [11] and [12].

References

[1] Yu. Yu. Berest, Construction of fundamental solutions for Huygen's equations as invariant solutions, *Soviet Math. Dokl.*, Vol. **43**, No. 2, (1991) 496-499.

[2] Yu. Yu. Berest, Weak invariants of local groups of transformations, *Diff. Equ.*, **29**, No. 10 (1993), 1561-1567.

[3] YU. YU. BEREST, Group Analysis of linear Differential Equations in Distributions and the Construction of Fundamental Solutions, *Diff. Equ.*, **29**, No. 11 (1993), 1700-1711.

[4] YU. YU. BEREST, N. H. IBRAGIMOV, Group Theoretic Determination of Fundamental Solutions, *Lie Groups Appl.*, Vol **1**, No. 2, (1994), 65-80.

[5] J. F. COLOMBEAU, Elementary Introduction to New Generalized Functions, *North Holland*, Amsterdam 1985.

[6] N. ĐAPIĆ, S. PILIPOVIĆ, Lie groups of symmetries for $u_t + f(u)u_x = 0$, *preprint*, 1998.

[7] R. HERMANN, M. OBERGUGGENBERGER, Ordinary Differential Equations and Generalized Functions, this volume.

[8] N. H. IBRAGIMOV (Ed.), CRC Handbook of Lie Group Analysis of Differential Equations, Vol. **1 - 3**, *CRC Press*, Florida 1994 - 1996.

[9] I. S. KRASIL'SHCHIK, V. V. LYCHAGIN, A. M. VINOGRADOV Geometry of Jet Spaces and Nonlinear Partial Differential Equations, Gordon and Breach 1986.

[10] M. KUNZINGER, Lie-Transformation Groups in Colombeau Algebras, doctoral thesis, University of Vienna, 1996.

[11] M. KUNZINGER, M. OBERGUGGENBERGER, Symmetries of Differential Equations in Colombeau Algebras, N. H. Ibragimov, F. M. Mahomed (Eds.), Modern Group Analysis VI, 9-20, *New Age Int. Publ.* 1997.

[12] M. KUNZINGER, M. OBERGUGGENBERGER, Group Analysis of Differential Equations and Generalized Functions, submitted (1998).

[13] P. D. METHÉE, Sur les distributions invariantes dans le groupe des rotations de Lorentz, *Comment. Math. Helv.* **28** (1954), 224-269.

[14] M. OBERGUGGENBERGER, Multiplication of Distributions and Applications to Partial Differential Equations, *Pitman Research Notes in Mathematics* **259**, Longman, Harlow, 1992.

[15] M. OBERGUGGENBERGER, M. KUNZINGER, Characterization of Colombeau Generalized Functions by their Pointvalues, to appear in *Math. Nachr.*

[16] M. OBERGUGGENBERGER, E. E. ROSINGER, Solution of Continuous Nonlinear PDEs through Order Completion, *North Holland*, Amsterdam 1994.

[17] P. J. OLVER, Applications of Lie Groups to Differential Equations, 2nd ed., *Springer*, New York 1993.

[18] L. V. OVSIANNIKOV, Group Analysis of Differential Equations, Academic Press, New York, 1982.

[19] E. E. ROSINGER, Non-Linear Partial Differential Equations. An Algebraic View of Generalized Solutions, *North Holland*, Amsterdam, 1990.

[20] E. E. ROSINGER, M. RUDOLPH, Group invariance of global generalized solutions of smooth nonlinear PDEs: a Dedekind order completion method, *Lie Groups Appl.* **1** (1994), no. 1, 203-215.

[21] E. E. ROSINGER, Y. E. WALUS, Group invariance of generalized solutions obtained through the algebraic method, *Nonlinearity* **7** (1994), 837 - 859.

[22] Z. SZMYDT, On homogeneous rotation invariant distributions and the Laplace operator, *Ann. Pol. Math.* **6** (1979), 249-259.

[23] Z. SZMYDT, B. ZIEMIAN, Invariant fundamental solutions of the wave operator, *Demonstr. Math.* **19**, 371-386 (1986).

[24] A. TENGSTRAND, Distributions invariant under an orthogonal group of arbitrary signature, *Math. Scand.* **8** (1960), 201-218.

[25] B. ZIEMIAN, On distributions invariant with respect to some linear transformations, *Ann. Pol. Math.* **6** (1979), 261-276.

M. KUNZINGER
Universität Wien, Institut für Mathematik, Strudlhofgasse 4, A-1090 Wien, Austria,
E-mail: Michael.Kunzinger@univie.ac.at

Arbitrary Global Lie Group Actions on Generalised Solutions of Nonlinear PDEs and an Answer to Hilbert's Fifth Problem

ELEMÉR E. ROSINGER

1 Introduction

Let us consider nonlinear PDEs of the general smooth form - here and in the rest of this paper 'smooth' will mean $\mathcal{C}^\infty$-smooth - namely,

(1) $\quad T(x, D)U(x) = F(x, U(x), \ldots, D_x^p U(x), \ldots) = 0, \quad x \in \Omega \subseteq \mathbf{R}^n$

where Ω is nonvoid open, F is smooth in all its arguments, while $p \in \mathbf{N}^n$, $|p| \leq m$, with $m \in \mathbf{N}$ given, and finally, $U : \Omega \longrightarrow \mathbf{R}$ is the unknown or solution function, be it classical or generalised. In such a case, the relevant Lie group G of symmetries of solutions will act on the manifold $M = \Omega \times \mathbf{R}$, according to

(2) $\quad G \times M \ni (g, (x, u)) \longmapsto g(x, u) = (g_1(x, u), g_2(x, u)) \in M$

where $g \in G$ is the group element acting, while $x \in \Omega, u \in \mathbf{R}$ are the independent and dependent variables, respectively. In other words $x' = g_1(x, u) \in \Omega, u' = g_2(x, u) \in \mathbf{R}$, for $x \in \Omega, u \in \mathbf{R}$. Moreover, g_1, g_2 are smooth. We note that, given $g \in G$, then in view of the Lie group axioms, we obtain the smooth diffeomorphism

(3) $\quad M \ni \ (x, u) \overset{g}{\longmapsto} g(x, u) \in \ M$

The two problems we solve in this paper are: a) to use *parametrisation* for defining *globally arbitrary* Lie group actions (2) both on classical and generalised functions U, some of them solutions of nonlinear PDEs (1); and b) to prove the corresponding basic property, namely, that even in the parametric case, usual Lie group symmetries of the PDEs in (1) will transform their solutions into solutions, both in the classical and in the generalised case.

These results can be applied to cases of nonclassical solutions of interest, such as for instance, the rarefaction waves and Riemann solvers of the nonlinear shock wave equation, when these solutions are acted upon globally by the nonprojectable Lie group symmetries of that equation, see chapters 6 - 9 in Rosinger [22].

Before we go further, first let us recall here that a convenient particular case of Lie group actions (2), called *projectable* or *fibre preserving*, Olver [8, 9], is when $x' = g_1(x,u) = g_1(x)$, for $g \in G, x \in \Omega, u \in \mathbf{R}$. In this case (3) gives for each $g \in G$ the smooth diffeomorphism $\Omega \ni x \longmapsto g_1(x) \in \Omega$. Therefore, given any smooth function $U : \Delta \longrightarrow \mathbf{R}$, where $\Delta \subseteq \Omega$ is nonvoid open, we can easily define the global action of (2) on U, by $gU = U' : \Delta' \longrightarrow \mathbf{R}$, where $\Delta' = g_1(\Delta)$ and $U'(x') = g_2(g_1^{-1}(x'), U(g_1^{-1}(x')))$, for $g \in G, x' \in \Delta'$.

Furthermore, as seen in Oberguggenberger & Rosinger [7], Rosinger & Rudolph [23], and Rosinger & Walus [24, 25], this well-known procedure which defines globally projectable Lie group actions on smooth functions can quite easily be extended to such actions on large classes of generalised functions. Unfortunately, in the case of arbitrary Lie group actions (2) which are not projectable, such a global definition will typically fail even in the simplest cases of Lie group actions and smooth functions. For instance, one cannot rotate globally a parabola in the plane.

It should be pointed out, however, that in a certain limited sense, the parametric approach to the definition of arbitrary Lie group actions has, in fact, been in use for a long time by now. Indeed, given a smooth function $U : \Omega \longrightarrow \mathbf{R}$, with $\Omega \subseteq \mathbf{R}^n$ nonvoid open, and given an arbitrary Lie group action $G \times M \longrightarrow M$, where $M = \Omega \times \mathbf{R}$, then the usual way this Lie group action on M is extended to U is as follows. We consider the graph of U, that is, the set $\gamma_U = \{ (x, U(x)) \mid x \in \Omega \}$, which is obviously a subset of M. Therefore, for any $g \in G$, we can define pointwise the action $g\gamma_U$ and obtain again a subset of M. Unfortunately, however, in general, it will *not* be true that $g\gamma_U = \gamma_V$, for a certain smooth function $V : \Omega \longrightarrow \mathbf{R}$, which function V, if it existed, it would obviously correspond to the *global* action of g on U, that is, we would have $gU = V$. And then, the usual way in dealing with arbitrary Lie group actions on functions will overcome this difficulty simply by a *double* localization : g is taken near the identity $e \in G$, and in addition, U is restricted to suitable nonvoid, open subsets Δ of Ω.

It is clear, however, that the consideration of the graph γ_U of U amounts to replacing $U : \Omega \longrightarrow \mathbf{R}$ by the following special *parametric* form of it, namely $U_* : \Omega \longrightarrow M$, where $\Omega \ni x \longmapsto U_*(x) = (x, U(x)) \in M$. Furthermore, in this case $g\gamma_U$ is nothing else but gU_*, which can *always* be defined globally, irrespective of the smooth function U, or of the Lie group action $g \in G$ on M.

And thus it becomes clear that the *only* difficulty we have ever faced with defining globally arbitrary Lie group actions on functions is *not* at all related to Lie groups or functions, but solely to our rather unformulated, and yet quite implacable *intent* to have gU_* retranslated into a usual, *nonparametric* function $V : \Omega \longrightarrow \mathbf{R}$.
On the other hand, in this paper, the parametric approach to Lie group actions is adopted and pursued in its *full* extent, this being the simple and fundamental reason for the fact that *arbitrary* Lie group actions can now be defined *globally* on classical and generalised functions.

2 The parametric approach

In its essence, the parametric method amounts to an *embedding* of the usual definition of a function into a larger concept. A usual function $A \ni x \overset{f}{\longmapsto} y = f(x) \in B$ is actually *constrained* to be a correspondence from the set A of its independent variable x, to the set B of its dependent variable y. A parametric representation of f can be given by *any pull back* type mapping $P \ni p \overset{h}{\longmapsto} h(p) = (x(p), y(p)) \in A \times B$ which maps any suitably given parameter domain P into the graph of f, under the following two conditions, namely, $y(p) = f(x(p))$, for $p \in P$, and the mapping $P \ni p \longmapsto x(p) \in A$ is surjective. With respect to P, this, in general, only implies that its cardinal is not smaller than that of A. However, in this paper, further suitable topological conditions will be required on the parameter domains P, while the parametrisations h will be assumed smooth.
In the usual nonparametric approach, when we want to define the Lie group action on a function, and obtain again a function, we cannot do so, unless at the end we are able to *separate* the independent and dependent variables, by expressing the latter as a function of the former. And in the nonprojectable case of Lie group actions, this will often not be possible, except locally in the independent variable, and also, near to the trivial, identical Lie group transformation.
On the other hand, if we start, and end, with parametrically given functions, then we shall have no difficulties at all. And when in addition we make use of the mentioned, and rather neglected, calculus advantage of parametric representation with respect to the explicit computation of derivatives, we can obtain in a natural manner the results in this paper.
The idea to dispense with our usual attachment to the distinction between indepen-

dent and dependent variables was suggested in a different context, see Olver [8], p. 222, where quotient manifolds were defined, in order to provide a rigorous formulation of the basic method for finding group invariant classical solutions for partial differential equations. That approach does not involve parametrisation, and it does not make use of the functorial advantage. In this paper, however, the functorial advantage of parametric representation is used in order to define, and then employ in several nontrivial applications, arbitrary nonlinear Lie group actions on usual and generalised functions.

And now, to the details of the parametric approach. Let us denote by

(4) $\quad \mathcal{C}_n^\infty(M)$

the set of all smooth functions $V : \Lambda \longrightarrow M$, where $\Lambda \subseteq \mathbf{R}^n$ is nonvoid open, and call them *n-dimensional parametric functions with values in M*. Given now a usual smooth function $U : \Delta \longrightarrow \mathbf{R}$, where $\Delta \subseteq \Omega$ is nonvoid open, we shall associate with it its *canonical* parametrisation $U_* : \Delta \longrightarrow M$, defined by $U_*(x) = (x, U(x))$, for $x \in \Delta$. Then clearly $U_* \in \mathcal{C}_n^\infty(M)$. However, U can have many other parametrisations. For instance, if $V : \Lambda \longrightarrow M$ is any function in $\mathcal{C}_n^\infty(M)$, then we say that V is a *parametrisation* of U, if and only if

(5) $\quad V(\Lambda) = \gamma_U$

where we recall that $\gamma_U = \{\ (x, U(x)) \mid x \in \Delta\ \}$ is the graph of U. Now we can use the functorial advantage of parametric representations and *extend* the arbitrary global Lie group actions (2) as follows

(6) $\quad G \times \mathcal{C}_n^\infty(M) \ni (g, V) \longmapsto gV = g \circ V \in \mathcal{C}_n^\infty(M)$

which clearly uses nothing else but the commutative diagram

(7) $$\Lambda \xrightarrow{\ V\ } M \xrightarrow{\ g\ } M \qquad (\Lambda \xrightarrow{\ gV\ } M)$$

The facts which are important for us and which result easily from this construction in (6), (7), are the following. Given $g \in G$ and a usual smooth function $U : \Delta \longrightarrow \mathbf{R}$, with $\Delta \subseteq \Omega$ nonvoid open, we *cannot* in general define globally the Lie group action gU. On the other hand, the canonical parametrisation U_* can always be defined, and then, applying to it (6), (7), we can *always* define *globally* the *parametric* Lie group action gU_*. However, in general, even when gU may exist globally, we need *not* have the commutative diagram

$$(8) \qquad \begin{array}{ccc} \mathcal{C}^{\infty}_{par}(\Omega) \ni U & \xrightarrow{\quad g \quad} & gU \in \mathcal{C}^{\infty}_{par}(\Omega) \\ \Big\downarrow (\ )_* & & \Big\downarrow (\ )_* \\ \mathcal{C}^{\infty}_{n}(M) \ni U_* & \xrightarrow[\quad g \quad]{} & gU_* \neq (gU)_* \in \mathcal{C}^{\infty}_{n}(M) \end{array}$$

We can deal with this difficulty by introducing a *preorder*, that is, a reflexive and transitive binary relation $\leq$ on the set $\mathcal{C}^{\infty}_{n}(M)$ of parametric functions, as follows. Given $V : \Lambda \longrightarrow M$ and $V' : \Lambda' \longrightarrow M$ in $\mathcal{C}^{\infty}_{n}(M)$, we say that $V \leq V'$, if and only if there exists a smooth *surjection* $\varphi : \Lambda \longrightarrow \Lambda'$, such that

(9) $\quad V = V' \circ \varphi$

In this case it is obvious that we shall also have $V(\Lambda) = V'(\Lambda')$, and it is easy to see that a natural interpretation of this preorder $V \leq V'$ is that the parametric function V' is *simpler* than the parametric function V. In particular, one can show directly that, given any usual smooth function $U : \Delta \longrightarrow \mathbf{R}$, with $\Delta \subseteq \Omega$ nonvoid open, as well as a parametric function $V : \Lambda \longrightarrow M$ in $\mathcal{C}^{\infty}_{n}(M)$, then the following four conditions are *equivalent*

$$(10) \qquad \begin{array}{l} V \text{ is a parametrisation of } U, \qquad V(\Lambda) = \gamma_U, \qquad V \leq U_*, \\ \\ V_1 \text{ is surjective and } V_2 = U \circ V_1 \end{array}$$

where we assumed that $\Lambda \ni y \longmapsto V(y) = (V_1(y), V_2(y)) \in M$, with

$$(11) \qquad \begin{array}{l} \Lambda \ni y \longmapsto V_1(y) \in \Omega \\ \Lambda \ni y \longmapsto V_2(y) \in \mathbf{R} \end{array}$$

Now we can return and improve on (8) by noting that, whenever the usual Lie group action gU exists globally, we shall also have

(12) $\quad gU_* \leq (gU) \leq gU_*$

In this way, we are led to replace the usual Lie group action gU, especially when it fails to exist globally, with the *parametric* Lie group action gU_* which always exists globally, see for further details Rosinger [22], chap. 4.

3 Derivatives of parametric functions

As in (11), let us take any parametric function $V : \Lambda \to M$ in $\mathcal{C}^\infty_n(M)$. Then $\Lambda \subseteq \mathbf{R}^n$ is nonvoid open, and we obtain the following system of $n+1$ equations

(13) $\quad x = V_1(y), \quad U = V_2(y), \quad y \in \Lambda, \quad x \in \Omega, \quad U \in \mathbf{R}$

Let $J_{V_1} : \Lambda \longrightarrow \mathbf{R}$ be the Jacobian of V_1 in (11). We assume that the open set

(14) $\quad D_V = \{ y \in \Lambda \mid J_{V_1}(y) \neq 0 \}$

is nonvoid, see (25) below. Given now any $y_0 \in D_V$, the inverse function theorem yields open neighbourhoods $\Lambda' \subseteq D_V$ and $\Delta' \subseteq \Omega$ of y_0 and $x_0 = V_1(y_0)$ respectively, such that $V_1 : \Lambda' \longrightarrow \Delta'$ is a smooth diffeomorphism, therefore the smooth inverse relation

(15) $\quad y = V_1^{-1}(x), \quad x \in \Delta'$

holds. Then using the last equation in (13), we obtain locally, that is, on Δ', the smooth function $U : \Delta' \longrightarrow \mathbf{R}$, given by

(16) $\quad U(x) = V_2(V_1^{-1}(x)), \quad x \in \Delta'$

Here it is important to note that U in (16) may depend on Δ', and not only on V_1, V_2, that is V, as may be suggested by that relation. Indeed, this is the consequence of the fact that the mapping $V_1 : \Lambda \longrightarrow \Omega$ need not have a global smooth inverse.

The fact of interest to us, however, is the following. When we want *effectively* to compute the partial derivatives of the function U in (16), we can do so both *globally* and *explicitly*, straight from the original parametric form (11), that is, without first having to go through the often only locally valid and difficult to compute steps leading to U in (16). Indeed, the *chain rule* will give the relation

$$(17) \quad \begin{array}{l} (\partial_1 U(x), ..., \partial_n U(x)) = (\partial_1 V_2(y), ..., \partial_n V_2(y))(\mathcal{J}_{V_1}(y))^{-1}, \\ y \in D_V \subseteq \Lambda, \quad x = V_1(y) \in V_1(D_V) \subseteq \Omega \end{array}$$

while ∂_i, with $1 \leq i \leq n$, denotes the i-th partial derivative in the respective variables. We recall here that in view of (14), $V_1(D_V)$ is an open subset of Ω. In (17), we used the notation $\mathcal{J}_W(z)$ for the differential at $z \in E$ of a mapping $W : E \longrightarrow \mathbf{R}^l$, where $E \subseteq \mathbf{R}^k$ is open. It follows that $\mathcal{J}_W(z)$ is the $l \times k$ matrix of first-order partial derivatives of W, thus in particular $J_{V_1}(y) = \det\mathcal{J}_{V_1}(y), \quad y \in \Lambda$, and also $\mathcal{J}_U(x) = (\partial_1 U(x), ..., \partial_n U(x))$, $\mathcal{J}_{V_2}(y) = (\partial_1 V_2(y), ..., \partial_n V_2(y))$, with x and y as in (17). Clearly, the right-hand side of the equation in (17) can be computed *directly* in terms of the parametric representation (11) alone. Furthermore, unlike (16), which in general is only valid locally, the relation (17) is *global*, see Section 7, and in particular (25), about the size of D_V. Finally, *higher order* partial derivatives of U in (16) can also be computed directly in the terms of the parametric representation (11) alone, simply by a further application of the chain rule to (17). And again, the respective relations will be valid *globally*.

4 Parametric functions as solutions

Given the general smooth nonlinear PDE in (1) and a parametric function $V : \Lambda \longrightarrow M$ in $\mathcal{C}_n^\infty(M)$ as in (11), then with the use of (13), (17), as well as the relations for higher order partial derivatives of U, which result from (17), we can define by *pull back* the nonlinear partial differential operator

(18) $\quad S(y,D)\, V(y) \;=\; F(\, V_1(y),\; V_2(y),\; \ldots\,), \quad y \in D_V \subseteq \Lambda \subseteq \mathbf{R}^n$

where in the right-hand term of the equation in (18), the places left free and marked by dots, correspond to the partial derivatives $D_x^p U$, with $p \in \mathbf{N}^n$, $1 \leq |p| \leq m$, in (1). As noted, in view of (17), these higher order partial derivatives of U can be computed globally on D_V and directly in terms of the partial derivatives of V_1 and V_2 alone. Clearly, if the open subset $D_V \subseteq \Lambda$ is nonvoid, then on this set D_V, the resulting $S(y,D)$ in (18) is of the same form as $T(x,D)$ in (1) is on Ω.

Now by definition, we shall say that V is a *parametric solution* of the PDE in (1), if and only if the relation holds

(19) $\quad S(y,D)V(y) \;=\; 0, \quad y \in D_V$

It is important to note here that the natural *domain* on which the parametric function $V : \Lambda \to M$ is a solution, is not necessarily the whole of its domain of definition Λ, but its open subset D_V. In view of (13), (14), we obviously have $V_1(D_V) \subseteq \Omega$, therefore in case $V_1(D_V) \neq \Omega$, the resulting concept of parametric solution given by (19) is weaker, that is, it is more general, than the usual one.

However, we should remember that the same solution may have different parametrisations. Let us, for instance, consider two such different parametrisations V, V', with V' being *simpler* than V, that is, see (9), $V \leq V'$. Then the chain rule gives $J_{V_1}(y) \;=\; J_{V_1'}(\varphi(y))\; J_\varphi(y), \quad y \in \Lambda$, hence, see (14), $\varphi(D_V) \subseteq D_{V'}$, and therefore

(20) $\quad V_1(D_V) \subseteq V_1'(D_{V'}) \subseteq \Omega$

and this means that, even in the case of parametric functions V, V' which are comparable in the sense of (9), the respective domains in (20) need *not* be the same. Moreover, a parametrisation V' which is *simpler* than V, will in view of (20) give a *less weak* concept of parametric solution. In particular, according to (12), the canonical parametrisation U_* of a solution given by a usual function U will yield the least weak concept of parametric solution, and in fact, it will give the usual concept.

5 Equivalent definition of parametric solution

Before we go further, it will be useful to recall the typically local, however, usual, that is, nonparametric *equivalent* formulation of the concept of parametric solution given in (16). This naturally centers around (16), and in view of (14), (18), it is clear that for $y_0 \in D_V$ and $x_0 = V_1(y_0) \in \Omega$, the following two relations

(21) $\quad S(y, D)V(y) \; = \; 0, \quad y \in \Lambda'$

(22) $\quad T(x, D)U(x) \; = \; 0, \quad x \in \Delta'$

are equivalent for suitable neighbourhoods $\Lambda' \subseteq D_V$ of y_0, respectively, $\Delta' \subseteq \Omega$ of x_0, which are such that the mapping, see (14), $\Lambda' \xrightarrow{V_1} V_1(\Lambda') = \Delta'$ is a smooth diffeomorphism.

6 Global symmetries of parametric solutions

Now, within the parametric context, we turn to the issue of fundamental interest related to Lie group symmetries of solutions of nonlinear PDEs, namely that such symmetries transform solutions into solutions. Let us assume that a parametric function V, see (11), is a parametric solution in the sense of (19) of the nonlinear PDE in (1), namely

(23) $\quad S(y, D)V(y) \; = \; 0, \quad y \in D_V$

Further, let us assume that the arbitrary Lie group G in (2) is a symmetry group of the PDE in (1). Given any $g \in G$, by using (7), we can always define the action gV of g on the parametric solution V.

Here we shall show that gV will again be a parametric solution of the PDE in (1). Indeed, in view of (6), we have $gV \; = \; g \circ V : \Lambda \longrightarrow M$, hence according to (19), we only have to try to show that $S(y, D)((gV)(y)) \; = \; 0, \quad y \in D_{gV}$. The precise result is the following, and it can be obtained by direct computation :

Theorem 1. Let $V : \Lambda \longrightarrow M$ be a parametric solution on the nonvoid open set $D_V \subseteq \Lambda$ of the nonlinear PDE in (1), and let G in (2) be the Lie group of symmetries of that equation. Given any $g \in G$, if

(24) $D_V \cap D_{gV} \neq \phi$

then the parametric function $gV : \Lambda \longrightarrow M$, which is the result of the action of g on V, will again be a parametric solution of the PDE in (1) on the nonvoid open set $D_V \cap D_{gV}$.

7 The nature of the domain condition (24)

At first sight, it may appear that the validity of Theorem 1 above would unduly be restricted by the *domain* condition (24). However, we can immediately note that such is not the case, and that the respective domain condition is natural, as well as minimal. Indeed, (24) *only* asks that the domain on which *both* V and its Lie group transformation gV are nonsingular, should be nonvoid.

In many applications, this domain condition (24) is actually satisfied in a much *stronger* form. Indeed, as we noted, according to (14), the open subsets D_V and D_{gV} are, respectively, the sets on which both V and its Lie group transformation gV are not singular. In particular, in case V happens to be the canonical parametric representation of an analytic function $U : \Delta \longrightarrow \mathbf{R}$, then we shall have, Olver [8]

(25) D_V open and dense in $\Lambda = \Delta$

Therefore, if the Lie group G is also analytic, then both D_V and D_{gV} will by necessity be open, as well as *dense* in $\Lambda = \Delta$. In this case, instead of (24), we shall actually have satisfied the *much stronger* condition

(26) $D_V \cap D_{gV}$ open and dense in $\Lambda = \Delta$

8 Global symmetries of generalised solutions

With Theorem 1, in Section 6, we obtained - in the case of classical, that is, smooth solutions - the desired maximal program of the parametric approach. Namely, we can now act *globally* with arbitrary Lie group symmetries on all classical solutions of the large class of smooth nonlinear PDEs in (1), and as a result of such actions, obtain again classical solutions of the respective equations.

Here we indicate the way the mentioned result in Theorem 1 can be extended to large classes of *generalised* solutions of the same type of nonlinear PDEs, as in (1). The detailed presentation can be found in Chapters 6 - 12, in Rosinger [22].

The generalised solutions, upon which the arbitrary Lie group symmetries will act globally, belong to a *parametrised* extension of the so-called *nowhere dense* differential algebras A_Ω of generalised functions, algebras which were first introduced and used in Rosinger [13] - [18].

Both the algebras A_Ω and their parametrised extensions Z_Ω have important properties. For instance, they can accommodate generalised functions with the *largest* class of *singularities* so far, namely, singularities given by arbitrary closed nowhere dense subsets Γ in Ω. Two important related fact are : first, such singularity sets Γ can have arbitrarily large positive Lebesgue measures, and second, in the neighbourhood of such singularity sets Γ, the generalised functions are not subjected to any conditions.

Here it should be mentioned that none of the other spaces of generalised functions known and used so far in the literature — be they of linear, distributional type, or of the various more recent nonlinear types — can contain generalised functions which have singularities on sets of positive Lebesgue measure, or which are not restricted in the neighbourhood of their singularities.

Further details about the properties of the algebras A_Ω and of their extensions Z_Ω can be found in Rosinger [15] - [21], respectively, Rosinger [22], chap. 1, 12.

Let us now present the construction of the spaces of parametric generalised functions Z_Ω. On the space $(\mathcal{C}_n^\infty(M))^{\mathbf{N}}$ of sequences of smooth functions, we define the symmetric and transitive binary relation $\approx_{nd}$ as follows. Given two sequences of smooth functions $s = (\psi_\nu | \nu \in \mathbf{N})$ and $t = (\chi_\nu | \nu \in \mathbf{N})$, then we denote $s \approx_{nd} t$, if and only if for a certain nonvoid open subset Λ in $\mathbf{R}^n$, we have $\psi_\nu,\ \chi_\nu : \Lambda \longrightarrow \mathbf{R},\quad \nu \in \mathbf{N}$, while at the same time

$$\begin{aligned}
&\exists\ \Gamma \subseteq \Lambda \ \text{ closed, nowhere dense } :\\
&\forall\ x \in \Lambda \setminus \Gamma\ :\\
(27)\qquad &\exists\ \Lambda' \subseteq \Lambda \setminus \Gamma \ \text{ neighbourhood of } \ x,\ \mu \in \mathbf{N}\ :\\
&\forall\ \nu \in \mathbf{N},\ \nu \geq \mu\ :\\
&\quad \psi_\nu = \chi_\nu
\end{aligned}$$

Now we can define the mentioned extensions of the algebras A_Ω by the space of parametric generalised functions

$$(28)\qquad Z_\Omega = ((\mathcal{C}_n^\infty(M))^{\mathbf{N}}) / \approx_{nd}$$

where we note that, since $\approx_{nd}$ is only symmetric and transitive, while it is not also reflexive, in the definition of the quotient in (28) we need a slight and rather obvious extension of the usual construction, see details Rosinger [22] pp. 71, 72.

The fact of importance to us is that, owing to their structure in (27), (28), the arbitrary global Lie group actions (6) on parametric functions, can easily be extended to similarly arbitrary and global Lie group actions on the parametric generalised functions, namely

(29) $\quad G \times Z_\Omega \xrightarrow{\mathcal{L}_{prm}} Z_\Omega$

Similarly, the nonlinear partial differential operator (18), can easily be extended to Z_Ω. In particular, given any $V \in Z_\Omega$, we can always check whether the relation holds

(30) $\quad S(y, D)V = 0 \;\text{ on }\; D_V$

where similar to (14), D_V is an open subset of $\mathbf{R}^n$.

Therefore, (30) gives the definition of a *parametric generalised solution* $V \in Z_\Omega$ of the nonlinear PDE in (1).

In this way, the following extension of Theorem 1, in Section 6, this time to parametric generalised solutions is obtained :

Theorem 2. Let $V \in (\mathcal{C}_n^\infty(M))^{\mathbf{N}} / \approx_{nd}$ be a parametric generalised solution of the nonlinear PDE in (1) on the open set D_V, and let G in (2) be the usual Lie group of symmetry of that equation. Given any $g \in G$, if

(31) $\quad D_V \cap D_{gV} \neq \phi$

then the parametric generalised function $g\,V \in (\mathcal{C}_n^\infty(M))^{\mathbf{N}} / \approx_{nd}$, which is the result of the action of g on V, will again be a parametric generalised solution of the PDE in (1), this time, on the open set $D_V \cap D_{gV}$. □

Related to condition (31), comments similar to those in Section 5 will apply.

9 Four concepts of solution

For clarification, let us note here that *four* concepts of solution for the general smooth nonlinear PDEs in (1) can be considered: two concepts for the usual nonparametric case, and two concepts for the parametric case, see the diagram below. The first two deal with the classical smooth, respectively, the generalised solutions. These two concepts have been dealt with quite extensively in the literature.

As a basic novelty in this paper, two more concepts of solution are introduced. Namely, in (19), we introduced the concept of *parametric classical solution*, while in (30), the concept of *parametric generalised solution* was presented.

Let us indicate here the relevant *connections* between these four concepts of solutions, with each arrow pointing to a more general and inclusive concept:

$$\begin{array}{ccc}
\text{CLASS NONPAR} & \xrightarrow{\text{(I)}} & \text{GEN NONPAR} \\
\text{(II)}\Big\downarrow & & \Big\downarrow\text{(III)} \\
\text{CLASS PARAM} & \xrightarrow[\text{(IV)}]{} & \text{GEN PARAM}
\end{array}$$

where CLASS NONPAR, GEN NONPAR, CLASS PARAM and GEN PARAM stand respectively for classical nonparametric, generalised nonparametric, classical parametric, and generalised parametric solutions.

10 Hilbert's fifth problem

We have seen in (29) that arbitrary Lie group actions can be defined on the spaces of parametric generalised functions Z_Ω. This result amounts to an answer to Hilbert's fifth problem, when this problem is considered in its original full generality, that is, of Lie groups acting on outside entities as well, and not only on themselves. Details in this regard can be found in Rosinger [22], Chap. 11, Olver [10], as well as in the basic references Hilbert [3], Browder [1], Yang [27].

11 Genuine Lie semigroup actions

As mentioned, one of the rather unexpected significant advantages of the parametric approach to Lie group actions is the natural and easy way such an approach can be extended to Lie *semigroup* actions. And unlike studied so far in the literature, see Hilgert & Hofmann & Lawson [5], Hilgert & Neeb [4], Weinstein [26]; now, these Lie semigroup actions can be *genuine* semigroup actions; that is, they need *not* be subsemigroup actions of some Lie group actions. Also, one should note that the mentioned Lie semigroup actions in the literature cited are not applied to PDEs or their solutions.

Let us note here that a significant interest in this kind of extension to genuine Lie semigroup actions can come from the following two facts.

First, the class of such Lie semigroup actions is far *larger* than that of Lie group actions. And as we have seen, one simple way to suggest the extent of such a difference between these two classes of actions is to compare it to the difference between the group of smooth diffeomorphisms of M, and on the other hand, the much larger semigroup of smooth mappings from M to M.
Second, following from the above, one can obtain a far more *powerful* tool in the study of solutions of nonlinear PDEs. Indeed, now one can apply to such solutions the far larger class of Lie semigroup actions, and find their invariance properties in this extended framework of what we shall call *semisymmetries*. Details in this respect can be found in Chapter 13 in Rosinger [22].

References

[1] F. E. Browder (Ed), Mathematical Developments Arising from Hilbert Problems. Proceedings of Symposia in Pure Mathematics, Vol. XXVIII, AMS Providence, 1976

[2] J.-F. Colombeau, Elementary Introduction to New Generalized Functions. North-Holland Mathematics Studies, Vol. 113, Amsterdam, 1985

[3] D. Hilbert, Mathematical Problems. Bull. AMS, Vol. 8, 1902, 437-479 (reprinted in F.E. Browder)

[4] J. Hilgert, K.H. Neeb, Lie Semigroups and their Applications. Springer Lecture Notes in Mathematics, Vol. 1552, Heidelberg, 1993

[5] J. Hilgert, K.H. Hofmann, J.D. Lawson, Lie Groups, Convex Cones, and Semigroups. Oxford Univ. Press, Oxford, 1989

[6] M.B. Oberguggenberger, Multiplication of Distributions and Applications to PDEs. Pitman Research Notes in Mathematics, Vol. 259. Longman, Harlow, 1992

[7] M.B. Oberguggenberger, E.E. Rosinger, Solution of Continuous Nonlinear PDEs through Order Completion. North-Holland Mathematics Studies, Vol. 181. North-Holland, Amsterdam, 1994

[8] P.J. OLVER, Applications of Lie Groups to Differential Equations. Springer, New York, 1986

[9] P.J. OLVER, Equivalence, Invariants and Symmetry. Cambridge Univ. Press, 1995

[10] P.J. OLVER, Nonassociative local Lie groups. J. Lie Theory, 6, 1996, 23-51

[11] E.E. ROSINGER, Embedding of the $\mathcal{D}'$ distributions into pseudo-topological algebras. Stud. Cerc. Math., Vol. 18, No. 5, 1966, 687-729.

[12] E.E. ROSINGER, Pseudotopological spaces, the embedding of the $\mathcal{D}'$ distributions into algebras. Stud. Cerc. Math., Vol. 20, No. 4, 1968, 553-582.

[13] E. E. ROSINGER, Division of Distributions. Pacif. J. Math., Vol. 66, No. 1, 1976, 257-263

[14] E.E. ROSINGER, Nonsymmetric Dirac distributions in scattering theory. In Springer Lecture Notes in Mathematics, Vol. 564, 391-399, Springer, New York, 1976

[15] E.E. ROSINGER, Distributions and Nonlinear Partial Differential Equations. Springer Lectures Notes in Mathematics, Vol. 684, Springer, New York, 1978.

[16] E.E. ROSINGER, Nonlinear Partial Differential Equations, Sequential and Weak Solutions, North Holland Mathematics Studies, Vol. 44, Amsterdam, 1980.

[17] E.E. ROSINGER, Generalized Solutions of Nonlinear Partial Differential Equations. North Holland Mathematics Studies, Vol. 146, Amsterdam, 1987.

[18] E.E. ROSINGER, Nonlinear Partial Differential Equations, An Algebraic View of Generalized Solutions. North Holland Mathematics Studies, Vol. 164, Amsterdam, 1990.

[19] E.E. ROSINGER, Global version of the Cauchy-Kovalevskaia theorem for nonlinear PDEs. Acta Appl. Math., Vol. 21, 1990, 331–343.

[20] E.E. ROSINGER, Characterization for the solvability of nonlinear PDEs, Trans. AMS, Vol. 330, 1992, 203–225.

[21] E.E. Rosinger, Nonprojectable Lie Symmetries of nonlinear PDEs and a negative solution to Hilbert's fifth problem. In (Eds. N.H. Ibragimov and F.M. Mahomed) Modern Group Analysis VI, Proceedings of the International Conference in the New South Africa, Johannesburg, January 1996, 21-30. New Age Inter. Publ., New Delhi, 1997

[22] E.E. Rosinger, Parametric Lie group Actions on Global Generalized Solutions of Nonlinear PDEs and an Answer to Hilbert's Fifth Problem. Kluwer Academic Publishers, Dordrecht, 1998

[23] E.E. Rosinger, M. Rudolph, Group invariance of global generalised solutions of nonlinear PDEs : A Dedekind order completion method. Lie Groups and their Applications, Vol. 1, No. 1, July-August 1994, 203-215

[24] E.E. Rosinger, E.Y. Walus, Group invariance of generalized solutions obtained through the algebraic method. Nonlinearity, Vol. 7, 1994, 837-859

[25] E.E. Rosinger, E.Y. Walus, Group invariance of global generalised solutions of nonlinear PDEs in nowhere dense algebras. Lie Groups and their Applications, Vol. 1, No. 1, July-August 1994, 216-225

[26] A. Weinstein, Grupoids : unifying internal and external symmetry, a tour through some examples. Notices of the AMS, Vol. 43, No. 7, July 1996, 744-752

[27] C.T. Yang, Hilbert's fifth problem and related problems on transformation groups. In F.E. Browder (Ed), Mathematical Developments Arising from Hilbert Problems, Proceedings of Symposia in Pure Mathematics, Vol. XXVIII, 142-146, AMS, Providence, 1976

E. E. Rosinger
Department of Mathematics and Applied Mathematics, University of Pretoria, 0002 Pretoria, South Africa,
E-mail: rosinger@scientia.up.ac.za or eerosinger@hotmail.com

Distributional Description of Impulsive Gravitational Waves

R. Steinbauer

In this contribution we present a recent application [13, 7] of the nonlinear theory of generalized functions to the theory of general relativity, i.e., we treat the geodesic and the geodesic deviation equation for impulsive pp-waves in the algebra of generalized functions. Formally these equations are ODEs nonlinear in the components of the metric tensor which is proportional to the Dirac δ-distribution; hence we have to deal with *nonlinear singular ODEs*. We derive existence and uniqueness results for these equations in Colombeau's algebra of generalized functions and calculate distributional limits of the unique solutions explicitly using the concept of association. These "distributional solutions" display the physically expected behavior, which shows that the geometry of impulsive pp-waves can be described consistently using distributions.

1 Impulsive pp-waves

In this section we briefly recall the basic features of impulsive plane fronted gravitational waves with parallel rays (impulsive pp-waves). PP-wave space times possess a covariantly constant null vector field, which in turn allows to write the metric in the form

$$ds^2 = H(u,x,y)du^2 - du\,dv + dx^2 + dy^2, \tag{1}$$

where u, v is a pair of null coordinates ($u = t - z$, $v = t + z$) and x, y are transverse (Cartesian) coordinates. We are especially interested in *impulsive* pp-waves where the profile function takes the form [10] $H(u,x,y)=f(x,y)\,\delta(u)$, and δ is the Dirac measure. Hence the spacetime is flat everywhere except for the (null) hypersurface $u = 0$, where it has a δ-like shape, modelling a gravitational shock wave.
Physically these geometries arise as ultrarelativistic limits of Kerr-Newman space times [1]. More precisely, a boost is applied to e.g., a Schwarzschild black hole with

respect to an asymptotic observer where the boost velocity approaches the speed of light. Hence impulsive pp-waves model the gravitational field of highly energetic particles and play an important role in scattering processes at the Planck scale [8]. Penrose [10] in his "scissor and paste" approach constructed impulsive pp-waves by glueing together two halves of Minkowski space time along the null hypersurface $u = 0$ with an (x, y)-dependent shift in the v-coordinate. Penrose also introduced a different coordinate system where the metric tensor is actually continuous. But, clearly, these new coordinates are not (even) continuously related to the ones used in (1) and therefore the differential structure of the manifold is changed.

In this work we shall use the distributional form of the metric, mainly for physical reasons (thereby following the approaches of [2, 4]), since in the ultrarelativistic limit (which is a mathematically delicate process, cf. [12]) the metric arises that way. We describe the geometry of impulsive pp-waves entirely in the distributional picture by solving the geodesic and the geodesic deviation equation in Colombeau's algebra of generalized functions. Using the concept of association we derive the distributional particle motion which, however, turns out to be equivalent to the particle motion derived by Podolsky and Vesely using the continuous form of the metric [11]. Hence from a physical point of view the different manifold structures are "identical".

2 The geodesic equation

We start with a pp-metric of the form

$$ds^2 = f(x^i)\,\delta(u)\,du^2 - du\,dv + (dx^i)^2\,, \tag{2}$$

where x^i $(i = 1, 2)$ denote the transverse coordinates and f is the *smooth* wave profile. It is straightforward to derive the geodesic equation which (using u as affine parameter, thereby excluding only trivial geodesics parallel to the shock) takes the form

$$\begin{aligned} \ddot{v}(u) &= f(x^j(u))\,\dot{\delta}(u) + 2\,\partial_i f(x^j(u))\,\dot{x}^i(u)\,\delta(u)\,, \\ \ddot{x}^i(u) &= \frac{1}{2}\partial_i f(x^j(u))\,\delta(u)\,, \end{aligned} \tag{3}$$

where denotes the derivative with respect to u.

Note that the first line involves a product of the δ-function with $\dot{x}^i$ which we cannot even expect to be continuous (and in fact turns out to be proportional to the

Heaviside function, cf. [13]). Therefore we cannot treat system (3) within Schwartz' linear theory of distributions and we proceed by transferring the equations into the framework of Colombeau's (full) algebra [3, 9].

First of all this amounts to a regularization of the δ-distribution to make it an element of $\mathcal{G}(\mathbb{R})$. Since we are interested in the most general result, we will not be satisfied with the canonical embedding $\iota(\delta)$, which corresponds to a "model δ-net" regularization, but use the largest class of "reasonable" (smooth) regularizations of the Dirac measure. Hence we define (cf. the definition of strict δ-nets in [9] ch. 2.7) a *generalized delta function* to be an element D of $\mathcal{G}(\mathbb{R}^n)$ possessing a representative $(D(\varphi, .))_{\varphi\in\mathcal{A}_0}$ such that $\exists p \in \mathbb{N}_0\ \forall\varphi \in \mathcal{A}_p(\mathbb{R}^n)\ \exists\eta = \eta(\varphi) > 0:$

$$
\begin{array}{ll}
\text{(i)} & \operatorname{supp}(D(\varphi_\varepsilon, .)) \subseteq [-\varepsilon, \varepsilon]\ \ \forall\varepsilon \in (0, \eta) \\
\text{(ii)} & \displaystyle\int D(\varphi_\varepsilon, x)\, dx \to 1\ (\varepsilon \to 0) \\
\text{(iii)} & \exists C = C(\varphi) > 0 \text{ such that } \displaystyle\int |D(\varphi_\varepsilon, x)|\, dx \le C\ \ \forall\varepsilon \in (0, \eta)
\end{array}
$$

Note that condition (i) has been chosen to avoid technicalities in the calculation of association relations which, of course, remain valid if (i) is replaced by the condition $\operatorname{supp}(R(\varphi_\varepsilon, .)) \to \{0\}\ (\varepsilon \to 0)$.

We proceed by transferring system (3) into the $\mathcal{G}$-setting. A main problem one has to face in this approach is that, since generalized solutions are constructed componentwise, i.e., from componentwise classical (smooth) solutions to (here) *nonlinear* equations, we have to ensure that the interval of existence does not shrink to a point in the limit $\epsilon \to 0$. Usually one puts strong restrictions on the right-hand side of the equation to guarantee this behavior (cf. [5, 6]). However, this would seriously affect physical applicability of our considerations and so we are forced to make use of the special form of the equations. Indeed we have the following technical result which guarantees global existence of solutions for small ϵ (cf. Appendix of [13]).

Lemma *Let $g : \mathbb{R}^n \to \mathbb{R}^n$ smooth and $(\rho_\varepsilon)_{\varepsilon\in(0,1)}$ be a net of smooth functions bounded uniformly in the L^1-norm and supp$(\rho_\epsilon) \in [-\epsilon, \epsilon]$. For any $x_0, \dot{x}_0 \in \mathbb{R}^n$ and any $\varepsilon \in (0, 1)$ consider the system*

$$
\begin{aligned}
\ddot{x}_\varepsilon(t) &= g(x_\varepsilon(t))\rho_\varepsilon(t) \\
x_\varepsilon(-1) &= x_0 \\
\dot{x}_\varepsilon(-1) &= \dot{x}_0
\end{aligned}
\tag{4}
$$

Let $b > 0$, $I = \{x \in \mathbb{R}^n : |x - x_0| \leq b + |\dot{x}_0|\}$ *and* $\alpha = \min\left\{\frac{b}{C\|g\|_{L^\infty(I)} + |\dot{x}_0|}, \frac{1}{2LC}, 1\right\}$ *with* L *a Lipschitz constant for* g *on* I*. Then (4) has a unique solution on* $J_\varepsilon = [-1, \alpha - \varepsilon]$*. Consequently, for* ε *sufficiently small* x_ε *is globally defined and both* x_ε *and* $\dot{x}_\varepsilon$ *are bounded, uniformly in* ε*, on compact sets.*

The proof uses the fact that the integral operator solving the equation is a contraction on the Banach space $\{f \in C(J_\varepsilon, \mathbb{R}^n) : |f(t) - x_0| \leq b + |\dot{x}_0|\}$. We have chosen initial conditions in $u = -1$, i.e., "before the shock" since geodesics starting at $u = 0$ do not allow any regularization independent treatment. Note that the bounds on x^i will also allow composition of the solutions constructed in Theorem 1 with arbitrary smooth functions f.

Now we are prepared to consider the geodesic equations in the Colombeau algebra. Denoting the $\mathcal{G}$-functions corresponding to x^i and v by capital letters we state the following

Theorem 1 *Let* $D \in \mathcal{G}(\mathbb{R})$ *be a generalized delta function,* $f \in C^\infty(\mathbb{R}^2)$ *and let* v_0, $\dot{v}_0$, x_0^i, $\dot{x}_0^i \in \mathbb{R}$ $(i = 1, 2)$*. The initial value problem*

$$\begin{aligned}
\ddot{V}(u) &= f(X^j(u))\,\dot{D}(u) + 2\,\partial_i f(X^j(u))\,\dot{X}^i(u)\,D(u) \\
\ddot{X}^i(u) &= \frac{1}{2}\partial_i f(X^j(u))\,D(u) \\
V(-1) &= v_0 \qquad X^i(-1) = x_0^i \\
\dot{V}(-1) &= \dot{v}_0 \qquad \dot{X}^i(-1) = \dot{x}_0^i
\end{aligned} \tag{5}$$

has a unique locally bounded solution $(V, X^1, X^2) \in \mathcal{G}(\mathbb{R})^3$.

To prove *existence* one has to verify that componentwise constructed classical solutions are moderate. This, however, is an easy consequence of the lemma above. To prove *uniqueness* observe that the lemma (with obvious modifications) remains valid if one adds a smooth function independent of ϵ to the right-hand side of the first line in (4). This allows to establish the $\mathcal{N}$-estimates on the difference of two locally bounded solutions of (5).

Note that we only had to use conditions (i) and (iii) of the generalized delta function whereas (ii) is the essential ingredient to prove the association relations stated in

Theorem 2 *The unique solution* (V, X^i) *of the geodesic equation (5) satisfies the following association relations:*

$$
\begin{aligned}
X^i &\approx x_0^i + \dot{x}_0^i(1+u) + \frac{1}{2}\partial_i f(x_0^j + \dot{x}_0^j)u_+ \\
V &\approx v_0 + \dot{v}_0(1+u) + f(x_0^j + \dot{x}_0^j)H(u) + \\
&\quad +\partial_i f(x_0^j + \dot{x}_0^i)\left(\dot{x}_0^i + \frac{1}{4}\partial_i f(x_0^j + \dot{x}_0^j)\right) u_+ \,,
\end{aligned} \tag{6}
$$

where H *and* u_+ *denote the Heaviside and kink function, respectively.*

Hence we see that even within the largest "reasonable" class of regularizations of the Dirac δ the geodesics of impulsive pp-waves (on the distributional level) are given by refracted straight lines. More precisely, the geodesics suffer a jump and a kink in the v- as well as a kink in the x^i-direction when crossing the shock hypersurface. The scale of the effect is entirely determined by the values of the profile function and its first derivatives at the shock hypersurface.

3 Jacobi equation

In this section we treat the Jacobi (or geodesic deviation) equation which describes the behavior of infinitesimally nearby geodesics, thereby using essentially the same method as above. Despite the fact that these equations involve terms proportional even to "$H^2\delta$" and "δ^2" (see [13]) we still manage to end up with a regularization independent and physically reasonable distributional interpretation of again unique generalized solutions.

To keep formulas transparent we shall restrict ourselves to the special case of axisymmetry, i.e., we demand that $f = f(r)$, where $r^2 = (x^1)^2 + (x^2)^2$. Furthermore, using initial conditions $x_0^2 = 0 = \dot{x}_0^2$ we will be able to work entirely in the $x^2 = 0$ hypersurface. Finally, we put $v_0 = 0 = \dot{x}_0^1$. Under these assumptions the Jacobi equation for the deviation vector field $N^a(u) = (N^u(u), N^v(u), N^x(u), N^y(u))$ over a geodesic takes the following explicit form in $\mathcal{G}$.

$$
\begin{aligned}
\ddot{N}^v &= 2[N^x f'(X)D]^{\cdot} - N^x f'(X)\dot{D} + [N^u f(X)D]^{\cdot\cdot} - \\
&\quad -N^u f''(X)\dot{X}^2 D - N^u f'(X)\ddot{X}D \\
\ddot{N}^x &= [\dot{N}^u f'(X) + \frac{1}{2}N^x f''(X)]D + \frac{1}{2}f'(X)N^u\dot{D} \\
\ddot{N}^y &= \ddot{N}^u = 0 \,,
\end{aligned} \tag{7}
$$

where D again denotes a generalized delta function and X is the (unique) solution to system (5) with simplification as discussed above. Since these equations are linear in the components of the Jacobi field, it is straightforward to derive the following result

Theorem 3 *Let n^a, $\dot{n}^a \in \mathbb{R}^4$. Then system (7) with initial conditions $N^a(-1) = n^a$ and $\dot{N}^a(-1) = n^a$ has a unique solution in $\mathcal{G}(\mathbb{R})^4$.*

The explicit calculation of association relations for the unique solutions to (7) is a long and tedious computation and we are only going to state a result for two special choices of initial conditions. We have the following

Theorem 4 *The unique solution of the geodesic deviation equation (7) with initial conditions $N^a(-1) = 0$ and $\dot{N}^a(-1) = (a, b, 0, 0)$ satisfies the following association relations:*

$$\begin{aligned} N^x &\approx \frac{1}{2} a f'(x_0)(u_+ + H(u)) \\ N^v &\approx b(1+u) + a[f(x_0)\delta(u) + \frac{1}{4} f'(x_0)^2 (H(u) + u_+)] \end{aligned} \tag{8}$$

For the initial conditions $N^a(-1) = 0$ and $\dot{N}^a(-1) = (0, a, b, 0)$ we have

$$\begin{aligned} N^x &\approx b(1+u) + \frac{1}{2} b\, f''(x_0)\, u_+ \,, \\ N^v &\approx a(1+u) + \frac{1}{4} b\, f'(x_0)\, f''(x_0)\, u_+ + b\, f'(x_0)\, H(u) \end{aligned} \tag{9}$$

Hence in the first case the Jacobi field suffers a kink, a jump and a δ-pulse across the shock overlapping the flat space behavior. The scale of the effect is proportional to the constant factor a which amounts to a "time advance" of nearby geodesics. In fact, the initial velocity a of the Jacobi field in u-direction causes nearby geodesics to cross the shock at different values of the parameter u. But as can be seen from (9), this "time advance" is not the only effect producing kinks and jumps. Equations (9) can be understood in terms of a (formal) Taylor expansion of the profile function near the point at which the reference geodesic crosses the shock, hence display the physically expected behavior. For further details see the discussion in [13].

We have described the geometry of impulsive pp-wave space times by an analysis of the geodesic and geodesic deviation equation in Colombaeu's algebra of generalized functions. Via the concept of association we derived a distributional picture, which perfectly coincides with physical expectations, thereby showing that a consistent distributional description of these geometries is indeed possible, as long as careful mathematical techniques are invoked to handle the nonlinear operations on the singular (i.e., distributional) wave profile instead of ad-hoc "multiplication rules" extending Schwartz linear theory.

Acknowledgment

The author wishes to thank M. Kunzinger and M. Oberguggenberger for many helpful discussions. This work was supported by Austrian Academy of Science, Ph.D. programme, grant #338 and by Research Grant P12023-MAT of the Austrian Science Foundation (FWF).

References

[1] AICHELBURG, P. C., SEXL, R. U.:, On the gravitational field of a massless particle, J. Gen. Rel. Grav. **2**, 303-312 (1971).

[2] BALASIN H.: Geodesics for impulsive gravitational waves and the multiplication of distributions, Class. Quant. Grav. **14**, 455-462 (1997).

[3] COLOMBEAU, J. F.: Elementary Introduction to New Generalized Functions, (North Holland, Amsterdam 1985).

[4] FERRARI V., PENDENZA P., VENEZIANO G.: Beam like gravitational waves and their geodesics, J. Gen. Rel. Grav. **20**, 1185-1191 (1988).

[5] HERMANN, R., OBERGUGGENBERGER, M.: Generalized functions, calculus of variations and nonlinear ODEs, preprint (1995).

[6] KUNZINGER, M.: Lie Transformation Groups in Colombeau Algebras, Ph.D.-thesis, University of Vienna (1996).

[7] KUNZINGER M., STEINBAUER, R.: A rigorous solution concept for geodesic and geodesic deviation equations in impusive gravitational waves, preprint, UWThPh-1998-25, (1998).

[8] LOUSTÓ, C. O., SÁNCHEZ, N.: The ultrarelativistic limit of the Kerr-Newman geometry and particle scattering at the Planck scale, Phys. Lett. B **232**, 462-466 (1989).

[9] OBERGUGGENBERGER, M.: Multiplication of Distributions and Applications to Partial Differential Equations, Longman Scientific and Technical, New York (1992).

[10] PENROSE, R.: The geometry of impulsive gravitational waves in *General Relativity, Papers in Honour of J. L. Synge,* edited by L. O'Raifeartaigh, Clarendon Press, Oxford (1972), pp. 101-115.

[11] PODOLSKY, J., VESELY, K.: Continuous coordinates for all impulsive pp-waves, Phys. Lett. A241, 145-147 (1998).

[12] STEINBAUER, R.: The ultrarelativistic Reissner-Nordstrøm field in the Colombeau algebra, J. Math. Phys. **38**, 1614-1622 (1997).

[13] STEINBAUER, R.: Geodesics and geodesic deviation for impulsive gravitational waves, J. Math. Phys. **39**, 2201-2212 (1998).

ROLAND STEINBAUER
Department of Mathematics and Department of Theoretical Physics, University of Vienna, Strudlhofg.4, A-1090 Wien/AUSTRIA
E-mail: Roland.Steinbauer@univie.ac.at

Nonlinear Generalised Functions in General Relativity

JAMES A. VICKERS

1 Introduction

In linear field theories distributions arise naturally in a number of different ways. They are used to describe idealised situations where the matter is confined to some lower dimensional submanifold – point particles, strings and shells of matter – they are used to describe junction conditions between matter and vacuum regions, and they are used in the description of shock waves. However general relativity is different from other field theories in that one does not have a fixed background metric, but instead the geometry is determined by the field equations. Spacetime is described by a manifold together with a Lorentz metric which is assumed to be sufficiently differentiable for Einstein's equations to be defined. One then detects the presence of singularities by showing that the spacetime is incomplete in some sense. The problem with this approach is that many physical spacetimes contain singularities according to this definition. For example, solutions to Einstein's equations representing impulsive gravitational waves are singular. What one would like to do is to lower the differentiability required of the metric to permit "distributional geometries" which represent physically acceptable weak singularities and are also mathematically tractable. Furthermore, as pointed out by Isham [1], distributional geometries are not only important in describing singularities but must also be included in a path integral description of quantum gravity. Owing to the non-linear nature of Einstein's equations such distributional geometries will in general require some non-linear theory of generalised functions. This review will show how a number of authors have used Colombeau's theory to provide a suitable mathematical framework for such calculations.

The curvature tensor is a non-linear function of the metric and its first two derivatives. Thus the metric needs to be at least C^2 to guarantee that the curvature is continuous. Mathematically one can weaken this condition to allow a C^{2-} metric and this is sufficient for most results in differential geometry to remain valid. However as far as the physics is concerned, one would also like to have conservation of the energy-momentum tensor which, because of Einstein's equations, is equivalent to the contracted Bianchi identities and therefore requires higher differentiability. Such problems would be avoided if one could give a distributional interpretation to the curvature, but this is in general not possible owing to the non-linearities. However in certain special situations one can formulate the field equations in a way which avoids ill-defined products of distributions. For example it is possible to describe shells of matter [2], [3] and gravitational radiation [4] within the context of classical distribution theory. Using the standard definition of the curvature

$$R^a{}_{bcd} = \Gamma^a_{db,c} - \Gamma^a_{cb,d} + \Gamma^a_{cf}\Gamma^f_{db} - \Gamma^a_{df}\Gamma^f_{cb}$$

one sees that one wants the connection to have a (weak) derivative and be square integrable. This lead Geroch and Traschen [5] to introduce a class of 'regular metrics' for which the components of the curvature are well defined as a distribution. They require that

1. the metric components g_{ab} and g^{ab} are locally bounded
2. the first weak derivative $g_{ab,c}$ exists and is locally square integrable.

Geroch and Traschen went on to show that such regular metrics can only have their singular support on a submanifold of co-dimension at most one. Thus for a 4-dimensional spacetime, a metric representing a shell of matter could belong to this class but a string or particle could not.

Although there has long been a desire to allow distributional geometries in general relativity it is only comparatively recently that any real progress has been made in realizing this aim. Many of the early attempts [6], [7], [8], [9] used methods which were specifically adapted to the particular problem being considered and whose general applicability was uncertain. More recently a number of different authors [10], [11], [12] have investigated distributional geometries using an approach based on Colombeau algebras. These were developed by Colombeau in the 1980s and contain the smooth functions as a subalgebra and the distributions as a linear subspace. The advantage of using this approach is that it gives a mathematically consistent way of multiplying distributions and provides a framework for understanding and justifying calculations involving various regularisation procedures. An important feature of these algebras is the notion of association which gives a correspondence between elements of the algebra and distributions. This allows one to use the power of the algebras to do mathematical calculations but then use the concept of association to interpret the final result in terms of distributions. However, most of the theory relating to Colombeau algebras has focused on the properties of *functions* on $\mathbb{R}^n$, whereas what is really required are generalised *tensor fields* on *manifolds*. As a result there does not exist at present a definitive method for treating distributional geometries within General Relativity, but instead a number of different approaches using different variants of the Colombeau algebras have been used. However, despite these technical differences, it is becoming clear that such algebras provide a fruitful method for investigating distributional geometries.

Historically one of the first singular solutions of Einstein's equations to be studied from a distributional viewpoint was the Schwarzschild solution which was considered by Parker in 1979 [6]. The key observation is that when written in Kerr-Schild coordinates the coefficients of the metric are locally integrable. The Schwarzschild and Kerr solutions have been studied in these coordinates by Balasin and Nachbagauer in a number of papers. Following a brief review of Colombeau theory in §2 we describe in §3 their work using the language of the simplified algebra. If one boosts the Schwarzschild solution with velocity v, then one obtains in the limit that the velocity is that of light the 'ultrarelativistic Schwarzschild solution'. This metric was investigated by Aichelburg and Sexl [13] who showed that it could be thought of as a gravitational shock wave. When one does a similar calculation with the ultrarelativistic limit of the Reissner-Nordstrøm solution one obtains a solution with vanishing electromagnetic field but δ-function energy density. In §4 we show how Steinbauer was able to explain this physically surprising result using Colombeau algebras. In §5 we consider geodesic equations for impulsive gravitational wave spacetimes and show how Colombeau algebras provide an appropriate way of obtaining the results expected on physical ground without the need to make use of ad hoc 'rules' for the multiplication of distributions. The study of conical singularities is another area where various authors had used a variety of regularisation procedures to obtain the physically plausible result that a cone has δ-function curvature at the vertex. However, this result appeared to be at odds with the results of Geroch and Traschen. In §6 we review the analysis and resolution of the problem by Clarke, Vickers and Wilson using the full Colombeau algebra. Finally in §7 we discuss some of the difficulties with using Colombeau algebras and attempts to develop a theory of generalised tensor fields on manifolds.

2 A brief review of Colombeau theory

In this section we give a brief introduction to the Colombeau algebras. For more details see [14], [15]. The basic idea is to consider generalised functions as 1-parameter families of smooth functions $\{f_\epsilon\}$. Our basic space will thus be

$$\mathcal{E}(\mathbb{R}^n) = \{\{f_\epsilon\} : \ 0 < \epsilon < 1, \ \ f_\epsilon \in C^\infty\}.$$

We may represent a smooth function by the constant family $f_\epsilon(x) = f(x)$, but for a function of finite differentiability we may obtain a family of smooth functions by taking the convolution with a suitable smoothing kernel or mollifier Φ

$$f_\epsilon(x) = \frac{1}{\epsilon^n}\int f(y)\Phi((y-x)/\epsilon)\mathrm{d}^n y.$$

However, for the case of a smooth function, we would like both representations to be equivalent. Using a Taylor series expansion to compare the difference between these expressions we are lead to define two representations to be equivalent if they differ by a 'negligible function' which is defined as a 1-parameter family of functions which on any compact set vanishes faster than any given positive power of ϵ. Since we are trying to construct a *differential algebra*, we also require the derivatives of f to satisfy this property and the resulting set $\mathcal{N}$ to be an ideal. Unfortunately $\mathcal{N}$ is not an ideal in $\mathcal{E}(\mathbb{R}^n)$, but by restricting the space to 'moderate functions' $\mathcal{E}_M(\mathbb{R}^n)$ which grow no faster than some inverse power of ϵ, one does have an ideal and we may define the differential algebra $\mathcal{G}$ as the quotient.

Definition (Moderate functions)

$$\mathcal{E}_M(\mathbb{R}^n) = \{\{f_\epsilon\} : \ \forall K \subset\subset \mathbb{R}^n, \forall \alpha \in \mathbb{N}_0^n, \exists p \in \mathbb{N}, \exists \eta > 0, \exists C > 0 \text{ such that} \\ \mathrm{Sup}|D^\alpha f_\epsilon(x)| \leqslant C\epsilon^{-p} \text{ for } 0 < \epsilon < \eta\}.$$

Note $K \subset\subset \mathbb{R}^n$ indicates that K is *compact* and we have also employed the standard multi-index notation for $D^\alpha f$.

Definition (Negligible functions)

$$\mathcal{N}(\mathbb{R}^n) = \{\{f_\epsilon\} : \forall K \subset\subset \mathbb{R}^n, \forall \alpha \in \mathbb{N}_0^n, \forall q \in \mathbb{N}, \exists \eta > 0, \exists C > 0 \text{ such that } \\ \mathrm{Sup}|D^\alpha f_\epsilon(x)| \leqslant C\epsilon^q \text{ for } 0 < \epsilon < \eta\}.$$

Definition (Simplified algebra)

$$\mathcal{G}(\mathbb{R}^n) = \mathcal{E}_M(\mathbb{R}^n)/\mathcal{N}(\mathbb{R}^n)$$

Thus a non-linear generalised function is represented by a moderate sequence of smooth functions modulo a negligible sequence. The space $\mathcal{E}_M(\mathbb{R}^n)$ is a differential algebra with pointwise operations and since the space of negligible functions is a differential ideal, $\mathcal{G}$ is also a commutative differential algebra. As remarked earlier, one of the advantages of the Colombeau approach is that one may frequently interpret the results in terms of distributions using the concept of association or weak equivalence. A representative $\{f_\epsilon\}$ of an element of $\mathcal{G}$ is said to be associated to a distribution $T \in \mathcal{D}'$ if

$$\forall \phi \in \mathcal{D}, \quad \lim_{\epsilon \to 0} \int f_\epsilon(x)\phi(x)\mathrm{d}^n x = \langle T, \phi \rangle$$

and we then write $\{f_\epsilon\} \approx T$.

Note that not all elements of $\mathcal{G}$ are associated to distributions.

More generally we say $\{f_\epsilon\} \approx \{g_\epsilon\}$ if

$$\forall \phi \in \mathcal{D}, \quad \lim_{\epsilon \to 0} \int f_\epsilon(x)\phi(x)\mathrm{d}^n x = \lim_{\epsilon \to 0} \int g_\epsilon(x)\phi(x)\mathrm{d}^n x.$$

Association is an equivalence relation which respects addition and differentiation. It also respects multiplication by *smooth* functions, but by the Schwarz impossibility results cannot respect multiplication in general.

The algebra presented above is the simplest of the Colombeau algebras. Furthermore the notion of smooth function and the growth conditions in the definitions of moderate and negligible functions only depend upon the differential structure. This allows the simplified algebra to be readily generalised to arbitrary manifolds. However it does suffer from the disadvantage that the embedding of C^p functions and distributions is not canonical. Thus one has to appeal to mathematical or physical arguments outside the theory to justify a particular representation of a non-smooth function. In some physical theories there is a natural parameter such as a coupling constant that can be used as the Colombeau parameter; in other cases one can embed by taking the convolution with an element of some special function space and only use general properties of elements of the space in proving results. However this is not always satisfactory and in §6 the full Colombeau algebra will be introduced. This involves enlarging the algebra substantially by making the generalised functions depend upon the mollifier. This allows one to construct a canonical embedding of distributions but at the expense of problems with coordinate invariance. Ways of circumventing these difficulties are considered in §7.

3 The Schwarzschild and Kerr spacetimes

In this section we review the work of Balasin and Nachbagauer [8], [9] who consider rotating, charged, black hole Kerr-Newman solutions in a number of papers. We will present the work using the language of the simplified algebra (cf. Balasin [10]). The solutions considered have the feature that they are all examples of Kerr-Schild geometries. Such metrics may be written in the form

$$g_{ab} = \eta_{ab} + f k_a k_b$$

where η_{ab} is the flat Minkowski metric, f is an arbitrary function, $k_a = \eta_{ab}k^b$ and k^a is null and geodetic with respect to η_{ab}.
The simplest example of such a solution is the Schwarzschild solution which in the standard Minkowski coordinates has Kerr-Schild form given by

$$f = \frac{2m}{r}, \quad k^a = (1, x^i).$$

The expression for the Ricci tensor of a Kerr-Schild metric takes the surprisingly simple form

$$R^a_b = \tfrac{1}{2}\eta^{cd}\eta^{ea}[\partial_e\partial_c(fk_dk_b) + \partial_b\partial_c(fk_dk_e) + \partial_c\partial_d(fk_ek_b)].$$

The energy momentum tensor is then given by Einstein's equations as

$$T^a_b = R^a_b - \tfrac{1}{2}\delta^a_b R.$$

One may then calculate the energy momentum tensor as follows. One regards the Kerr-Schild form as being valid on the *whole* of Minkowski space with $\frac{2m}{r}$ replaced by some suitable regularised function. One now regards this as an element of $\mathcal{G}$ and computes the components of T^a_b (in Minkowski coordinates) as elements of $\mathcal{G}$. Finally one can show that

$$T^0_0 \approx -m\delta^{(3)} \tag{1}$$

$$T^a_b \approx 0 \quad \text{otherwise.} \tag{2}$$

The calculation of the energy-momentum tensor for the Kerr solution is significantly harder. Unlike the Schwarzschild case we cannot easily write the Kerr solution as the limit of a one parameter family of regular Kerr-Schild metrics. The problem arises because of the topology of the maximum analytic extension of the Kerr solution which leads to a branch singularity when the metric is written using the standard 'flat' Kerr-Schild decomposition. Balasin [10] avoided this problem by considering metrics of generalised Kerr-Schild form. These are metrics which can be written

$$g_{ab} = \hat{g}_{ab} + fk_a k_b$$

where $\hat{g}_{ab}$ is now a background metric, $k_a = \hat{g}_{ab}k^b$ and k^a is null and geodetic with respect to $\hat{g}_{ab}$ (and also g_{ab} because of the form of the metric). One now has

$$R^a_b = \hat{R}^a_b + \hat{g}^{cd}\hat{g}^{ef}\hat{R}^a{}_{ceb}fk_dk_f + \tfrac{1}{2}\hat{g}^{cd}\hat{g}^{ea}[\partial_e\partial_c(fk_dk_b) + \partial_b\partial_c(fk_dk_e) + \partial_c\partial_d(fk_ek_b)]$$

where $\hat{R}^a{}_{bcd}$ is the curvature of $\hat{g}_{ab}$.

We may write the Kerr metric in generalised Kerr-Schild form by taking the background metric to have the form

$$\hat{g}_{ab}dx^a dx^b = -dt^2 + \frac{\Sigma}{r^2 + a^2}dr^2 + \Sigma d\theta^2 + (r^2 + a^2)d\phi^2$$

where $\Sigma = r^2 + a^2\cos^2\theta$ and $(t, r, \theta, \phi) \in \mathbb{R}^2 \times S^2$. Performing all the calculations in $\mathcal{G}(\mathbb{R}^2 \times S^2)$ one may compute $\sqrt{\hat{g}}R^a_b$ which is found to be given by

$$\sqrt{\hat{g}}R^a_b \approx 2\pi\delta(\cos\theta)(-a\delta(u)\partial^a_u du_b + \partial^a_\theta d\theta_b + m\delta'(u)(\partial^a_t - (1/a)\partial^a_\phi)(dt_b + ad\phi_b).$$

Note that the density $\sqrt{\hat{g}}R^a_b$ rather than R^a_b is calculated since on manifolds we want to regard distributions as transforming like densities.

The above calculation was carried out by Balasin using the simplified algebra. This is probably the only practical way of doing the calculation given the topology of the manifold and the complexity of the metric. It suffers from the usual problem when using the simplified algebra of a non-canonical embedding. Rather than embed using a convolution, the embedding has been chosen to preserve the generalised Kerr-Schild form. However the embedding used is not unique within this class and it would be desirable to show that any reasonable embedding which preserved the form of the decomposition gave the same result. It is even less clear that a 'natural regularisation' in some other coordinate system would give the same result. Nevertheless the calculation is an impressive example of the complicated calculations that can be performed in general relativity using the simplified algebra.

4 Ultrarelativistic black holes

In 1971 Aichelburg and Sexl [13] derived the ultrarelativistic limit of the Schwarzschild geometry. Below we give a description of the limit using the language of the simplified algebra. We start by considering the Kerr-Schild form of the Schwarzschild metric written in double null coordinates $u = t - x$ and $w = t = x$,

$$ds^2 = dudw - dy^2 - dz^2 + \frac{2m}{r} k_a k_b dx^a dx^b \tag{3}$$

where in these coordinates $k^a = ((r-x)/r, (r+x)/r, y/r, z/r)$.
The Minkowski background enables us to have a well-defined concept of boost and we may therefore boost the solution by velocity v along the x-axis. We therefore write

$$u = \sqrt{\frac{1+v}{1-v}}\bar{u} \qquad w = \sqrt{\frac{1-v}{1+v}}\bar{w}$$

and to keep the energy of the 'particle' finite we rescale the mass according to the special relativistic formula

$$m = (1-v^2)^{\frac{1}{2}} p.$$

Substituting this into (3) gives a 1-parameter family of metrics depending on the boost velocity v. We are interested in the ultrarelativistic limit in which v reaches the speed of light (ie $v \to 1$), so we replace v by $1-\epsilon$ and regard $g_{(\epsilon)ab}$ as elements of the simplified algebra. It is readily shown that most of the terms in the perturbation $\frac{2m}{r} k_a k_b$ are associated to zero. The only surviving term is $\frac{2m}{r} k_0 k_0$. Using calculations very similar to Steinbauer [11] we find this is associated to $8p\ln\rho\delta(u)$ and hence $g_{(\epsilon)ab} \approx g_{(0)ab}$ where

$$g_{(0)ab}dx^a dx^b = 8p\ln\rho\delta(u)du^2 + dudw - dy^2 - dz^2.$$

This metric describes a pp-wave and is flat everywhere except on the null plane $u = 0$ which contains the 'particle'. This calculation was first derived by Aichelburg and Sexl [13], who started with Schwarzschild written in isotropic coordinates and simultaneously boosted the solution and made a v-dependent coordinate transformation to compute the limiting metric.

It should be pointed out that this result is entirely consistent with the calculation of the energy-momentum tensor of the Schwarzschild solution given in the previous section. The ultrarelativistic limit of (1), (2) in the (u, w, x, y) coordinate system gives

$$T^a_b = \delta(u)\delta^{(2)}(y, z)p_a p^b \quad \text{where } p_a = (1, 0, 0, 0),$$

which is just the energy-momentum tensor of $g_{(0)ab}$. Indeed this observation was used by Balasin and Nachbagauer [16] to *derive* the ultrarelativistic limit of the Schwarzschild and Kerr geometries.

It is also possible to calculate the ultrarelativistic limit of the Reissner-Nordstrøm solution. However in this case it is also necessary to rescale the charge according to the formula

$$e^2 = (1 - v^2)^{\frac{1}{2}} f^2, \quad f \text{ a constant.}$$

The limiting metric was found by Loustó and Sánchez [17] using a rather ad-hoc regularisation to be

$$ds^2 = \left\{ 8p\ln\rho + \frac{3\pi f^2}{2\rho} \right\} \delta(u)du^2 + dudw - dy^2 - dz^2. \tag{4}$$

This was confirmed using a calculation in $\mathcal{G}$ by Steinbauer [18]. The solution obtained again represents a pp-wave and is flat everywhere except on the null plane $u = 0$. The ultrarelativistic limit of the electromagnetic energy-momentum tensor of Reissner-Nordstrøm is also found to be compatible with the energy-momentum tensor of (4). However Steinbauer [11] also noted the rather surprising result that all the components of the electromagnetic field were associated to zero. One thus appeared to have the rather unexpected result that the electromagnetic field vanished, but the associated energy-momentum tensor did not. However it must be stressed that regarded as elements of $\mathcal{G}$ the electromagnetic field components are *non-zero* (even though they are *associated* to zero). Calculating the energy-momentum tensor of this field within $\mathcal{G}$ gives the correct result. There is nothing unusual within Colombeau theory in having objects in $\mathcal{G}$ which are associated to zero having products which are not associated to zero. This is simply a reflection of the fact that association does not respect multiplication in general.

5 Geodesics for impulsive gravitational wave spacetimes

In the previous section we showed how the ultrarelativistic limit of the Schwarzschild solution lead to an impulsive gravitational wave spacetime. We now consider geodesics in such spacetimes with metric taking the form

$$ds^2 = f(x^A)\delta(u)du^2 + dudw - \delta_{AB}dx^A dx^B$$

where $A,\ B = 2,\ 3$ denote the transverse coordinates. Geodesics for these impulsive gravitational wave spacetimes have been considered by Steinbauer [19], Ferrari, Pendenza and Veneziano [20] and by Balasin [21]. As one might expect, one can regularise the geodesic equations, and show that the geodesics consist of broken and refracted straight lines. A more rigorous derivation of these results using existence and uniqueness theorems within the Colombeau algebra has recently been given by Kunzinger and Steinbauer [22]. From a physical point of view these results are important because they show that using the Colombeau algebra gives the same results as viewing an impulsive wave as the limiting case of a sandwich wave.

6 Conical singularities and cosmic strings

An important example of a two-dimensional weak singularity is provided by the conical spacetime

$$ds^2 = dt^2 - dr^2 - A^2r^2d\phi^2 - dz^2.$$

This spacetime is locally flat for $r \neq 0$ and heuristic arguments suggest that it has delta-function like curvature on the $r = 0$ axis. The corresponding energy-momentum tensor then describes a string with stress equal to the mass per unit length μ, where $\mu = 2\pi(1-A)/A$. This is precisely the form of the energy-momentum tensor of a cosmic string in the weak field thin string limit. Such spacetimes are also important mathematically as they provide simple examples of quasi-regular singularities (i.e., singularities for which the components of the Riemann tensor measured in a parallelly propagated frame tend to a well-defined limit, see e.g. Vickers [23] for further details).

Unfortunately such spacetimes do not fall into the Geroch-Traschen class of 'regular metrics' [5], so one cannot expect the curvature to be well defined using conventional distribution theory. Furthermore Geroch and Traschen showed how it was possible to obtain different values of the mass per unit length by taking different regularisations. Because of this Clarke, Vickers and Wilson [12] chose to investigate the curvature of the cone using the 'full' Colombeau algebra where one has a *canonical* embedding of distributions. This is achieved by enlarging the algebra so that generalised functions depend explicitly upon the mollifier. The mollifiers belong to certain function spaces $\mathcal{A}_k$ with vanishing moments up to order k. The space of moderate and negligible functions are again defined by power law behavior in ϵ where now the power depends upon k. See [14] or [12] for further details.

The singular part of the curvature arises from the conical singularity in the 2-cone

$$ds^2 = dr^2 + A^2r^2d\phi^2$$

and for simplicity we will describe the calculation of the curvature (density) of this metric. One must first write the metric in coordinates which are regular at the origin. In Cartesian coordinates one has

$$g_{ab} = \tfrac{1}{2}(1 + A^2)\delta_{ab} + \tfrac{1}{2}(1 - A^2)h_{ab},$$

where

$$h_{ab} = \begin{pmatrix} \frac{x^2-y^2}{x^2+y^2} & \frac{2xy}{x^2+y^2} \\ \frac{2xy}{x^2+y^2} & \frac{y^2-x^2}{x^2+y^2} \end{pmatrix}.$$

We now regard the metric as an element of $\mathcal{E}_M(\mathbb{R}^2)$ by taking the convolution of the components g_{ab} with a suitable mollifier. To find these we need to calculate

$$\tilde{h}_\epsilon(x, y) = \frac{1}{\epsilon^2}\int_{\mathbb{R}^2} h(u + x, v + y)\Phi((u/\epsilon, (v/\epsilon))dudv$$

where $h(x, y) := e^{2i\phi}$.

By expanding in circular harmonics, using the compactness of the support of the mollifier and using the residue theorem, one can obtain a Fourier series for $\tilde{h}$ and hence for $\tilde{g}_{(\epsilon)ab}$. One may now estimate the curvature $\tilde{R}_\epsilon$ of $\tilde{g}_{(\epsilon)ab}$, and use the Gauss-Bonnet theorem to show that

$$[\tilde{R}\sqrt{\tilde{g}}] \approx 4\pi(1-A)\delta^{(2)},$$

the densitised version of the curvature being chosen since we want to regard the delta function as transforming like a density.

It is important to stress that the methods described here may be used to calculate the curvature of the full four-dimensional cone, although one may no longer use the Gauss-Bonnet theorem to calculate the curvature and therefore requires more delicate estimates (see Wilson [24] for details). One then obtains the heuristically expected energy-momentum tensor. Furthermore it is not hard to modify the results to deal with a metric which is not exactly conical, but approaches one quadratically as $r \to 0$. A second point to note is that the result does not depend upon the coordinates in which the calculation is carried out so long as they are smoothly related to Cartesians. This may be shown by explicitly transforming to new coordinates $X = X(x, y)$ and $Y = Y(x, y)$ and doing the whole calculation in the new coordinates. Again more delicate estimates are required but one can show that the resulting curvature density transforms in exactly the same way as a delta-function (see Vickers and Wilson [25] for details).

A significant generalisation of this result was obtained by Wilson [26] who extended the results to a four-dimensional time dependent cosmic string. He considered cylindrically symmetric pure radiation solutions of Einstein's equations. These metrics may be written in the form

$$ds^2 = e^{2\gamma(t-r)}(dt^2 - dr^2) + r^2 d\phi^2 + dz^2$$

and naively one would expect a delta-function contribution to the curvature due to the angular deficit of $2\pi(1 - e^{-\gamma(t)})$. However, because of the time dependence of the angular deficit, the singularity cannot be quasi-regular but must be stronger. In fact, it is an example of an 'intermediate singularity' and the components of the Riemann tensor have a limit in a special choice of frame. Wilson calculated the energy-momentum tensor of this spacetime by first writing the metric in null Cartesian coordinates and then proceeded to obtain a smooth metric by convolution, estimate the curvature of the smooth metric and show that it is associated to a distribution. This result shows that even in the radiating case the mass per unit length is given by the expected formula

$$\mu(u,z) \approx 2\pi(1 - e^{-\gamma(u)}).$$

Furthermore, by applying the methods of Ashtekar et al. one may show that this result agrees with the asymptotically measured mass per unit length. However, unlike the case of the static string it is unclear whether the calculation is coordinate invariant. The (u, x, y, z) coordinates used are natural for investigating radiating systems but have the disadvantage that Minkowski space appears to be singular on the axis when written in these coordinates. However, the correct mass per unit length for a static cosmic string is obtained, and this vanishes for Minkowski space.

7 Colombeau theory on manifolds

An underlying principle of General Relativity is that the measurement of physical quantities should be independent of the coordinate system used. Mathematically this is reflected in the fact that the the theory is formulated in terms of tensor fields on manifolds. The minimum differentiability of the coordinate transformations and the dependence on the differential structure of the manifold is quite subtle, but at the very least the theory should be invariant under smooth diffeomorphisms.
Although the definition of the simplified algebra may be generalised to $\mathcal{G}(M)$, an algebra on the manifold, it is hard to see how to embed distributions and functions of finite differentiability in a natural manner. In some applications there will be a natural physical parameter, such as a coupling constant, that may be used to provide the ϵ index. However in many situations the embedding is achieved using a convolution which depends upon the linear structure of $\mathbb{R}^n$. A second problem relates to the tensorial nature of the objects that one wants to embed, whereas the Colombeau algebra is formulated for functions. Thus even if one could develop an invariant embedding of functions, one would not have an invariant embedding of tensors. Embedding the components of a tensor and then transforming would in general give a different answer from transforming and then embedding since multiplication by a smooth function does not in general commute with the embedding.

One possibility is to regard the Colombeau algebra as an intermediate calculational tool for obtaining distributional answers. One picks some coordinate system, embeds the components of the metric into the Colombeau algebra (in the given coordinates) and calculates the components of the curvature and energy-momentum tensor, etc. One then tries to show that these objects are associated to distributions. Finally one repeats the entire calculation in some other (smoothly related) coordinate system and tries to show that the answer transforms in the expected way for a distribution. Although such calculations can be useful in some situations (for example in calculating the curvature of a conical spacetime [25]), it would be preferable to make the embedding into the algebra coordinate invariant.

An invariant embedding of a function of finite differentiability may be achieved by demanding that the mollifiers Φ transform in an appropriate way. This suggests that one works in the full algebra where this can be incorporated into the transformation law for a generalised function. Unfortunately the transformed mollifiers are no longer elements of the $\mathcal{A}_k$ spaces. However Colombeau and Meril [27] showed how weakening the moment conditions and enlarging $\mathcal{A}_k$ to a much larger space (of bounded paths Φ^{ϵ}) enabled one to generalise the full algebra and obtain an embedding that commutes with diffeomorphisms. The price that one has to pay for this (apart from the more complicated definition of the algebra) is the stricter definition of association, which comes from the enlargement of the space of mollifiers.

However, as remarked earlier, one cannot simply apply the invariant embedding of a scalar to the components of a tensor and obtain an invariant embedding of the tensor. The problem really stems from the way the convolution integrates the components of the tensor at *different* points of the manifold. The solution to this is to introduce some additional structure which enables one to first transport the tensor fields to the same point x in M so that one can then do the integration in a meaningful way. Such a transport is naturally provided by specifying a background connection, and in keeping with the spirit of the full algebra this is made an argument of the generalised tensor field. Thus a generalised tensor field depends upon a background connection γ, a mollifier Φ and the point x. The background connection also provides a background covariant derivative which commutes with the embedding. This may be used to specify other geometric objects such as a generalised Lie derivative or generalised connection (see Vickers and Wilson [28] for details).

References

[1] ISHAM, C.J.: Proc. R. Soc. **A 351** 209 (1976)

[2] ISRAEL, W.: Nouovo Cimento **B 41** 1–14 (1966)

[3] CLARKE, C.J.S. AND DRAY T.: Class. Quant. Grav. **4** 265–275 (1987)

[4] KHAN, K.A. AND PENROSE, R.: Nature, **229** 185 (1971)

[5] GEROCH, R. AND TRASCHEN, J.: Phys. Rev. D **15** 1017–1031 (1987)

[6] PARKER, P.E.: J. Math. Phys. **20** 1423–1426 (1979)

[7] RAJU, C.: J. Phys. A: Math. Gen. **15** 1785–1797 (1982)

[8] BALASIN, H. AND NACHBAGAUER, H: Class. Quant. Grav. **10** 2271–2278 (1993)

[9] BALASIN, H. AND NACHBAGAUER, H: Class. Quant. Grav. **11** 1453–1461 (1994)

[10] BALASIN, H.: Class. Quant. Grav. **14** 455–462 (1997)

[11] STEINBAUER, R.: J. Math. Phys. **38** 1614–1622 (1997)

[12] CLARKE, C.J.S., VICKERS, J.A. AND WILSON J.P.: Class. Quant. Grav. **13** 2485–2498 (1996)

[13] AICHELBURG, P.C. AND SEXL, R.U.: J. Gen. Rel. Grav. **2** 303-312 (1971)

[14] COLOMBEAU, J.F.: Elementary introduction to new generalized functions, North Holland, 1984.

[15] BIAGIONI, H.A.: A nonlinear theory of generalised functions, Lecture Notes in mathematics 1421, Springer (1990)

[16] BALASIN, H. AND NACHBAGAUER, H: Class. Quant. Grav. **12** 707-713 (1995)

[17] LOUSTÓ, C.O. AND SÁNCHEZ, N.: Nucl. Phys. B **232** 462–466 (1989)

[18] STEINBAUER, R.: MSc University of Vienna (1996)

[19] Steinbauer, R.: J. Math. Phys. **39** 2201–2212 (1998)

[20] Ferrari, V. Pendenza, P and Veneziano, G: J. Gen. Rel. Grav. **20** 1185–1191 (1988)

[21] Balasin, H.: Class. Quant. Grav. **14** 455–462 (1997)

[22] Kunzinger, M. and Steinbauer, R.: preprint gr-qc/9806009

[23] Vickers, J.A.: Class. Quant. Grav. **7** 731–741 (1990)

[24] Wilson, J.P.: PhD Thesis, University of Southampton (1997)

[25] Vickers, J.A. and Wilson, J.P.: ESI preprint 565 (1998)

[26] Wilson, J.P.: Class. Quant. Grav. **14** 3337–3351 (1997)

[27] Colombeau, J.F. and Meril, A.: J. Math. Anal. Appl. **186** 357–364 (1994)

[28] Vickers, J.A. and Wilson, J.P.: ESI preprint 566 (1998)

J.A.Vickers
University of Southampton, Department of Mathematics, Southampton, SO17 1BJ,
E-mail: jav@maths.soton.ac.uk

STOCHASTIC ANALYSIS

A White Noise Approach to Stochastic Differential Equations Driven by Wiener and Poisson Processes

HELGE HOLDEN AND BERNT ØKSENDAL

1 Introduction

The main idea behind the white noise approach to stochastic partial differential equations (SPDEs) can be described as follows:
Recall the fact that even quite simple SPDEs, like the stochastic Poisson equation

$$\begin{aligned} \Delta F(x) &= -W(x)\,; \qquad x \in D \subset \mathbb{R}^d\,, \quad d > 3 \\ F(x) &= 0\,; \qquad x \in \partial D \end{aligned} \tag{1.1}$$

(where $\Delta = \sum_{i=1}^{d} \partial^2/\partial x_i^2$ is the Laplacian and W is Gaussian white noise), do not allow solutions which are ordinary stochastic processes/fields $F(x,\omega)\colon \overline{D} \times \Omega \to \mathbb{R}$. However, Walsh [9] proved that solutions exist in the distribution sense:

(1.2) For almost all ω the map $x \mapsto F(x,\omega)$ is a distribution in x.

This approach works fine for linear equations like (1.1), but requires more work in the nonlinear case. It is, however, possible also to handle some such cases by introducing Colombeau random generalized functions. See [7] and the references therein.
The *white noise approach* looks for generalized solutions in the following sense:

(1.3) For all x the map $\omega \mapsto F(x,\omega)$ is a *stochastic* distribution in ω.

Since there is a natural product in the stochastic distribution spaces (the Wick product), this approach has the advantage of extending easily to many nonlinear SPDEs, provided they are interpreted in the Wick sense. Moreover, there is an efficient solution algorithm for such equations, based on the Hermite transform (see, e.g., [5, Theorem 4.1.1], and Section 3 in this paper).

The theory of *Gaussian* white noise and its applications to stochastic differential equations (SDEs), both ordinary and partial, are now well-known. See for example the presentations in [4] and [5]. It is also known—although not so widely—that a similar type of theory exists for Poissonian white noise. Moreover, this Poissonian white noise theory has similar applications to SDEs driven by the Poisson process. In fact, as described in Section 2 there is even a unitary correspondence $\mathcal{U}$ between the solutions of Gaussian and Poissonian white noise equations, provided they both are of Wick-type. See the exposition in [1] and the references therein. A brief survey of this connection can be found in [5, Section 4.9].

However, this connection does not immediately explain how the white noise theory can be modified to cover SDEs driven by *combinations* of Gaussian and Poissonian white noise. The purpose of this paper is to outline how this can be done and at the same time review the separate Gaussian and Poissonian cases and the relation between them in more detail.

In Section 2 we briefly review the white noise theory in the Gaussian and Poissonian cases separately and some relations between them.

Then in Section 3 we outline how to combine the two white noise theories into one theory that cover equations driven by both Gaussian and Poissonian noise.

In Section 4 we illustrate the theory by considering some examples.

2 A review of the Gaussian and Poissonian white noise theories – and relations between them

Let $\mathcal{S}(\mathbb{R}^d)$ denote the Schwartz space of rapidly decreasing smooth functions on $\mathbb{R}^d$ and let $\mathcal{S}'(\mathbb{R}^d)$ be its dual (the space of tempered distributions). Then by the Bochner-Minlos theorem there exists a unique probability measure μ on $\Omega_1 := \mathcal{S}'(\mathbb{R}^d)$ defined by the property that

$$\int_{\Omega_1} e^{i\langle\omega_1,\phi\rangle} d\mu(\omega_1) = \exp(-\tfrac{1}{2}\|\phi\|^2)$$

for all $\phi \in \mathcal{S}(\mathbb{R}^d)$, where $\|\phi\|^2 = \int_{\mathbb{R}^d} \phi(x)^2 dx$ and $\langle\omega_1, \phi\rangle$ is the action of $\omega_1 \in \Omega_1$ on $\phi \in \mathcal{S}(\mathbb{R}^d)$. (See, e.g., [5, Section 2.1] for more details.) The measure μ is called the

Gaussian white noise probability measure.
Similarly, again by the Bochner-Minlos theorem, there exists a unique probability measure ν on $\Omega_2 := \mathcal{S}'(\mathbb{R}^d)$ defined by the property that

$$\int_{\Omega_2} e^{i\langle\omega_2,\phi\rangle} d\nu(\omega_2) = \exp\left(\int_{\mathbb{R}^d} (e^{i\phi(x)} - 1)dx\right)$$

for all $\phi \in \mathcal{S}(\mathbb{R}^d)$, where (again) $\langle\omega_2, \phi\rangle$ is the action of $\omega_2 \in \Omega_2$ on $\phi \in \mathcal{S}(\mathbb{R}^d)$. The measure ν is called the *Poissonian white noise probability measure.*
In the *Gaussian* case one exploits the isometry

$$E_\mu[\langle\cdot,\phi\rangle^2] = \|\phi\|^2 \, ; \qquad \phi \in \mathcal{S}(\mathbb{R}^d)$$

to define

$$\langle\omega_1, \psi\rangle = \lim_{n\to\infty} \langle\omega_1, \phi_n\rangle,$$

with limit in $L^2(\mu)$, for all $\psi \in L^2(\mathbb{R}^d)$, where ϕ_n is a sequence in $\mathcal{S}(\mathbb{R}^d)$ such that $\phi_n \to \psi$ in $L^2(\mathbb{R}^d)$. In particular, this defines *d-parameter Brownian motion* as (the x-continuous version of)

$$B(x,\omega_1) := \langle\omega_1, \mathcal{X}_{[0,x]}\rangle \, ; \qquad x \in \mathbb{R}^d, \ \omega_1 \in \Omega_1 \, ,$$

where

$$\mathcal{X}_{[0,x]}(t_1, \ldots, t_d) = \mathcal{X}_{[0,x_1]}(t_1) \cdots \mathcal{X}_{[0,x_d]}(t_d)$$

if $x = (x_1, \ldots, x_d)$ with $x_j \geq 0$ for $j = 1, \ldots, d$. (If $x_j < 0$, we put $\mathcal{X}_{[0,x_j]} = -\mathcal{X}_{[-x_j,0]}$.)
Similarly, in the *Poissonian* case one exploits the isometry

$$E_\nu\left[\left(\langle\cdot,\phi\rangle - \int_{\mathbb{R}^d} \phi(x)dx\right)^2\right] = \|\phi\|^2 \, ; \qquad \phi \in \mathcal{S}(\mathbb{R}^d)$$

to define

$$\langle\omega_2, \psi\rangle = \lim_{n\to\infty} \langle\omega_2, \phi_n\rangle$$

with limit in $L^2(\nu)$, for all $\psi \in L^1(\mathbb{R}^d) \cap L^2(\mathbb{R}^d)$, where ϕ_n is a sequence in $\mathcal{S}(\mathbb{R}^d)$ converging to ψ in $L^1(\mathbb{R}^d) \cap L^2(\mathbb{R}^d)$. In particular, this defines the d-parameter Poisson process as (the right-continuous version of)

$$P(x,\omega_2) := \langle\omega_2, \mathcal{X}_{[0,x]}\rangle \, ; \qquad x \in \mathbb{R}^d, \ \omega_2 \in \Omega_2.$$

The *compensated Poisson process* is then defined by

$$Q(x,\omega_2) := P(x,\omega_2) - \prod_{j=1}^{d} x_j\ ; \qquad x = (x_1,\dots,x_d).$$

From here on one can construct stochastic integrals with respect to B and Q, respectively and the corresponding stochastic polynomials and so on. We summarize the corresponding concepts in the two cases in the list below. Unless otherwise mentioned we use the terminology of [5]. In particular, recall that $\{\eta_k\}_{k=1}^{\infty}$ is the basis of $L^2(\mathbb{R}^d)$ consisting of tensor products of the *Hermite functions*

$$\xi_n(x) = \pi^{-1/4}((n-1)!)^{-1/2}e^{-x^2/2}h_{n-1}(\sqrt{2}\,x)\ ; \qquad n\in\mathbb{N};\quad x\in\mathbb{R},$$

where

$$h_m(y) = (-1)^m e^{y^2/2}\frac{d^m}{dy^m}(e^{-y^2/2})\ ; \qquad m\in\mathbb{N}\cup\{0\};\quad y\in\mathbb{R}$$

are the *Hermite polynomials.* The symbol $\mathcal{J}$ denotes the set of all multi-indices $\alpha=(\alpha_1,\dots,\alpha_n)$ where the α_i's are nonnegative integers and $n\in\mathbb{N}$.

In the following list we use the following notation. $x=(x_1,\dots,x_d)\in\mathbb{R}^d$, $\alpha=(\alpha_1,\dots,\alpha_n)\in\mathcal{J}$, $\varepsilon^{(k)}$ denotes the kth unit multi-index, namely $\varepsilon^{(k)}=(0,0,\dots,1)$ with 1 on kth place. (In [5] $\varepsilon^{(k)}$ is denoted by ε_k.) Furthermore, for any function $\xi(x)$ we define $\overline{\xi}=\int_{\mathbb{R}}\xi(x)dx$. Finally, $\hat{\otimes}$ denotes the symmetric tensor product and $\widehat{L}^2(\mathbb{R}^m)$ denotes the symmetric functions in $L^2(\mathbb{R}^m)$.

(2.1) **White noise probability measure**

GAUSSIAN: $\int_{\mathcal{S}'(\mathbb{R}^d)} e^{i\langle\omega_1,\phi\rangle}d\mu(\omega_1) = \exp(-\frac{1}{2}\|\phi\|^2)$; $\phi\in\mathcal{S}(\mathbb{R}^d)$.

POISSONIAN: $\int_{\mathcal{S}'(\mathbb{R}^d)} e^{i\langle\omega_2,\phi\rangle}d\nu(\omega_2) = \exp(\int_{\mathbb{R}^d}(e^{i\phi(x)}-1)dx)$; $\phi\in\mathcal{S}(\mathbb{R}^d)$.

(2.2) **Processes**[1]

GAUSSIAN: Brownian motion B: $B(x,\omega_1)=\langle\omega_1,\mathcal{X}_{[0,x]}\rangle$.

POISSONIAN: Poisson process P: $P(x,\omega_2)=\langle\omega_2,\mathcal{X}_{[0,x]}\rangle$. Compensated Poisson process Q: $Q(x,\omega_2)=P(x,\omega_2)-\prod_{k=1}^{d}x_k$.

(2.3) **Stochastic integrals**

[1]Note that both $\langle\omega_1,\mathcal{X}_{[0,x]}\rangle$ and $\langle\omega_2,\mathcal{X}_{[0,x]}\rangle$ are only defined almost everywhere with respect to μ and ν, respectively. Since μ and ν are not absolutely continuous with respect to each other, the two processes $B(x)$ and $P(x)$ are not the same even though they are defined by the same expression.

GAUSSIAN: $\langle \omega_1, \phi \rangle = \int_{\mathbb{R}^d} \phi(x) dB(x)$ a.s. μ for all $\phi \in L^2(\mathbb{R}^d)$.
POISSONIAN: $\langle \omega_2, \psi \rangle = \int_{\mathbb{R}^d} \psi(x) dP(x)$ a.s. ν, and
$\int_{\mathbb{R}^d} \psi(x) dQ(x) = \int_{\mathbb{R}^d} \psi(x) dP(x) - \overline{\psi}$ for all $\psi \in L^1(\mathbb{R}^d) \cap L^2(\mathbb{R}^d)$.

(2.4) **Stochastic polynomials**
GAUSSIAN: Hermite polynomials $H_\alpha(\omega_1) = \prod_j h_{\alpha_j}(\langle \omega_1, \eta_j \rangle)$,
POISSONIAN: Charlier polynomials $C_\alpha(\omega_2) = C_{|\alpha|}(\omega_2; \overbrace{\eta_1, \ldots, \eta_1}^{\alpha_1}, \ldots, \overbrace{\eta_m, \ldots, \eta_m}^{\alpha_m})$,
where

$$C_n(\omega_2; \psi_1, \ldots, \psi_n)$$
$$= \frac{\partial^n}{\partial u_1 \cdots \partial u_n} \exp\Big(\langle \omega_2, \ln(1 + \sum_{j=1}^n u_j \psi_j) \rangle - \sum_{j=1}^n u_j \int_{\mathbb{R}^d} \psi_j(x) dx \Big)|_{u_1 = \cdots = u_n = 0},$$

for $\psi_j \in S(\mathbb{R}^d)$; $1 \leq j \leq n$.

(2.4a) **Example**
GAUSSIAN: $H_{\varepsilon^{(k)}}(\omega_1) = h_1(\langle \omega_1, \eta_k \rangle) = \langle \omega_1, \eta_k \rangle = \int_{\mathbb{R}^d} \eta_k(x) dB(x)$.
POISSONIAN: $C_{\varepsilon^{(\ell)}}(\omega_2) = \langle \omega_2, \eta_\ell \rangle - \overline{\eta}_\ell = \int_{\mathbb{R}^d} \eta_\ell(x) dQ(x)$.

(2.5) **Expansion in terms of iterated integrals**
GAUSSIAN: If $f \in L^2(\mu)$, then $f = \sum_{n=0}^\infty \int_{\mathbb{R}^{nd}} f_n(x) dB^{\otimes n}(x)$, where
$f_n \in \widehat{L}^2(\mathbb{R}^{nd})$. Furthermore $\|f\|^2_{L^2(\mu)} = \sum_{n=0}^\infty n! \|f_n\|^2_{L^2(\mathbb{R}^{nd})}$.
POISSONIAN: If $g \in L^2(\mu)$, then $g = \sum_{n=0}^\infty \int_{\mathbb{R}^{nd}} g_n(x)\, dQ^{\otimes n}(x)$, where
$g_n \in \widehat{L}^2(\mathbb{R}^{nd})$. Furthermore $\|g\|^2_{L^2(\nu)} = \sum_{n=0}^\infty n! \|g_n\|^2_{L^2(\mathbb{R}^{nd})}$.

(2.6) **Expansion in terms of polynomials**
GAUSSIAN: If $f \in L^2(\mu)$, then $f(\omega_1) = \sum_{\alpha \in \mathcal{J}} a_\alpha H_\alpha(\omega_1)$, where $a_\alpha \in \mathbb{R}$.
Furthermore $\|f\|^2_{L^2(\mu)} = \sum_{\alpha \in \mathcal{J}} a_\alpha^2 \alpha!$.
POISSONIAN: If $g \in L^2(\nu)$, then $g(\omega_2) = \sum_{\beta \in \mathcal{J}} b_\beta C_\beta(\omega_2)$, where $b_\beta \in \mathbb{R}$.
Furthermore $\|g\|^2_{L^2(\nu)} = \sum_{\beta \in \mathcal{J}} b_\beta^2 \beta!$.

(2.7) **Relation between the two expansions**
GAUSSIAN: $\int_{\mathbb{R}^{|\alpha|d}} \eta^{\widehat{\otimes} \alpha} dB^{\otimes |\alpha|} = H_\alpha(\omega_1)$.
POISSONIAN: $\int_{\mathbb{R}^{|\alpha|d}} \eta^{\widehat{\otimes} \alpha} dQ^{\otimes |\alpha|} = C_\alpha(\omega_2)$.

(2.8) **Kondratiev norms**, $-1 \leq \rho \leq 1$, $q \in \mathbb{Z}$
GAUSSIAN: If $F(\omega_1) = \sum_{\alpha \in \mathcal{J}} a_\alpha H_\alpha(\omega_1)$, then $\|F\|^2_{\rho,q,\mu} = \sum_\alpha a_\alpha^2 (\alpha!)^{1+\rho} (2\mathbb{N})^{q\alpha}$.
POISSONIAN: if $G(\omega_2) = \sum_{\beta \in \mathcal{J}} b_\beta C_\beta(\omega_2)$, then $\|G\|^2_{\rho,q,\nu} = \sum_\beta b_\beta^2 (\beta!)^{1+\rho} (2\mathbb{N})^{q\beta}$.

(2.9) **Kondratiev Hilbert spaces**

GAUSSIAN: $(\mathcal{S})_{\rho,q,\mu} := \{F; \|F\|_{\rho,q,\mu} < \infty\}$.

POISSONIAN: $(\mathcal{S})_{\rho,q,\nu} := \{G; \|G\|_{\rho,q,\nu} < \infty\}$.

(2.10) **Kondratiev stochastic test function spaces,** $0 \leq \rho \leq 1$

GAUSSIAN: $(\mathcal{S})_{\rho,\mu} = \bigcap_{r\in\mathbf{N}}(\mathcal{S})_{\rho,r,\mu}$, with projective topology.

POISSONIAN: $(\mathcal{S})_{\rho,\nu} = \bigcap_{r\in\mathbf{N}}(\mathcal{S})_{\rho,r,\nu}$, with projective topology.

(2.11) **Kondratiev stochastic distribution spaces,** $0 \leq \rho \leq 1$

GAUSSIAN: $(\mathcal{S})_{-\rho,\mu} = \bigcup_{r\in\mathbf{N}}(\mathcal{S})_{-\rho,-r,\mu}$, with inductive topology.

POISSONIAN: $(\mathcal{S})_{-\rho,\nu} = \bigcup_{r\in\mathbf{N}}(\mathcal{S})_{-\rho,-r,\nu}$, with inductive topology.

(2.12) **Hida spaces**

GAUSSIAN: $(\mathcal{S})_{\mu} := (\mathcal{S})_{0,\mu}$, $\quad (\mathcal{S})^*_{\mu} := (\mathcal{S})_{-0,\mu}$.

POISSONIAN: $(\mathcal{S})_{\nu} := (\mathcal{S})_{0,\nu}$, $\quad (\mathcal{S})^*_{\nu} := (\mathcal{S})_{-0,\nu}$.

(2.13) **White noise**

GAUSSIAN: $W(x,\omega_1) = \sum_{k=1}^{\infty} \eta_k(x) H_{\varepsilon^{(k)}}(\omega_1)$, $\quad W(x,\cdot) \in (\mathcal{S})^*_{\mu}$ for all x.

POISSONIAN: $V(x,\omega_2) = \sum_{k=1}^{\infty} \eta_k(x) C_{\varepsilon^{(k)}}(\omega_2)$, $\quad V(x,\cdot) \in (\mathcal{S})^*_{\nu}$ for all x.

(2.14) **Wick product**

GAUSSIAN: $(\sum_{\alpha} a_{\alpha} H_{\alpha}) \diamond (\sum_{\beta} b_{\beta} H_{\beta}) = \sum_{\alpha,\beta} a_{\alpha} b_{\beta} H_{\alpha+\beta} = \sum_{\gamma}(\sum_{\alpha+\beta=\gamma} a_{\alpha} b_{\beta}) H_{\gamma}$.

POISSONIAN: $(\sum_{\alpha} a_{\alpha} C_{\alpha}) \diamond (\sum_{\beta} b_{\beta} C_{\beta}) = \sum_{\alpha,\beta} a_{\alpha} b_{\beta} C_{\alpha+\beta} = \sum_{\gamma}(\sum_{\alpha+\beta=\gamma} a_{\alpha} b_{\beta}) C_{\gamma}$.

(2.15) **Properties of the Wick product**

GAUSSIAN: If $f, g \in (\mathcal{S})_{1,\mu}$, then $f \diamond g \in (\mathcal{S})_{1,\mu}$, and if $F, G \in (\mathcal{S})_{-1,\mu}$, then $F \diamond G \in (\mathcal{S})_{-1,\mu}$.

POISSONIAN: If $f, g \in (\mathcal{S})_{1,\nu}$, then $f \diamond g \in (\mathcal{S})_{1,\nu}$, and if $F, G \in (\mathcal{S})_{-1,\nu}$, then $F \diamond G \in (\mathcal{S})_{-1,\nu}$.

(2.16) **Wick products in terms of iterated integrals**

GAUSSIAN: $(\int_{\mathbb{R}^{nd}} f_n dB^{\otimes n}) \diamond (\int_{\mathbb{R}^{md}} f_m dB^{\otimes m}) = \int_{\mathbb{R}^{(n+m)d}} (f_n \widehat{\otimes} f_m) dB^{\otimes(n+m)}$.

POISSONIAN: $(\int_{\mathbb{R}^{nd}} g_n dQ^{\otimes n}) \diamond (\int_{\mathbb{R}^{md}} g_m dQ^{\otimes m}) = \int_{\mathbb{R}^{(n+m)d}} (g_n \widehat{\otimes} g_m) dQ^{\otimes(n+m)}$.

(2.17) **Wick powers and iterated integrals**[2]

GAUSSIAN: $\langle \omega_1, \phi \rangle^{\diamond n} = \int_{\mathbb{R}^{nd}} \phi^{\otimes n} dB^{\otimes n}$ for $\phi \in L^2(\mathbb{R}^d)$.

POISSONIAN: $(\langle \omega_2, \psi \rangle - \overline{\psi})^{\diamond n} = \int_{\mathbb{R}^{nd}} \psi^{\otimes n} dQ^{\otimes n}$ for $\psi \in L^1(\mathbb{R}^d) \cap L^2(\mathbb{R}^d)$.

[2]These formulas follow by combining (2.3) and (2.16).

(2.18) **Wick exponentials**[3]

GAUSSIAN: $\exp^\diamond(\langle\omega_1,\phi\rangle) := \sum_{n=0}^\infty \frac{1}{n!}\langle\omega_1,\phi\rangle^{\diamond n} = \exp(\langle\omega_1,\phi\rangle - \frac{1}{2}\|\phi\|^2)$ for all $\phi \in L^2(\mathbb{R}^d)$.

POISSONIAN: $\exp^\diamond(\langle\omega_2,\psi\rangle) := \sum_{n=0}^\infty \frac{1}{n!}\langle\omega_2,\psi\rangle^{\diamond n} = \exp(\langle\omega_2,\ln(1+\psi)\rangle)$ for all $\psi \in L^1(\mathbb{R}^d)\cap L^2(\mathbb{R}^d)$, $\psi \geq -1+\varepsilon$ for some $\varepsilon > 0$.

(2.19) **Wick powers and Wick exponentials**[4]

GAUSSIAN: $\langle\omega_1,\phi\rangle^{\diamond n} = \frac{\partial^n}{\partial u^n}\exp^\diamond(\langle\omega_1,u\phi\rangle)|_{u=0}$
$= \frac{\partial^n}{\partial u^n}\exp(\langle\omega_1,u\phi\rangle - \frac{1}{2}u^2\|\phi\|^2)|_{u=0}$ for all $\phi \in L^2(\mathbb{R}^d)$.

POISSONIAN: $\langle\omega_2,\psi\rangle^{\diamond n} = \frac{\partial^n}{\partial u^n}\exp^\diamond(\langle\omega_2,u\psi\rangle)|_{u=0}$
$= \frac{\partial^n}{\partial u^n}\exp(\langle\omega_2,\ln(1+u\psi)\rangle)|_{u=0}$ for all $\psi \in L^1(\mathbb{R}^d)\cap L^2(\mathbb{R}^d)$.

(2.20) **Wick polynomials and Wick exponentials**[4]

GAUSSIAN: $\langle\omega_1,\phi_1\rangle \diamond\cdots\diamond \langle\omega_1,\phi_n\rangle = \frac{\partial^n}{\partial u_1\cdots\partial u_n}\exp^\diamond(\langle\omega_1,\sum_{j=1}^n u_j\phi_j\rangle)|_{u_1=\cdots=u_n=0}$
$= \frac{\partial^n}{\partial u_1\cdots\partial u_n}\exp(\langle\omega_1,\sum_{j=1}^n u_j\phi_j\rangle - \frac{1}{2}\|\sum_{j=1}^n u_j\phi_j\|^2)|_{u_1=\cdots=u_n=0}$
for $\phi_j \in L^2(\mathbb{R}^d)$ for $1 \leq j \leq n$.

POISSONIAN: $\langle\omega_2,\psi_1\rangle \diamond\cdots\diamond \langle\omega_2,\psi_n\rangle = \frac{\partial^n}{\partial u_1\cdots\partial u_n}\exp^\diamond(\langle\omega_2,\sum_{j=1}^n u_j\psi_j\rangle)|_{u_1=\cdots=u_n=0}$
$= \frac{\partial^n}{\partial u_1\cdots\partial u_n}\exp(\langle\omega_2,\ln(1+\sum_{j=1}^n u_j\psi_j)\rangle)|_{u_1=\cdots=u_n=0}$
for $\psi_j \in L^1(\mathbb{R}^d)\cap L^2(\mathbb{R}^d)$ and $1 \leq j \leq n$.

(2.21) **Representation of stochastic polynomials in terms of Wick exponentials**[5]

GAUSSIAN: $H_\alpha(\omega_1) = \partial^\alpha \exp^\diamond(\int_{\mathbb{R}^d}\sum_{j=1}^n u_j\eta_j(x)dB(x))|_{u_1=\cdots=u_n=0}$,
where $\partial^\alpha = \frac{\partial^{|\alpha|}}{\partial u_1^{\alpha_1}\cdots\partial u_n^{\alpha_n}}$.

POISSONIAN: $C_\alpha(\omega_2) = \partial^\alpha \exp^\diamond(\int_{\mathbb{R}^d}\sum_{j=1}^n u_j\eta_j(x)dQ(x))|_{u_1=\cdots=u_n=0}$.

(2.22) **Skorohod integrals**

GAUSSIAN: If $Y(t,\omega_1) = \sum_{n=0}^\infty \int_{\mathbb{R}^n} f_n(t,\cdot)dB^{\otimes n}$, where $\sum_{n=0}^\infty (n+1)!\|\widetilde{f}_n\|_{L^2(\mathbb{R}^{n+1})} < \infty$, then $\int_{\mathbb{R}} Y(t)\delta B(t) := \sum_{n=0}^\infty \int_{\mathbb{R}^{n+1}} \widetilde{f}_n dB^{\otimes(n+1)}$. Here

[3] The Gaussian case can be found, e.g., in [5, Lemma 2.6.16]. For a proof in the Poissonian case and when $\alpha = 1$ see Corollary 2.3 below.

[4] This follows from (2.18).

[5] The Gaussian case follows from (2.4), the fact that $h_{\alpha_j}(\langle\omega_1,\eta_j\rangle) = \langle\omega_1,\eta_j\rangle^{\diamond\alpha_j}$ (which is seen from (2.7) and (2.17)) plus (2.20). To prove the Poissonian case note that by (2.3) we have, for $\psi \in L^1(\mathbb{R}^d)\cap L^2(\mathbb{R}^d)$, $\int_{\mathbb{R}^d}\psi(x)dQ(x) = \int_{\mathbb{R}^d}\psi(x)dP(x) - \overline{\psi} = \langle\omega_2,\psi\rangle - \overline{\psi}$. Therefore, by (2.18) we get $\exp^\diamond(\int_{\mathbb{R}^d}\psi(x)dQ(x)) = \exp^\diamond(\langle\omega_2,\psi\rangle - \overline{\psi}) = \exp(\langle\omega_2,\ln(1+\psi)\rangle - \overline{\psi})$, assuming that $\psi \geq -1+\varepsilon$ for some $\varepsilon > 0$. Applying this to appropriate approximations to $\psi_0 = \sum_{j=1}^n u_j\eta_j$ we see that (2.21) follows from the definition of C_α in (2.4).

$\widetilde{f}_n$ is the symmetrization of $f_n(t, x_1, \ldots, x_n)$ with respect to all its $n+1$ variables.
POISSONIAN: If $Z(t, \omega_2) = \sum_{n=0}^{\infty} \int_{\mathbb{R}^n} g_n(t, \cdot) dQ^{\otimes n}$, where $\sum_{n=0}^{\infty} (n+1)! \|\widetilde{g}_n\|_{L^2(\mathbb{R}^{n+1})} < \infty$, then $\int_{\mathbb{R}} Z(t) \delta Q(t) := \sum_{n=0}^{\infty} \int_{\mathbb{R}^{n+1}} \widetilde{g}_n dQ^{\otimes(n+1)}$.

(2.23) **Connection between Skorohod integrals and Ito-Poisson integrals**[6]
GAUSSIAN: $\int_{\mathbb{R}} Y(t) \delta B(t) = \int_{\mathbb{R}} Y(t) dB(t)$ if $Y(t)$ is $\mathcal{F}_t$-adapted.
POISSONIAN: $\int_{\mathbb{R}} Z(t) \delta Q(t) = \int_{\mathbb{R}} Z(t) dQ(t)$ if $Z(t)$ is adapted to the filtration $\mathcal{G}_t$ generated by $\{P(s, \cdot)\}_{s \leq t}$.

(2.24) **White noise description of Skorohod integrals**[6]
GAUSSIAN: If $Y(t)$ is Skorohod integrable, see (2.22), then $\int_{\mathbb{R}} Y(t) \delta B(t) = \int_{\mathbb{R}} Y(t) \diamond W(t) dt$.
POISSONIAN: If $Z(t)$ is Skorohod integrable, see (2.22), then $\int_{\mathbb{R}} Z(t) \delta Q(t) = \int_{\mathbb{R}} Z(t) \diamond V(t) dt$.

(2.25) **Hermite transform**
GAUSSIAN: If $F(\omega_1) = \sum_{\alpha \in \mathcal{J}} a_\alpha H_\alpha(\omega_1)$, then $(\mathcal{H}F)(z) = \sum_{\alpha \in \mathcal{J}} c_\alpha z^\alpha$ for $z = (z_1, z_2, \ldots)$, where $z^\alpha = z_1^{\alpha_1} z_2^{\alpha_2} \cdots$.
POISSONIAN: If $G(\omega_2) = \sum_{\beta \in \mathcal{J}} b_\beta C_\beta(\omega_2)$, then $(\mathcal{H}G)(z) = \sum_{\beta \in \mathcal{J}} b_\beta z^\beta$.

(2.26) S**-transform (definition)**, $\rho < 1$
GAUSSIAN: If $F \in (\mathcal{S})_{-\rho,\mu}$, then $(SF)(\phi) = \langle F, \exp^\diamond \int_{\mathbb{R}^d} \phi \, dB \rangle$ for all $\phi \in \mathcal{S}(\mathbb{R}^d)$.
POISSONIAN: If $G \in (\mathcal{S})_{-\rho,\nu}$, then $(SG)(\phi) = \langle G, \exp^\diamond \int_{\mathbb{R}^d} \phi dQ \rangle$ for all $\phi \in \mathcal{S}(\mathbb{R}^d)$.

(2.27) S**-transform (properties)**, $\rho < 1$
GAUSSIAN: If $F = \sum_\alpha a_\alpha H_\alpha(\omega_1) \in (\mathcal{S})_{-\rho,\mu}$, then $(SF)(\phi) = \sum_\alpha a_\alpha (\phi, \eta)^\alpha$ for all $\phi \in S(\mathbb{R}^d)$.
POISSONIAN: If $G = \sum_\beta b_\beta C_\beta(\omega_2) \in (\mathcal{S})_{-\rho,\nu}$, then $(SG)(\phi) = \sum_\beta b_\beta (\phi, \eta)^\beta$ for all $\phi \in S(\mathbb{R}^d)$.

The unitary correspondence $\mathcal{U}$
We review some other concepts and results from [1]: Define $\mathcal{U} : (\mathcal{S})_{-1,\mu} \to (\mathcal{S})_{-1,\nu}$ by

$$\mathcal{U}(\sum_\alpha a_\alpha H_\alpha) = \sum_\alpha a_\alpha C_\alpha.$$

[6]For proofs and more details in the Gaussian case see, e.g., [5, Theorem 2.5.9] and the references therein. For a proof in the Poissonian case, see [5, Theorem 5.1]. This proof is valid in the more general setting of *colored* noise.

Then $\mathcal{U}$ is a continuous bijection and $\mathcal{U}|_{(\mathcal{S})_{\rho,q,\mu}}$ is an isometry onto $(\mathcal{S})_{\rho,q,\nu}$ for all $\rho \in [-1,1]$ and all $q \in \mathbb{Z}$. Moreover, $\mathcal{U}$ preserves the Wick product:

$$\mathcal{U}(F \diamond G) = (\mathcal{U}F) \diamond (\mathcal{U}G). \tag{2.28}$$

Some examples: We have that

$$\mathcal{U}(\langle \omega_1, \psi \rangle) = \langle \omega_2, \psi \rangle - \overline{\psi}\,; \qquad \psi \in L^1(\mathbb{R}^d) \cap L^2(\mathbb{R}) \tag{2.29}$$

which follows from (2.4a). In particular,

$$\mathcal{U}(B(x,\omega_1)) = Q(x,\omega_2), \tag{2.30}$$

$$\mathcal{U}(F^{\diamond n}(\omega_1)) = (\mathcal{U}F)^{\diamond n}(\omega_2). \tag{2.31}$$

by using induction on (2.28). From the Wick-Taylor expansions and (2.31)

$$\mathcal{U}(\exp^{\diamond} F(\omega_1)) = \exp^{\diamond} \mathcal{U}F(\omega_2). \tag{2.32}$$

Recall the following fundamental result from [1]: (For notation see Corollary 4.9.7 in [5].)

Theorem 2.1 *Suppose $F(t,x) : \mathbb{R}^{d+1} \to (\mathcal{S})_{-1,\mu}$ is a solution of a Gaussian Wick-type stochastic differential equation of the form*

$$A^{\diamond}(t, x, \partial_t, \nabla_x, F, W) = 0.$$

Then $G := \mathcal{U}F$ is a solution of the corresponding Poissonian Wick-type equation

$$A^{\diamond}(t, x, \partial_t, \nabla_x, G, V) = 0,$$

and conversely. Here $A^{\diamond}$ denotes the Gaussian (resp. Poissonian) Wick version of a given function A.

This elegant result shows that when considering Wick-type stochastic differential equations it is—at least in principle—enough to know the solution in the Gaussian case. The solution in the Poissonian case then follows by applying the map $\mathcal{U}$. We illustrate this by considering a simple example:

Example 2.2 Let $X(t)$ be the solution of the (1-parameter) SDE

$$dX(t) = \psi(t)X(t)dQ(t)\,; \qquad t > 0 \tag{2.33}$$

where $Q(t) = Q(t, \omega_2)$ is the compensated Poisson process and $\psi(t)$ is a continuous, deterministic function, $\psi(t) > -1+\varepsilon$ for all t, for some $\varepsilon > 0$. According to the Poissonian white noise theory, the equation can be regarded as a Wick-type stochastic differential equation in the (Poisson) Hida space $(\mathcal{S})^*_\nu$, namely the equation

$$\frac{dX(t)}{dt} = \psi(t)X(t) \diamond V(t) \tag{2.34}$$

where $\diamond$ denotes the Poissonian Wick product and $V(t) = dQ(t)/dt \in (\mathcal{S})^*_\nu$ is Poissonian white noise.

The corresponding Gaussian white noise differential equation is

$$\frac{dY(t)}{dt} = \psi(t)Y(t) \diamond W(t) \tag{2.35}$$

where $\diamond$ denotes the Gaussian Wick product and $W(t) = dB(t)/dt \in (\mathcal{S})^*_\mu$ is Gaussian white noise.

Using Wick calculus on equation (2.35), we get the solution

$$Y(t) = Y(0) \diamond \exp^\diamond \Big(\int_0^t \psi(s)dB(s)\Big) \tag{2.36}$$

(see, e.g., [5, Theorem 3.1.2] for a more general calculation).

If $Y(0) = X(0)$ is a constant, then we get

$$Y(t) = X(0) \exp^\diamond\Big(\int_0^t \psi(s)dB(s)\Big).$$

According to the correspondence $\mathcal{U}$ the solution $X(t)$ of (2.34) is therefore

$$X(t) = \mathcal{U}(Y(t)) = X(0) \exp^\diamond\Big(\int_0^t \psi(s)dQ(s)\Big). \tag{2.37}$$

In the Gaussian case we know that

$$\exp^\diamond\Big(\int_0^t \psi(s)dB(s)\Big) = \exp\Big(\int_0^t \psi(s)dB(s) - \tfrac{1}{2}\int_0^t \psi(s)^2 ds\Big),$$

but what is $\exp^\diamond(\int_0^t \psi(s)dQ(s))$ in the Poissonian case? The answer can be found using (2.18):

$$
\begin{aligned}
\exp^\diamond\Big(\int_0^t \psi(s)dQ(s)\Big) &= \exp^\diamond\Big(\int_0^t \psi(s)dP(s) - \int_0^t \psi(s)ds\Big)\\
&= \exp^\diamond\big(\langle\omega_2, \mathcal{X}_{[0,t]}\psi\rangle - \int_0^t \psi(s)ds\big)\\
&= \exp\big(\langle\omega_2, \ln(1+\mathcal{X}_{[0,t]}\psi)\rangle - \int_0^t \psi(s)ds\big)\\
&= \exp\Big(\int_0^t \ln(1+\psi(s))dP(s) - \int_0^t \psi(s)ds\Big).
\end{aligned}
\tag{2.38}
$$

Combining (2.37) and (2.38) we get the following expression for the solution $X(t)$ of (2.33):

$$
X(t) = X(0)\exp\Big(\int_0^t \ln(1+\psi(s))dP(s) - \int_0^t \psi(s)ds\Big).
\tag{2.39}
$$

It may be instructive to see that we can verify the formula (2.39) by using (2.34) directly as follows:

The Poisson process is described by a sequence of stopping times $0 < \tau_1 < \tau_2 < \cdots$ at which P jumps with size 1. Between these stopping times $X(t)$ satisfies the equation

$$
dX(t) = -\psi(t)X(t)dt\,; \qquad \tau_k \le t < \tau_{k+1},
$$

since $dP(t) = dQ(t) - dt$. Hence

$$
X(t) = X(\tau_k)\exp\Big(-\int_{\tau_k}^t \psi(s)ds\Big)\,; \qquad \tau_k \le t < \tau_{k+1}.
\tag{2.40}
$$

At the jumping times τ_k the value of X increases by

$$
\Delta X(\tau_k) := X(\tau_k) - X(\tau_k^-) = \psi(\tau_k)X(\tau_k^-).
$$

Hence

$$
X(\tau_k) = X(\tau_k^-)(1+\psi(\tau_k)) = X(\tau_k^-)\exp(\ln(1+\psi(\tau_k))).
\tag{2.41}
$$

Combining (2.40) and (2.41) we get

$$
\begin{aligned}
X(t) &= X(0)\prod_{\tau_k\le t}\exp(\ln(1+\psi(\tau_k)))\exp\Big(-\int_{\tau_k}^{t\wedge\tau_{k+1}} \psi(s)ds\Big)\\
&= X(0)\exp\Big(\sum_{\tau_k\le t}\ln(1+\psi(\tau_k)) - \int_{\tau_k}^{t\wedge\tau_{k+1}} \psi(s)ds\Big)\\
&= X(0)\exp\Big(\int_0^t \ln(1+\psi(s))dP(s) - \int_0^t \psi(s)ds\Big),
\end{aligned}
$$

which is (2.39). This ends Example 2.2.

Since (2.39) is equivalent to (2.37), the above argument proves the following result:

Corollary 2.3 *Suppose* ψ *is continuous,* $\psi > -1$ *and*

$$\int_0^t (\ln(1+\psi(s)))^2 ds < \infty. \tag{2.42}$$

Then

$$\exp^\diamond\Big(\int_0^t \psi(s)dQ(s)\Big) = \exp\Big(\int_0^t \ln(1+\psi(s))dP(s) - \int_0^t \psi(s)ds\Big). \tag{2.43}$$

Remark The above argument was carried out under the assumption that $\psi \geq -1+\varepsilon$ for some $\varepsilon > 0$. However, the identity (2.43) extends to the case (2.42) because

$$E[(\int_0^t \ln(1+\psi(s))dP(s))^2] = E[(\int_0^t \ln(1+\psi(s))dQ(s) + \int_0^t \ln(1+\psi(s))ds)^2]$$
$$= \int_0^t \ln(1+\psi(s))^2 ds + (\int_0^t \ln(1+\psi(s))ds)^2.$$

3 The mixed Gaussian-Poissonian case

Now suppose there are m noise sources, of which k are Gaussian white noises and the remaining $m-k$ are Poissonian. Then we define the *combined Gaussian-Poissonian white noise probability measure* θ on $\Omega := \mathcal{S}'(\mathbb{R}^d) \times \cdots \times \mathcal{S}'(\mathbb{R}^d)$ (m factors) by

$$d\theta(\omega_1,\dots,\omega_m) = d\mu(\omega_1)\cdots d\mu(\omega_k)d\nu(\omega_{k+1})\cdots d\nu(\omega_m) \tag{3.1}$$

where $\omega = (\omega_1,\dots,\omega_m) \in \Omega$.
For notational simplicity we will from now on assume that $m = 2$ and $k = 1$. For $\omega = (\omega_1,\omega_2) \in \Omega = \mathcal{S}'(\mathbb{R}^d) \times \mathcal{S}'(\mathbb{R}^d)$ and $\phi = (\phi_1,\phi_2) \in \mathcal{S}(\mathbb{R}^d) \times \mathcal{S}(\mathbb{R}^d)$ we define

$$\langle \omega, \phi\rangle = \langle \omega_1,\phi_1\rangle + \langle \omega_2,\phi_2\rangle. \tag{3.2}$$

Then by (2.1) we see that

$$\int_\Omega e^{i\langle\omega,\phi\rangle} d\theta(\omega) = \exp\Big(-\tfrac{1}{2}\|\phi_1\|^2 + \int_{\mathbb{R}^d} (e^{i\phi_2(x)} - 1)dx\Big). \tag{3.3}$$

Let $\Gamma = \mathcal{J} \times \mathcal{J}$ be the set of all pairs $\gamma = (\alpha, \beta)$ of multi-indices $\alpha, \beta \in \mathcal{J}$, and define

$$M_\gamma(\omega_1, \omega_2) = H_\alpha(\omega_1) G_\beta(\omega_2) \; ; \qquad \gamma = (\alpha, \beta) \in \Gamma. \tag{3.4}$$

For example, this gives (using (2.4a))

$$M_{(\varepsilon^{(k)},0)}(\omega_1, \omega_2) = H_{\varepsilon^{(k)}}(\omega_1) = \langle \omega_1, \eta_k \rangle, \tag{3.5}$$

$$M_{(0,\varepsilon^{(\ell)})}(\omega_1, \omega_2) = C_{\varepsilon^{(\ell)}}(\omega_2) = \langle \omega_2, \eta_\ell \rangle - \overline{\eta}_\ell. \tag{3.6}$$

From tensor product considerations, we see that we have the following analogue of (2.6): Any $f \in L^2(\theta)$ can be written (uniquely)

$$f(\omega_1, \omega_2) = \sum_{\gamma \in \Gamma} c_\gamma M_\gamma(\omega_1, \omega_2) \tag{3.7}$$

where $c_\gamma \in \mathbb{R}$. Moreover, we have the isometry

$$\|f\|^2_{L^2(\theta)} = \sum_{\gamma \in \Gamma} c_\gamma^2 \gamma! \tag{3.8}$$

where $\gamma! = \alpha! \, \beta!$ if $\gamma = (\alpha, \beta) \in \Gamma$.
We proceed as in the list in Section 2. For $\rho \in [-1, 1]$ and $q \in \mathbb{Z}$ define the *Kondratiev norms*

$$\|F\|^2_{\rho,q,\theta} = \sum_{\gamma \in \Gamma} c_\gamma^2 (\gamma!)^{1+\rho} (2\mathbb{N})^{q\gamma} \text{ if } F = \sum_{\gamma \in \Gamma} c_\gamma M_\gamma(\omega) \, , \tag{3.9}$$

where

$$(2\mathbb{N})^{q\gamma} = (2\mathbb{N})^{q\alpha} (2\mathbb{N})^{q\beta}, \qquad \gamma = (\alpha, \beta) \in \Gamma$$

and the *Kondratiev Hilbert spaces*

$$(\mathcal{S})_{\rho,q,\theta} = \{F ; \|F\|_{\rho,q,\theta} < \infty\}. \tag{3.10}$$

For $0 \leq \rho \leq 1$ we define the *Kondratiev stochastic test function spaces*

$$(\mathcal{S})_{\rho,\theta} = \bigcap_{r \in \mathbb{N}} (\mathcal{S})_{\rho,r,\theta} \qquad \text{with projective topology} \tag{3.11}$$

and the *Kondratiev stochastic distribution spaces*

$$(\mathcal{S})_{-\rho,\theta} = \bigcup_{r \in \mathbb{N}} (\mathcal{S})_{-\rho,-r,\theta} \qquad \text{with inductive topology.} \tag{3.12}$$

Furthermore, we define the *Hida test function space*

$$(\mathcal{S})_\theta := (\mathcal{S})_{0,\theta}, \tag{3.13}$$

and the *Hida distribution space*

$$(\mathcal{S})^*_\theta := (\mathcal{S})_{-0,\theta}. \tag{3.14}$$

The *combined Gaussian-Poissonian white noise* $U(x,\omega)$ is defined by

$$U(x,\omega) = (W(x,\omega_1), V(x,\omega_2)), \tag{3.15}$$

where

$$W(x,\omega_1) = \sum_{k=1}^{\infty} \eta_k(x) H_{\varepsilon^{(k)}}(\omega_1), \qquad V(x,\omega_2) = \sum_{\ell=1}^{\infty} \eta_\ell(x) C_{\varepsilon^{(\ell)}}(\omega_2)$$

is Gaussian and Poissonian white noise, respectively. If

$$F(\omega) = \sum_{\lambda\in\Gamma} a_\lambda M_\lambda(\omega) \in (\mathcal{S})_{-1,\theta},$$

and

$$G(\omega) = \sum_{\delta\in\Gamma} b_\delta M_\delta(\omega) \in (\mathcal{S})_{-1,\theta},$$

we define their *Wick product* $F \diamond G$ by

$$(F\diamond G)(\omega) = \sum_{\lambda,\delta\in\Gamma} a_\lambda b_\delta M_{\lambda+\delta}(\omega) = \sum_{\gamma\in\Gamma}\Big(\sum_{\lambda+\delta=\gamma} a_\lambda b_\delta\Big) M_\gamma(\omega). \tag{3.16}$$

As in the Gaussian case one can prove that

$$F, G \in (\mathcal{S})_{-1,\theta} \Rightarrow F\diamond G \in (\mathcal{S})_{-1,\theta}, \tag{3.17}$$

$$f, g \in (\mathcal{S})_{1,\theta} \Rightarrow f\diamond g \in (\mathcal{S})_{1,\theta} \tag{3.18}$$

Examples

a) Using (3.5) and (3.6) we get

$$\begin{aligned}
\langle\omega_1,\eta_k\rangle \diamond \langle\omega_2,\eta_\ell\rangle &= M_{(\varepsilon^{(k)},0)}(\omega) \diamond (M_{(0,\varepsilon^{(\ell)})}(\omega) + \overline{\eta}_\ell) \\
&= M_{(\varepsilon^{(k)},\varepsilon^{(\ell)})}(\omega) + \overline{\eta_\ell} M_{(\varepsilon^{(k)},0)}(\omega) \\
&= H_{\varepsilon^{(k)}}(\omega_1) C_{\varepsilon^{(\ell)}}(\omega_2) + \overline{\eta}_\ell H_{\varepsilon^{(k)}}(\omega_1) \\
&= \langle\omega_1,\eta_k\rangle(\langle\omega_2,\eta_\ell\rangle - \overline{\eta}_\ell) + \overline{\eta}_\ell\langle\omega,\eta_k\rangle = \langle\omega_1,\eta_k\rangle\cdot\langle\omega_2,\eta_\ell\rangle.
\end{aligned} \tag{3.19}$$

b) The Wick product of Gaussian and Poissonian white noise reads

$$\begin{aligned} W(x,\omega_1)\diamond V(y,\omega_2) &= \Big(\sum_k \eta_k(x) H_{\varepsilon^{(k)}}(\omega_1)\Big) \diamond \Big(\sum_\ell \eta_\ell(y) C_{\varepsilon^{(\ell)}}(\omega_2)\Big) \\ &= \sum_{k,\ell} \eta_k(x)\eta_\ell(y) H_{\varepsilon^{(k)}}(\omega_1) C_{\varepsilon^{(\ell)}}(\omega_2), \end{aligned} \tag{3.20}$$

which can be written

$$\sum_{k,\ell} \eta_k(x)\eta_\ell(y) M_{(\varepsilon^{(k)},\varepsilon^{(\ell)})}(\omega)$$

and therefore

$$\begin{aligned} \|W(x)\diamond V(y)\|^2_{0,-r,\theta} &= \sum_{k,\ell} \eta_k(x)^2 \eta_\ell(y)^2 \varepsilon^{(k)}!\varepsilon^{(\ell)}!(2\mathbb{N})^{-r(\varepsilon^{(k)},\varepsilon^{(\ell)})} \\ &= \sum_{k,\ell} \eta_k(x)^2\eta_\ell(y)^2 (2k)^{-r}(2\ell)^{-r} \\ &= \Big(\sum_k \eta_k(x)^2 (2k)^{-r}\Big)\Big(\sum_\ell \eta_\ell(y)^2(2\ell)^{-r}\Big) < \infty \qquad \text{for } r > 1 \end{aligned}$$

because

$$\sup_{x,k} |\eta_k(x)| < \infty. \tag{3.21}$$

Hence

$$W(x)\diamond V(y) \in (\mathcal{S})^*_\theta \qquad \text{for all } x, y.$$

c) The preceding example illustrates a general result: If

$$F(\omega) = F(\omega_1) = \sum_{\alpha\in\mathcal{J}} a_\alpha H_\alpha(\omega_1) \in (\mathcal{S})_{-1,\theta},$$

$$G(\omega) = G(\omega_2) = \sum_{\beta\in\mathcal{J}} b_\beta C_\beta(\omega_2) \in (\mathcal{S})_{-1,\theta},$$

then

$$(F\diamond G)(\omega_1,\omega_2) = F(\omega_1)\cdot G(\omega_2). \tag{3.22}$$

The proof is an immediate consequence of the definition (3.16):

$$\begin{aligned} (F\diamond G)(\omega_1,\omega_2) &= \sum_{\alpha,\beta\in\mathcal{J}} a_\alpha b_\beta M_{(\alpha,\beta)}(\omega_1,\omega_2) = \sum_{\alpha,\beta\in\mathcal{J}} a_\alpha b_\beta H_\alpha(\omega_1) C_\beta(\omega_2) \\ &= \Big(\sum_{\alpha\in\mathcal{J}} a_\alpha H_\alpha(\omega_1)\Big)\cdot\Big(\sum_{\beta\in\mathcal{J}} b_\beta C_\beta(\omega_2)\Big) = F(\omega_1)\cdot G(\omega_2). \end{aligned}$$

The *Hermite transform* of $F(\omega)=\sum_{\gamma\in\Gamma} c_\gamma M_\gamma(\omega) \in (\mathcal{S})_{-1,\theta}$ is the function $\mathcal{H}F\colon \mathbb{C}_0^{\mathbf{N}} \times \mathbb{C}_0^{\mathbf{N}} \to \mathbb{C}$ defined by

$$\widetilde{F}(z,w) := \mathcal{H}F(z,w) := \sum_{\gamma\in\Gamma} c_\gamma z^\alpha w^\beta\,; \qquad \gamma=(\alpha,\beta), \tag{3.23}$$

where

$$z=(z_1,z_2,\dots)\in\mathbb{C}_0^{\mathbf{N}}, \qquad w=(w_1,w_2,\dots)\in\mathbb{C}_0^{\mathbf{N}},$$

and

$$z^\alpha w^\beta = z_1^{\alpha_1} z_2^{\alpha_2}\cdots w_1^{\beta_1} w_2^{\beta_2}\cdots$$

if $\alpha=(\alpha_1,\alpha_2\dots)$, $\beta=(\beta_1,\beta_2,\dots)$.

The Hermite transform solution scheme

Consider a Gaussian-Poissonian white noise differential equation of Wick-type in the unknown $F\colon \mathbb{R}^d \to (\mathcal{S})_{-1,\theta}$:

$$A^\diamond(t,x,\partial_t,\nabla_x,F,U)=0 \tag{3.24}$$

where $A^\diamond$ is the Wick version of a given function A. We apply the Hermite transform to this equation and obtain

$$\widetilde{A}(t,x,\partial_t,\nabla_x,f(z,w),\widetilde{U}(z,w))=0, \tag{3.25}$$

where $f=\widetilde{F}$. Suppose that for each z,w we can solve the deterministic differential equation (3.25) for $f=f(t,x,z,w)$. Then under certain conditions we get that $f(t,x,z,w)$ is the Hermite transform of some $F:\mathbb{R}^d\to(\mathcal{S})_{-1,\theta}$. Moreover, F solves (3.24). Sufficient conditions for this are given in [5, Theorem 4.1.1] in the Gaussian case and a similar result can easily be proved in the mixed Gaussian-Poissonian case. We omit the details.

4 Applications to Gaussian-Poissonian SDEs

We now illustrate the theory by applying it to some stochastic (ordinary and partial) differential equations driven by a Brownian motion and (an independent) Poisson process.

Example 4.1 (Geometric Brownian-Poissonian motion) Consider the equation

$$dX(t)=a(t)X(t)dt+\phi(t)X(t)dB(t)+\psi(t)X(t)dQ(t) \tag{4.1}$$

where $a(t)$, $\phi(t)$, and $\psi(t)$ are deterministic functions, ψ is continuous, $\psi > -1$ and

$$\int_0^T (|a(t)| + \phi^2(t) + (\ln(1+\psi(t)))^2)dt < \infty .$$

In view of (2.24) we can rewrite the equation as

$$\begin{aligned} \frac{dX(t)}{dt} &= X(t) \diamond [a(t) + \phi(t)W(t) + \psi(t)V(t)] \\ &= X(t) \diamond [a(t) + (\phi(t), \psi(t)) \cdot U(t)]. \end{aligned}$$

Applying the usual trick of Wick-multiplying the equation by the integrating factor

$$\exp^\diamond \left[-\int_0^t (a(s) + \phi(s)W(s) + \psi(s)V(s))ds \right]$$

gives us the solution

$$X(t) = X(0) \diamond \exp^\diamond \left[\int_0^t a(s)ds + \int_0^t \phi(s)dB(s) + \int_0^t \psi(s)dQ(s) \right]. \tag{4.2}$$

If $X(0)$ is constant, this may be written (using (2.18) and Corollary 2.3)

$$\begin{aligned} X(t) = X(0) \exp \Bigg[&\int_0^t (a(s) - \tfrac{1}{2}\phi(s)^2 - \psi(s))ds \\ &+ \int_0^t \phi(s)dB(s) + \int_0^t \ln(1+\psi(s))dP(s) \Bigg]. \end{aligned} \tag{4.3}$$

Example 4.2 Consider the Poisson equation in a Gaussian-Poissonian stochastic medium

$$\begin{aligned} \Delta F(x) &= -\phi(s)W(x) - \psi(s)V(x) = -(\phi(x), \psi(x)) \cdot U(x) \,; \quad x \in D \\ F(x) &= 0 \,; \qquad x \in \partial D \end{aligned} \tag{4.4}$$

where $\phi(x)$, $\psi(x)$ are bounded, continuous, deterministic functions on a bounded domain $D \subset \mathbb{R}^d$ with regular boundary ∂D.

To solve this equation we proceed as in the purely Gaussian case (see [5, Section 4.2]): We first take the Hermite transform and obtain the equation

$$\begin{aligned} \Delta \widetilde{F}(x,z,w) &= -\phi(s)\widetilde{W}(x,z) - \psi(x)\widetilde{V}(x,w) \,; \quad x \in D \\ \widetilde{F}(x,z,w) &= 0 \,; \quad x \in \partial D \end{aligned} \tag{4.5}$$

with $(z,w) \in \mathbb{C}_0^{\mathbb{N}} \times \mathbb{C}_0^{\mathbb{N}}$ as parameters. For each fixed (z,w) this equation has the solution

$$\widetilde{F}(x,z,w) = \int_D G(x,y)[\phi(y)\widetilde{W}(y,z) + \psi(y)\widetilde{V}(y,w)]dy \tag{4.6}$$

where G is the classical Green function in D. We recognize (4.6) as the Hermite transform of

$$F(x) = F(x,\omega) = \int_D G(x,y)[\phi(y)W(y,\omega_1) + \psi(y)V(y,\omega_2)]dy\ ; \qquad \omega = (\omega_1,\omega_2). \tag{4.7}$$

This the solution of (4.4), in virtue of the Hermite transform solution scheme. Note that $F(x,\cdot) \in (\mathcal{S})_\theta^*$ for all x, for all dimensions d. Moreover, for $d \le 3$ we actually have $F(x,\cdot) \in L^2(\theta)$ because then $G(x,\cdot) \in L^2(dy)$ (see [5, p. 5]). And then the solution can be written in the more familiar form

$$F(x,\omega) = \int_D G(x,y)[\phi(y)dB(y,\omega_1) + \psi(y)dQ(y,\omega_2)] \qquad (d \le 3). \tag{4.8}$$

Similarly the reader can verify that the discussions of the other Wick-type stochastic partial differential equations in [5, Chapter 4] carry over to the mixed Gaussian-Poissonian case.

The Gjessing Formula and the Benth-Gjessing Method in the Mixed Gaussian-Poissonian Case

By inspecting the proof of the Gjessing formula [3] for Gaussian white noise (see, e.g., [5, Theorem 2.10.7], we see that it carries over to the mixed Gaussian-Poissonian case as follows:

Theorem 4.4 (The Gjessing formula, mixed Gaussian-Poissonian case) *Suppose* $\int_{\mathbb{R}^d}(\phi(x)^2 + |\psi(x)| + \ln(1+\psi(x))^2)dx < \infty$ *and* $X \in L^p(\theta)$ *for some* $p > 1$. *Then* $X \diamond \exp^\diamond(\int_{\mathbb{R}^d}\phi dB + \int_{\mathbb{R}^d}\psi dQ) \in L^\rho(\theta)$ *for all* $\rho < p$ *and almost surely we have*

$$\Big(X \diamond \exp^\diamond\big(\int_{\mathbb{R}^d}\phi dB + \int_{\mathbb{R}^d}\psi dQ\big)\Big)(\omega_1,\omega_2) = X(\omega_1-\phi,\omega_2-\psi)\cdot\exp^\diamond\Big(\int_{\mathbb{R}^d}\phi dB + \int_{\mathbb{R}^d}\psi dQ\Big)(\omega_1,\omega_2). \tag{4.9}$$

In the Gaussian case Benth and Gjessing [2] used the Gjessing formula to solve a class of quasi-linear stochastic differential equations. (See also the exposition in [5, Section 4.8].) This method carries over to the mixed Gaussian-Poissonian case. To illustrate this we discuss the following example:

Example 4.5 Consider the stochastic differential equation

$$dX(t) = a(t)X(t)^{\lambda}dt + \phi(t)X(t)dB(t) + \psi(t)X(t)dQ(t) \tag{4.10}$$

with $a(t) \geq 0$, $\phi(t), \psi(t)$ as in Example 4.1, where $\lambda \in \mathbb{R}$ is a constant and we assume $X(0) > 0$. The case $\lambda = 1$ is covered in Example 4.1, so we assume $\lambda \neq 1$. The noise interpretation of this equation is (see (2.24))

$$\frac{dX(t)}{dt} = a(t)X(t)^{\lambda} + X(t) \diamond [\phi(t)W(t) + \psi(t)V(t)].$$

This equation can be written

$$\frac{d}{dt}[J(t) \diamond X(t)] = a(t)J(t) \diamond X(t)^{\lambda}, \tag{4.11}$$

where

$$J(t) = \exp^{\diamond}\Big(-\int_0^t \phi(s)W(s)ds - \int_0^t \psi(s)V(s)ds\Big). \tag{4.12}$$

Define

$$Y(t,\omega) = (J(t) \diamond X(t))(\omega) = J(t,\omega) \cdot X(t, \omega_1 + \phi^{(t)}, \omega_2 + \psi^{(t)}) \tag{4.13}$$

where $\phi^{(t)}(s) = \mathcal{X}_{[0,t]}(s)\phi(s)$, $\psi^{(t)}(s) = \mathcal{X}_{[0,t]}(s)\psi(s)$. Then by (4.11), (4.9) and (4.13)

$$\begin{aligned} \frac{dY(t,\omega)}{dt} &= a(t)(J(t) \diamond X(t)^{\lambda})(\omega) \\ &= a(t)(J(t,\omega) \cdot X(t, \omega_1 + \phi^{(t)}, \omega_2 + \psi^{(t)})^{\lambda}) \\ &= a(t)J(t,\omega)[J^{-1}(t,\omega)Y(t,\omega)]^{\lambda} \\ &= a(t)J(t,\omega)^{1-\lambda}Y(t,\omega)^{\lambda}. \end{aligned} \tag{4.14}$$

For fixed ω this is a (deterministic) separable differential equation in $Y(t,\omega)$, with solution

$$Y(t,\omega) = [Y(0)^{1-\lambda} + (1-\lambda)\int_0^t a(s)J(s,\omega)^{1-\lambda}ds]^{\frac{1}{1-\lambda}}. \tag{4.15}$$

Finally, we use (4.13) to find X:

$$X(t,\omega_1,\omega_2) = J(t,\omega_1-\phi^{(t)},\omega_2-\psi^{(t)})Y(t,\omega_1-\phi^{(t)},\omega_2-\psi^{(t)}), \tag{4.16}$$

where, by (4.12), (2.18) and (2.3),

$$J(t,\omega_1,\omega_2) = \exp\Big(-\int_0^t \phi dB - \int_0^t \ln(1+\psi)dP - \int_0^t (\tfrac{1}{2}\phi^2-\psi)ds\Big). \tag{4.17}$$

From (4.15) we conclude that the solution $X(t)$ exists for all $t \geq 0$ if $\lambda < 1$ (since $a \geq 0$). However, if $\lambda > 1$ we may have explosions: $Y(t,\omega)$ only exists up to the first time t such that

$$X(0)^{1-\lambda} + (1-\lambda)\int_0^t a(s)J(s,\omega)^{1-\lambda}ds = 0.$$

For extensions of white noise theory beyond the mixed Gaussian-Poissonian case we refer to [6] and [8].

Acknowledgments
We wish to thank the Erwin Schrödinger Institute, Vienna and Michael Oberguggenberger for their kind invitation to the Workshop on Nonlinear Theory of Generalized Functions in November/December 1997 and for their hospitality during our stay there.

References

[1] Benth, F.E., and Gjerde, J.: A remark on the equivalence between Poissonian and Gaussian stochastic partial differential equations, Potential Analysis 8 (1998), 179–193.

[2] Benth, F.E., and Gjessing, H.: A non-linear parabolic equation with noise. A reduction method, Manuscript Nr. 179 (1994), Universität Mannheim.

[3] Gjessing, H.: Anticipating Calculus and Wick Products, Dr. Scient. Thesis, University of Bergen, Norway, 1995.

[4] Hida, T., Kuo, H.-H., Potthoff, J., and Streit, L.: White Noise Analysis, Kluwer 1993.

[5] Holden, H., Øksendal, B., Ubøe, J., and Zhang, T.: Stochastic Partial Differential Equations, Birkhäuser 1996.

[6] Kondratiev, Y., Streit, L., Westerkamp, W., and Yan, J.: Generalized functions in infinite dimensional analysis, Manuscript, Universität Bielefeld 1995.

[7] Oberguggenberger, M., Russo, F.: Nonlinear SPDEs: Colombeau solutions and pathwise limits, In Decreusefond et al. (eds.): Stochastic Analysis and Related Topics VI, Birkhäuser 1998, pp. 319–332.

[8] Us, G.: Towards a coloured noise analysis, Manuscript, Kiev University 1997.

[9] Walsh, J.: An introduction to stochastic partial differential equations, In R. Carmona et al. (eds.): École d'Été de Probabilités de Saint-Flour XIX–1984, Springer Lecture Notes in Mathematics, vol. 1180 (1986), 265–437.

H.Holden
Department of Mathematical Sciences, Norwegian University of Science and Technology, N–7034 Trondheim, Norway, E-mail: holden@math.ntnu.no

B.Øksendal
Department of Mathematics, University of Oslo, Box 1053 Blindern, N–0316 Oslo, Norway and Norwegian School of Economics and Business Administration, Helleveien 30, N–5035 Bergen-Sandviken, Norway, E-mail: oksendal@math.uio.no

White Noise Driven Stochastic Partial Differential Equations: Triviality and Non-Triviality

FRANCESCO RUSSO AND MICHAEL OBERGUGGENBERGER

1 Introduction and notations

In this paper a complete probability space $(\Omega, \mathcal{A}, P)$ will be fixed; d will be a positive integer. $\mathcal{S}(\mathbb{R}^d)$ will be the space of fast decreasing functions and $\mathcal{S}'(\mathbb{R}^d)$ the space of tempered distributions. $C_b^\infty(\mathbb{R}^d)$ is the space of smooth functions on $\mathbb{R}^d$ whose derivatives of each order are bounded; $\mathcal{O}_M(\mathbb{R}^d)$ will be the space of smooth functions having at most polynomial growth, together with all their derivatives. $\mathcal{F}$ will symbolize Fourier transform. $C(\mathbb{R}_+)$ will denote the space of continuous functions on $\mathbb{R}_+$; it is a Fréchet space if equipped with the topology of uniform convergence on each compact set. $\mathbb{C}$ is identified with $\mathbb{R}^2$.
The object of study is a semilinear heat equation which is driven by a space-time Gaussian white noise

$$(1.1) \qquad \begin{cases} (\partial_t - L)\, U(t,x) = \lambda\, F(U(t,x))\, \chi(x) + \dot{W}(t,x) & \text{on } \mathbb{R}_+ \times \mathbb{R}^d \\ U(0,.) = U_0 \end{cases}$$

where L is a uniformly elliptic symmetric partial differential operator with C_b^∞-coefficients, F is a globally Lipschitz real valued function, $U_0 \in \mathcal{S}'(\mathbb{R}^d), \chi$ is smooth with compact support, λ is a real constant, and $\dot{W}(t,x)$ is a Gaussian space-time white noise.
For $\lambda = 0$, the formally written equation (1.1) has a function valued solution if and only if $d = 1$, see for instance [34]. If $d > 1$, (1.1) with $\lambda = 0$ can be stated and solved but the solution is in this case a random Gaussian $\mathcal{S}'(\mathbb{R}^d)$-valued process.
If $\lambda \neq 0, d = 1$, there are no problems to give a meaning to (1.1); the solution exists as a continuous random field. If $\lambda \neq 0, d > 1$ it is possible to handle (1.1) using the theory of generalized functions of Colombeau type. For this theory we

refer for instance to [5, 6, 7, 24, 30, 31]. In this paper, our principal aim is to show that, at the macroscopic level, the solution is associated with a Gaussian random field for a large class of non-linearities F. This will be called the *triviality effect* and it translates the fact that F only acts partially. This phenomenon has already been observed in [1] when F is the Fourier transform of a complex measure and in [26, 27, 28] when F is a Lipschitz function having some limit ℓ at infinity.

In this paper we extend these results to a larger class of functions F producing the triviality effect. This happens when $F(X) = ax + b + G(x)$, where $a, b \in \mathbb{R}$ and G is Lipschitz such that its Fourier transform has no mass at zero; for the precise definition, see Section 4. In order to obtain nontriviality, one possibility is to replace in (1.1) F with a Wick reordered non-linearity, at least in the case when F is the Fourier transform of a complex measure with compact support.

The occurrence of triviality effects is not specific of parabolic equations. Similar phenomena can be observed for the stochastic wave equation ($d \geq 2$), elliptic equations on $\mathbb{R}^d (d \geq 4)$ or the Schrödinger equation ($d \geq 1$). However, (1.1) is interesting because it is motivated by quantum field theory. If $F(y) = \cos y$, equation (1.1) is an analogue of stochastic quantization of the Sine-Gordon field. The stimulations to studying (1.1) are essentially the following:

1) Most of the literature on stochastic quantization concerns the case where $d = 2$ and F is a polynomial (stochastic quantization of $P(\Phi)_2$). For this there is a huge list of references, see for instance [3, 10, 13, 15, 18, 22].

2) In the stochastic quantization studies, generally (1.1) is replaced with an equation which has formally the same invariant measure as (1.1), but the noise is regularized, generally by acting $L^{-\varepsilon}, \varepsilon > 0$.

3) In stochastic quantization, only solutions in law appear. We are interested in *strong* probabilistic solutions.

Concerning other references on generalized functions in stochastic analysis, we can quote [20, 21, 25, 32]. In particular, [32] explores the relation between Colombeau generalized functions and stochastic integrals and equations of Stratonovich type. Other approaches concerning strong solutions which make use of Wiener distributions can be found in [4, 12, 17].

We conclude this introduction by saying that in the case of $d > 1$, it is possible to get function valued solutions provided the white noise is replaced by a Gaussian homogeneous noise (with colored covariance). There is of course the case of nuclear covariance, see for instance [11]. Recently there has been a lot of interest concerning the study of limiting conditions on the noise covariance for having function valued solutions. Above all we quote [9], but also [8, 19, 23]. These papers concern either the case of the stochastic heat equation or the wave equation, which are similar with respect to this feature.

2 The free (Gaussian) case

The expression $(\dot{W}(t,x), t \geq 0, x \in \mathbb{R}^d)$ for white noise is a formal one; however, we can give a meaning to

$$W(A) = \int_A \dot{W}(s,y)\, dsdy \tag{2.1}$$

by introducing white noise as an orthogonal Hilbert measure $W : \mathcal{B}(\mathbb{R}^{d+1}) \to L^2(\Omega, \mathcal{A}, P)$ where $\mathcal{B}(\mathbb{R}^{d+1})$ is the δ-ring of the subsets of $\mathbb{R}^{d+1}$ whose Lebesgue measure is finite. We require moreover that

- $W(A)$ is a mean-zero Gaussian distributed random variable whose variance is equal to the Lebesgue measure of A.

- If A and B are disjoint, then $W(A)$ and $W(B)$ are independent.

The only stochastic integral needed in this paper is with respect to deterministic integrands. Using very elementary techniques of integration theory we can define

$$\int_{\mathbb{R}^{d+1}} g\, dW \tag{2.2}$$

for every $G \in L^2(\mathbb{R}^{d+1}, dsdy)$. This is possible using the "measure" property of white noise.

Remark 2.1 In spite of the fact that white noise is a Hilbert-valued measure, the situation is not so simple at the pathwise level where it is only a random tempered distribution. More precisely, there is a measurable version $\mathcal{W} : \Omega \to C(\mathbb{R}_+; \mathcal{S}'(\mathbb{R}^d))$ such that

$$\mathcal{W}(\omega)(t,\varphi) = \int_{[0,t]\times\mathbb{R}^d} \varphi(y)\, dW(s,y) \quad \text{almost surely}\ . \tag{2.3}$$

For more details, the reader can consult [34]. (2.3) is in fact a cylindrical Brownian motion, see also [11].

Let L be a symmetric uniformly elliptic partial differential operator acting in the variables $x \in \mathbb{R}^d$ with C_b^∞-coefficients. For $f \in \mathcal{S}'(\mathbb{R}^d)$, the problem

$$(2.4) \qquad \begin{cases} (\partial_t u - L u)(t,x) = 0 & \text{on } \mathbb{R}_+ \times \mathbb{R}^d \\ u(0,.) = f \end{cases}$$

has a unique solution $(t,x) \to P_t f(x) \in \mathcal{C}(\mathbb{R}_+; \mathcal{S}(\mathbb{R}^d))$ with the properties

- $(P_t)_{t \geq 0}$ is a semigroup on $\mathcal{S}'(\mathbb{R}^d)$;
- $P_t f \in \mathcal{S}(\mathbb{R}^d)$ if $f \in \mathcal{S}(\mathbb{R}^d)$;
- $P_t f \in C^\infty(\mathbb{R}^d)$ with polynomial growth for every $t > 0$;
- for $t > 0$, P_t has a density $p_t(x,y)$, that is to say

$$P_t f(x) = \int p_t(x,y)\, f(y)\, dy, \ t > 0\,.$$

In particular, $(t,x) \to p_t(x,y)$ solves problem (2.4) with f being the Dirac measure at point y.

Remark 2.2 There are constants $C_i, \delta_i > 0, i = 1,2$, such that (see for instance [34])

$$\frac{C_1}{t^{d/2}} \exp\left(-\frac{|x-y|^2}{\delta_1 t}\right) \leq p_t(x,y) \leq \frac{C_2}{t^{d/2}} \exp\left(-\frac{|x-y|^2}{\delta_2 t}\right).$$

For $a, b \in \mathbb{R}$, we denote by $\mathcal{L}^{a,b}$ the partial differential operator $\mathcal{L}^{a,b}\varphi = L\varphi + a\varphi + b$. The problem

$$(2.5) \qquad \begin{cases} (\partial_t v - \mathcal{L}^{a,b} v)(t,x) = 0 & \text{on } \mathbb{R}_+ \times \mathbb{R}^d \\ v(0,.) = f \end{cases}$$

has a unique solution in $\mathcal{C}(\mathbb{R}_+; \mathcal{S}'(\mathbb{R}^d))$ as well. It is given by

$$v(t,x) = e^{at} u(t,x) + \ell(t)$$

where

$$\ell(t) = \begin{cases} (e^{at} - 1)\,\frac{b}{a}, & a \neq 0 \\ bt, & a = 0 \end{cases}$$

and u is the solution of (2.4). In particular, for $f = \delta_y, b = 0$ we obtain the *fundamental solution*

$$q_t(x,y) = e^{at} p_t(x,y) \tag{2.6}$$

of the operator $\partial_t - \mathcal{L}^{a,0}$.

Remark 2.3 For $f \in \mathcal{S}'(\mathbb{R}^d)$ we set $(Q_t f)(x) = \langle f, q_t(x,.)\rangle$. Q_t defines again a semigroup on $\mathcal{S}'(\mathbb{R}^d)$ with density given by $q_t(x,y)$.

Remark 2.4

a) An easy consequence of Remark 2.2 and (2.6) is that for every $x \in \mathbb{R}^d$ and $t > 0$

$$\int_{[0,t]\times\mathbb{R}^d} q^2_{t-s}(x,y)\,dsdy < \infty \text{ iff } d = 1. \tag{2.7}$$

b) Further,

$$\int_{[0,t]\times\mathbb{R}^d} q^r_{t-s}(x,y)\,dsdy < \infty \text{ for } d = 2 \text{ and all } r < 2\,.$$

Remark 2.5 The problem

$$\begin{cases} \partial_t X - \mathcal{L}^{a,b} X = \dot{W} \\ X(0,.) = U_0 \end{cases} \tag{2.8}$$

has a meaning after integration in time:

$$X(t,\varphi) = U_0(\varphi) + \int_0^t ds\, X(s, \mathcal{L}^{a,b}\varphi) + \mathcal{W}(t,\varphi)\,. \tag{2.9}$$

It has a unique solution in $\mathcal{C}(\mathbb{R}_+; \mathcal{S}'(\mathbb{R}^d))$ since the paths of $\mathcal{W}$ are in $\mathcal{C}(\mathbb{R}_+; \mathcal{S}'(\mathbb{R}^d))$. By inspection, an $L^2(\Omega)$-valued version of it will be given by

$$X^{a,b}(t,\varphi) = U_0(Q_t\varphi) + \int_{[0,t]\times\mathbb{R}^d} dW(s,y)\, Q_{t-s}\varphi(y) + \ell(t) \int_{\mathbb{R}^d} \varphi(y)\,dy\,. \tag{2.10}$$

Remark 2.6 If $d = 1, X^{a,b}(t,\varphi) = \int_{\mathbb{R}^d} dx\, \varphi(x) X^{a,b}(t,x)$ where

$$X^{a,b}(t,x) = Q_t U_0(x) + \int_{\mathbb{R}^d} q_{t-s}(x,y)\,dW(s,y) + \ell(t)\,. \tag{2.11}$$

The stochastic integral makes sense because of Remark 2.4.

If $d > 1$, the right-hand side in (2.7) is infinite and it is a real problem to give a meaning to (2.8) perturbed by a non-linearity. For this reason we shall have to proceed via regularization.

Let us consider $\phi \in \mathcal{S}(\mathbb{R}^d)$ such that $\int \phi(x)dx = 1$. For $\varepsilon > 0$ we define

$$\mathcal{W}^\varepsilon(t,x) = \mathcal{W}(t, \phi^\varepsilon(x - .)) \tag{2.12}$$

where $\mathcal{W}$ is the cylindrical Brownian motion defined in (2.3) and $\phi^\varepsilon(x) = \varepsilon^{-d}\phi(x/\varepsilon)$. This naturally defines the random distributional process with paths in $\mathcal{C}(\mathbb{R}_+; \mathcal{S}'(\mathbb{R}^d))$ given by

$$\mathcal{W}^\varepsilon(t,\varphi) = \mathcal{W}(t, \phi^\varepsilon * \varphi).$$

At the $L^2(\Omega)$-level, $\mathcal{W}^\varepsilon$ can be identified with the (Gaussian) martingale measure (see [34])

$$W^\varepsilon(t,\varphi) = \int_{[0,t]\times\mathbf{R}^d} dW(s,y)(\phi^\varepsilon * \varphi)(y). \tag{2.13}$$

We recall that for a Borel function g defined on $\mathbb{R}_+ \times \mathbb{R}^d$,

$$\int_{[0,t]\times\mathbf{R}^d} g(s,y)\, dW^\varepsilon(s,y) = \int_{[0,t]\times\mathbf{R}^d} dW(s,z) \int_{\mathbf{R}^d} \phi^\varepsilon(y-z)\, g(s,y)$$

exists if and only if

$$\int_{[0,t]\times\mathbf{R}^d} dsdz \left(\int_{\mathbf{R}^d} \phi^\varepsilon(y-z)\, g(s,y)\right)^2 < \infty.$$

Let X_ε be the solution to

$$X_\varepsilon(t,\varphi) = U_0(\varphi) + \int_0^t ds\, X_\varepsilon(s, \mathcal{L}^{a,b}\varphi) + \mathcal{W}^\varepsilon(t,\varphi). \tag{2.14}$$

It has paths in $\mathcal{C}(\mathbb{R}_+; \mathcal{S}'(\mathbb{R}^d))$ and can be written as $X_\varepsilon(t,\varphi) = \int_{\mathbb{R}^d} dx\, \varphi(x) X_\varepsilon(t,x)$ where

$$X_\varepsilon(t,x) = Q_t U_0(x) + \int_{[0,t]\times\mathbf{R}^d} dW^\varepsilon(s,y)\, q_{t-s}(x,y) + \ell(t). \tag{2.15}$$

Then we can write

$$X_\varepsilon(t,x) = Q_t U_0(x) + \int_{[0,t]\times\mathbf{R}^d} dW(s,z) \int_{\mathbf{R}^d} dy\, \phi(y)\, q_{t-s}(x, z+\varepsilon y) + \ell(t). \tag{2.16}$$

Proposition 2.7 Let $x_1, x_2 \in \mathbb{R}^d, t > 0$. Then

$$\lim_{\varepsilon\to 0} \operatorname{Cov}\left(X^\varepsilon(t,x_1), X^\varepsilon(t,x_2)\right) = \int_0^t ds\, e^{2as} p_{2(t-s)}(x_1,x_2). \tag{2.17}$$

Proof. We have

$$\begin{aligned}&\mathrm{Cov}\,(X^\varepsilon(t,x_1),X^\varepsilon(t,x_2))\\&=\int_{[0,t]\times\mathbb{R}^d} dsdz \int dy_1dy_2\,\phi(y_1)\phi(y_2)q_{t-s}(x_1,z+\varepsilon y_1)\,q_{t-s}(x_2,z+\varepsilon y_2)\,.\end{aligned}$$

Using Fatou's lemma, Remark 2.4a) in case $x_1 = x_2$, the Lebesgue convergence theorem and Remark 2.2, the previous quantity is seen to converge to

$$\int_0^t ds \int dz\, q_{t-s}(x_1,z)q_{t-s}(x_2,z) = \int_0^t ds\, q_{2t-2s}(x_1,x_2)$$

because of the symmetry and the semigroup property of $(Q_t)_{t\geq 0}$; see also [2].
Remark 2.8 So (2.17) is finite if and only if $x_1 \neq x_2$, see Remark 2.4.

3 The nonlinear case; massless distributions

As we pointed out in the Introduction, we are interested in equation (1.1). After integrating formally in time (1.1) becomes

$$(3.1)\qquad U(t,x) = U_0(x) + \int_0^t LU(s,x)\,ds + \lambda\,\chi(x)\int_0^t F(U(s,x))\,ds + \mathcal{W}(t,x)\,.$$

If $\lambda = 0$ the equation is called *free* and the solution $X^{0,0}$ has been discussed in the previous section. If $\lambda \neq 0$ we are forced to deal with non-linear functions of distributions. It is possible to state an existence and uniqueness theorem for (3.1) in the framework of random tempered generalized functions as described in [2]. In that paper we used a class of Colombeau type tempered generalized functions in the spatial variable x only. A closer look suggests that the regularization has to take place only in this variable. For defining such a class of generalized functions we work with a formulation proposed in [5, 24] where the mollifier is fixed.
As in Section 2, we take a mollifier $\phi \in \mathcal{S}(\mathbb{R}^d)$ such that $\int_{\mathbb{R}^d}\phi(x)dx = 1$. Again, we set $\phi^\varepsilon(x) = \varepsilon^{-d}\phi(x/\varepsilon)$. For the purpose of generalized function theory, ϕ will be moreover supposed to satisfy $\int_{\mathbb{R}^d} x^m\phi(x)\,dx = 0$ for any multi-index m such that $|m| \geq 1$.
We now briefly describe the algebra of generalized functions we use. The starting point is the differential algebra $\mathcal{E}(\mathbb{R}^d;\mathcal{C}(\mathbb{R}_+))$ of complex valued functions $(t,x,\varepsilon)\to R(t,x,\varepsilon), t\in\mathbb{R}_+, x\in\mathbb{R}^d, \varepsilon>0$ such that the maps $x\to R(.,x,\varepsilon):\mathbb{R}^d\to\mathcal{C}(\mathbb{R}_+)$ are $\mathcal{C}^\infty$ for any $\varepsilon>0$. An element $R\in\mathcal{E}(\mathbb{R}^d;\mathcal{C}(\mathbb{R}_+))$ is said to have a *tempered moderate*

bound if it enjoys the following property: For all $T \in \mathbb{R}_+$ there exists $n \in \mathbb{N}$ such that

$$\sup_{t \le T, x \in \mathbb{R}^d} \frac{|R(t,x,\varepsilon)|}{1+|x|^n} = \mathcal{O}(\varepsilon^{-n}) \text{ as } \varepsilon \to 0\,. \tag{3.2}$$

In particular, the left-hand quantity in the previous expression is finite for small $\varepsilon > 0$; $\mathcal{O}$ is the usual Landau symbol. The family of all $R \in \mathcal{E}(\mathbb{R}^d; \mathcal{C}(\mathbb{R}_+))$ satisfying (3.2) constitutes a subalgebra.

Next, we consider the ideal of functions R such that for each $T \in \mathbb{R}_+, q \in \mathbb{N}$ there is $n \in \mathbb{N}$ with

$$\sup_{t \le T, x \in \mathbb{R}^d} \frac{|R(t,x,\varepsilon)|}{1+|x|^n} = \mathcal{O}(\varepsilon^{q}) \text{ as } \varepsilon \to 0\,. \tag{3.3}$$

$R \in \mathcal{E}(\mathbb{R}^d; \mathcal{C}(\mathbb{R}_+))$ is said to be *moderate* if for any partial differential operator D acting in the variable $X \in \mathbb{R}^d, DR$ has a tempered moderate bound (3.2). The family of moderate R constitutes a differential subalgebra of $\mathcal{E}(\mathbb{R}^d; \mathcal{C}(\mathbb{R}_+))$ and will be denoted by $\mathcal{E}_M(\mathbb{R}^d; \mathcal{C}(\mathbb{R}_+))$.

$R \in \mathcal{E}_M(\mathbb{R}^d; \mathcal{C}(\mathbb{R}_+))$ is said to be *null* if for any partial differential operator D acting on x, DR has a null tempered bound (3.3). The family of null elements R forms a differential ideal in $\mathcal{E}_M(\mathbb{R}^d; \mathcal{C}(\mathbb{R}_+))$, denoted by $\mathcal{N}(\mathbb{R}^d; \mathcal{C}(\mathbb{R}_+))$.

Finally, $\mathcal{G}(\mathbb{R}^d; \mathcal{C}(\mathbb{R}_+))$ is defined as the quotient $\mathcal{E}_M(\mathbb{R}^d; \mathcal{C}(\mathbb{R}_+)) \,/\, \mathcal{E}(\mathbb{R}^d; \mathcal{C}(\mathbb{R}_+))$ and this will be our algebra of generalized functions (or more precisely, $\mathcal{C}(\mathbb{R}_+)$-valued generalized functions).

A *random generalized function*, notation $G \in \mathcal{G}_\Omega(\mathbb{R}^d; \mathcal{C}(\mathbb{R}_+))$ will be an application $\Omega \to \mathcal{G}(\mathbb{R}^d; \mathcal{C}(\mathbb{R}_+))$ such that there is a map $R_G(\cdot,\cdot,\varepsilon) : \Omega \times \mathbb{R}^d \to \mathcal{C}(\mathbb{R}_+)$, measurable for each $\varepsilon > 0$, and such that $(R_G(\omega,\cdot,\varepsilon)$ is a representative of $G(\omega)$, P-almost surely.

The initial condition $U_0 \in \mathcal{S}'(\mathbb{R}^d)$ of (1.1) can be identified with the time independent $\mathcal{C}(\mathbb{R}_+)$-valued generalized function represented by

$$R_{U_0}(t,x,\varepsilon) = U_0 * \phi^\varepsilon(x - .)\,.$$

The cylindrical Brownian motion $\mathcal{W}$ defined in Section 2 can be identified with the random generalized function represented by

$$R_{\mathcal{W}}(t,x,\varepsilon) = \mathcal{W}^\varepsilon(t,x)\,.$$

We say that G is *associated* with a random $\mathcal{C}(\mathbb{R}_+)$-valued distribution $S : \Omega \times \mathcal{S}(\mathbb{R}^d) \to \mathcal{C}(\mathbb{R}_+)$ if for any $\varphi \in \mathcal{S}(\mathbb{R}^d), t \geq 0$

$$\lim_{\varepsilon \to 0} \int_{\mathbb{R}^d} R_G(\omega; t, x, \varepsilon)\, \varphi(x)\, dx = S(\omega, \varphi)(t)$$

in probability. Of course the association concept is weaker than the equality in the quotient $\mathcal{G}_\Omega(\mathbb{R}^d; \mathcal{C}(\mathbb{R}_+))$: the association is equality at a "macroscopic" level; the other one corresponds to identity at the "microscopic" level.

We come back to problem (3.1). Since it has no direct meaning, we concentrate on the regularized version of it:

$$(3.4) \qquad U^\varepsilon(t, \varphi) = U_0(\varphi) + \int_0^t ds\, U^\varepsilon(s, L\varphi) + \lambda \int_0^t ds\, F(U^\varepsilon)(s, \chi\varphi) + \mathcal{W}(t, \varphi)$$

where $U^\varepsilon(t, \varphi)$ denotes $\int_{\mathbb{R}^d} dx\, \varphi(x) U^\varepsilon(t, x)$ and $(U^\varepsilon(t, x))$ is a random field with paths in $\mathcal{C}(\mathbb{R}_+, \mathcal{O}_M(\mathbb{R}^d))$.

Remark 3.1 Using classical PDE-techniques, it is possible to show that (3.4) has a unique solution which moreover fulfills (as an $L^2(\Omega)$-valued function) the mild type equation

$$(3.5) \qquad U^\varepsilon(t, x) = P_t U_0(x) + \lambda \int_{[0,t] \times \mathbb{R}^d} ds dy\, p_{t-s}(x, y) F(U^\varepsilon(s, y)) \chi(y) + \int_{[0,t] \times \mathbb{R}^d} dW^\varepsilon(s, y) p_{t-s}(x, y)\,.$$

At the generalized function level, problem (3.1) makes sense and we can state the following theorem.

Theorem 3.2 Assuming that F is C_b^∞ and $U_0 \in \mathcal{G}(\mathbb{R}^d)$, there is a unique random generalized function solving equation (3.1) in $\mathcal{G}(\mathbb{R}^d; \mathcal{C}(\mathbb{R}_+))$. One representative is given by $R_U(t, x, \varepsilon) = U^\varepsilon(t, x)$ where U^ε is the solution to (3.5).

Proof. The techniques are classical in generalized function theory.

The questions we will try to answer in the sequel are the following:

a) Can one describe a reasonable large class of functions F for which $\lim_{\varepsilon \to 0} U^\varepsilon(t, \varphi)$ exists in probability?

b) If yes, what is the limit?

In fact we will exhibit an interesting class for which the limit exists and it is a Gaussian random field (the triviality effect).

Definition 3.3 A distribution $V \in \mathcal{S}'(\mathbb{R})$ is said to be *massless at zero* if

(3.6) $$\lim_{\varepsilon \to 0} \langle V, \psi(\cdot/\varepsilon)\rangle = 0\,,$$

for the function $\psi(y) = \exp(-y^2/2)$.

Remark 3.4 For our applications, we shall be mainly concerned with functions whose Fourier transform is massless at zero. For $G \in \mathcal{S}'(\mathbb{R})$ the Fourier transform $V = \mathcal{F}G$ is massless at zero, i.e., satisfies (3.6), if and only if

(3.7) $$\lim_{\varepsilon \to 0} \varepsilon \langle G, \psi(\varepsilon\, \cdot)\rangle = 0\,,$$

noting that ψ is identical with its Fourier transform up to a multiplicative factor. In particular, when G is a bounded function, condition (3.7) expresses the fact that the mean of the limiting occupation measure is zero.

Example 3.5 Let G be a continuous function such that the limits $\lim_{x \to -\infty} = a$ and $\lim_{x \to +\infty} = b$ exist. It follows from the symmetry of ψ that the limit as $\varepsilon \to 0$ in (3.7) equals $(a+b)/2$. Thus $\mathcal{F}G$ is massless at zero iff $a = -b$. In particular, this holds when G vanishes at infinity.

Example 3.6 Let G be a periodic, sufficiently regular function with period π_0. Expanding G in its Fourier series, we see that the limit in (3.7) equals $\int_0^{\pi_0} G(y)dy$ so that $\mathcal{F}G$ is massless at zero iff G has mean zero along its period.

Example 3.7 If $G \in L^p(\mathbb{R})$ for some $p \in [1,2]$ or if $x^{-q}G(x) \in L^1(\mathbb{R})$ then $\mathcal{F}G$ is massless at zero (direct computation).

Example 3.8 Every antisymmetric distribution $V \in \mathcal{S}'(\mathbb{R})$ is massless at zero. If $V = \mu$, an integrable measure with $\mu(\{0\}) = 0$, then V is massless at zero. If $V \in \mathcal{S}'(\mathbb{R})$ such that its support is contained in $\mathbb{R} \setminus \{0\}$, then V is massless at zero. For example, taking $a > 0, \chi$ smooth with compact support in $[a/2, 2a]$ and letting V be the principal value distribution $V(\xi) = \mathrm{vp}\,\frac{1}{\xi - a} \cdot \chi(\xi)$, we have that V is massless at zero. Its inverse Fourier transform G is a multiple of $(e^{-ax}\,\mathrm{sign}\,\mathrm{x}) * \check{\chi}$ and provides an example of a smooth bounded function which is not periodic, not antisymmetric, has no limits at infinity, but whose Fourier transform is massless at zero.

4 The triviality phenomenon

The main theorem of this section is the following; $T > 0$ will be fixed.

Theorem 4.1 Assume that F is a Lipschitz function of the form $F(x) = ax+b+G(x)$ where G is bounded and $\mathcal{F}G$ is massless at zero, $a, b \in \mathbb{R}$. Let U^ε be the solution to (3.4). Then, for every $p \geq 1, t \geq 0, \varphi \in \mathcal{S}(\mathbb{R}^d)$,

$$U^\varepsilon(t,\varphi) \to X^{a,b}(t,\varphi)$$

in $L^p(\Omega)$ as $\varepsilon \to 0$, where $X^{a,b}$ is the solution of (2.9).

Remark 4.2 We recall that $X = X^{a,b}$ solves the problem

$$\begin{cases} \partial_t X = LX + aX + b + \dot{W} \\ X(0) = U_0 \,. \end{cases}$$

Proof (of Theorem 4.1). For $\varepsilon > 0$, let us consider the random field $(X_\varepsilon(t,x))_{\varepsilon>0}$, the solution to (2.14). For simplicity of formulation, the proof will be written for $p = 1$. We will prove that

a) $X_\varepsilon(t,\varphi) \to X(t,\varphi)$ in $L^1(\Omega)$;

b) $\int_{\mathbb{R}^d} dx\, \varphi(x)\, (X_\varepsilon(t,x) - U^\varepsilon(t,x)) \to 0$ in $L^1(\Omega)$.

First, a) is immediate because of the explicit expression (2.16) and the property of stochastic integrals. Therefore we will concentrate on assertion b). For this we write

$$U^\varepsilon(t,\varphi) - X_\varepsilon(t,\varphi) = \int_0^t ds\, (U^\varepsilon - X_\varepsilon)(s, \mathcal{L}^{a,b}\varphi) + \int_0^t ds\, G(U^\varepsilon)(s, \varphi\chi)\,.$$

Consequently, using the mild form, we get

$$\begin{aligned} U^\varepsilon(t,x) - X_\varepsilon(t,x) = \int_0^t ds \int dy\, q_{t-s}(x,y)\, \chi(y)\, \left(G(U^\varepsilon(s,y)) - G(X_\varepsilon(s,y))\right) \\ + \int_0^t ds \int dy\, q_{t-s}(x,y)\, \chi(y)\, G(X_\varepsilon(s,y))\,. \end{aligned}$$

Let $K \in \mathbb{R}^d$ be a compact set containing $\operatorname{supp}\chi$. Taking the expectation of the absolute value and integrating over K, for

$$\Psi(t) = \int_K dx\, E\, |U^\varepsilon(t,x) - X_\varepsilon(t,x)|\, \chi(x)$$

we get

$$\begin{aligned} \Psi(t) \leq \int_0^t ds \int dy\, \chi(y)\, E\, |G(U^\varepsilon(s,y)) - G(X_\varepsilon(s,y))| \int_K dx\, q_{t-s}(x,y) \\ + \int ds \int_K dx\, E \left| \int dy\, q_{t-s}(x,y)\, \chi(y)\, G(X_\varepsilon(s,y)) \right|. \end{aligned}$$

Let k be a Lipschitz constant for G times $\sup|\chi|$. Using $\int_K dx q_{t-s}(x,y) \leq 1$, the previous expression implies

$$\Psi(t) \leq k \int_0^t ds\, \Psi(s) + \int_0^t ds \int_K dx\, E\left|\int dy\, q_{t-s}(x,y)\, \chi(y)\, G(X_\varepsilon(s,y))\right|$$

for every $t \leq T$. Using the Gronwall lemma, there is a constant C such that

$$\sup_{t\leq T} |\Psi(t)| \leq C \int_0^T ds \int_K dx\, E\left|\int dy\, q_{t-s}(x,y)\, \chi(y)\, G(X_\varepsilon(s,y))\right| \tag{4.1}$$

Since G is bounded and $\int_K dy q_{t-s}(x,y) \leq 1$, the Lebesgue convergence theorem allows us to arrive at the desired conclusion provided we show that for $\rho \in \mathcal{S}(\mathbb{R}^r)$

$$E\left|\int dy\, \rho(y)\, G(X_\varepsilon(s,y))\right| \to 0 \tag{4.2}$$

as $\varepsilon \to 0$, for all $s \in]0,T]$. In fact we can easily show the L^2-convergence to zero of

$$E\left|\int dy\, \rho(y)\, G(X_\varepsilon(s,y))\right|^2 = \int dy_1 \int dy_2\, \rho(y_1)\rho(y_2)\, E\left(G(X_\varepsilon(s,y_1))G(X_\varepsilon(s,y_2))\right).$$

To see this, we let $V_1(\varepsilon) = X_\varepsilon(s,y_1)$, $V_2(\varepsilon) = X_\varepsilon(s,y_2)$ for $y_1 \neq y_2 \in \mathbb{R}^d, t > 0$ and apply Proposition 4.3 below together with Proposition 2.7 and Remark 2.8. This will conclude the proof of the theorem.

Proposition 4.3 Let $(V_1(\varepsilon), V_2(\varepsilon))$ be a Gaussian vector such that $\sigma_2^2(\varepsilon) = \operatorname{Var} V_2(\varepsilon) \to \infty$ as $\varepsilon \to 0$. Let $G : \mathbb{R} \to \mathbb{R}$ be a bounded function such that $\mathcal{F}G$ is massless at zero. Then

$$E\left(G(V_1((\varepsilon))\, G(V_2(\varepsilon))\right) \to 0 \text{ as } \varepsilon \to 0.$$

Proof. The covariance matrix $\Sigma(\varepsilon)$ and its inverse are given by

$$\Sigma(\varepsilon) = \begin{pmatrix} \sigma_1^2 & \sigma_{12} \\ \sigma_{12} & \sigma_2^2 \end{pmatrix}, \quad \Sigma(\varepsilon)^{-1} = \begin{pmatrix} a_1^2 & a_{12} \\ a_{12} & a_2^2 \end{pmatrix},$$

Observing the relation

$$\det \Sigma^{-1} = a_1^2 a_2^2 - a_{12}^2 = \frac{a_1^2}{\sigma_2^2}$$

an easy algebraic computation gives that

$$\begin{aligned} &E\left(G(V_1)G(V_2)\right) \\ &= \int dx_1\, G(x_1) \int dx_2\, G(x_2)\, \det \Sigma^{-1/2} \exp\left(-\tfrac{1}{2}[a_1^2 x_1^2 + 2a_{12}x_1x_2 + a_2^2 x_2^2]\right) \\ &= \int dx_2\, \tfrac{1}{\sigma_2} G(x_2) \exp\left(\tfrac{-x_2^2}{2\sigma_2^2}\right) a_1 \int dx_1\, G(x_1) \exp\left(-\tfrac{1}{2}[a_1 x_1 + \tfrac{a_{12}}{a_1} x_2]^2\right). \end{aligned}$$

The second integral is bounded by the L^∞-norm of G, while the first converges to zero by assumption and (3.7).

5 Heat equation with Wick reordering

A way to overcome the triviality phenomenon is to introduce *Wick reordering* for F in equation (3.1). This has been the object of [2].
Wick reordering is a concept introduced by mathematical physicists for Gaussian random fields, see for instance [14, 16, 33], but the list of references is almost uncountable. This concept of non-linearity has been used to study the stochastic quantization equation. Such an equation has always been studied at the weak probabilistic level (solutions in law, martingale problems, ...); thereby, Wick renormalization with respect to free Gaussian fields suffices. The Wick non-linearity has also been introduced for studying stochastic partial differential equations in the strong probabilistic sense (when the probability space is given): in this case it is necessary to use a Wick reordering with respect to (not necessarily Gaussian) Wiener distributions, see [17].
The equation we study is also understood in the strong sense, but from a pathwise viewpoint. In [2], we introduced the Wick reordering of a random generalized function in the context of the stochastic heat equation. The idea is to counterbalance the triviality phenomena through the multiplication of $F(U)$ by an "infinite" correction. We discuss here the case $d = 2$ and without entering too much into details the object of study is

(5.1)
$$U(t,x) = U_0(x) + \int_0^t ds\, LU(s,x) + \lambda \int_0^t ds\ :F:(AU)(s,x)\,\chi(x) + \mathcal{W}(t,x),$$

where $t \geq 0, x \in \mathbb{R}^2, \lambda \in \mathbb{R}, F(y) = \int e^{iay} d\mu(a), \mu$ being a complex measure with compact support such that $\mu(\{0\}) = 0$ (see Example 3.8); $a \in \mathbb{R}$ will be a small parameter.
Using techniques of Wiener distributions, one equation of type (5.1) has been studied in [4]; there the techniques of [17] are implemented using special Wiener distributions of Kondratiev type. We conjecture that our approach and theirs are compatible.
In [2], we considered the case $U_0 = 0$, but the analysis carries over to the case $U_0 \in \mathcal{S}'(\mathbb{R}^2)$. The object $:F:(Au)(t,x)$ was essentially a random generalized

function represented by

$$R_{:F:(Au)}(t,x,\varepsilon) =: F : (Au)_\varepsilon(t,x,\varepsilon) \ \text{(by definition)} \tag{5.2}$$
$$= \int d\mu(a) \exp\left(iAau(t,x) + A^2a^2\, E(X_\varepsilon(t,x))^2\right)$$

where X_ε is the random field defined in Section 2. Here we have to deal with the problem that, because of the exponentials in (5.2), $R_{:F:(Au)}$ belongs almost surely to $\mathcal{E}(\mathbb{R}^2; \mathcal{C}(\mathbb{R}_+))$, but it is generally not moderate. We recall that X_ε solves

$$X_\varepsilon(t,\varphi) = U_0(\varphi) + \int ds\, X_\varepsilon(s, L\varphi) + \mathcal{W}^\varepsilon(t,\varphi)\,. \tag{5.3}$$

In order to give a meaning to (5.1) in the (Colombeau-) $\mathcal{C}(\mathbb{R}_+)$-valued generalized function sense, (5.2) must be changed into a moderate expression. One way of doing this is to replace the random generalized function defined by $X(t,\varphi)$ with the "indistinguishable" random generalized function $X^c(t,\varphi)$ where

$$R_{X^c}(t,x,\varepsilon) = R_X(t,x,h_c(\varepsilon)) \text{ with } h_c(\varepsilon) = (-1/\log\varepsilon)^{1/c}$$

for some suitably chosen $c > 0$. The object $: F : (Au)$ will then be replaced with $: F :_c (Au)$ where

$$R_{:F:_c(Au)}(t,x,\varepsilon) = \int d\mu(a) \exp\left(iAau(t,x) + A^2a^2\, E(X^c_\varepsilon(t,x))^2\right)\,. \tag{5.4}$$

With the assumptions above, we can state the following theorem:

Theorem 5.1 There is a unique random generalized function $U \in \mathcal{G}_\Omega(\mathbb{R}^2; \mathcal{C}(\mathbb{R}_+))$ which solves the equation

$$U(t,x) = U_0(x) + \int_0^t ds\, LU(s,x) + \lambda \int_0^t ds\ : F :_c (AU)(s,x) + \mathcal{W}_c(t,x) \tag{5.5}$$

where $R_{\mathcal{W}_c}(t,x,\varepsilon) = R_{\mathcal{W}}(t,x,h_c(\varepsilon))$.

Proof. See [2].

In the context of this section it is legitimate to ask whether U is in some sense associated with a random $\mathcal{C}(\mathbb{R}_+)$-valued distribution with paths in $\mathcal{C}(\mathbb{R}_+; \mathcal{S}'(\mathbb{R}^2))$. For this it is useful to study the behavior of the solution $U^\varepsilon(t,x)$ to the following equation (5.6) when $\varepsilon \to 0$:

$$U^\varepsilon(t,\varphi) = U_0(\varphi) + \int_0^t ds\, U^\varepsilon(s, L\varphi) \tag{5.6}$$
$$+\lambda \int_0^t ds \int_{\mathbb{R}^2} dy\ : F : (AU^\varepsilon)_\varepsilon(s,y)\, \chi(y)\varphi(y) + \mathcal{W}^\varepsilon(t,\varphi)\,.$$

As an L^2-valued function, U^ε solves the following mild equation;

$$(5.7)\quad U^\varepsilon(t,x) = P_t U_0(x) + \lambda \int_{[0,t]\times\mathbb{R}^2} dsdy\, p_{t-s}(x,y)\ :F:(AU^\varepsilon)_\varepsilon(s,y)\,\chi(y)$$
$$+ \int_{[0,t]\times\mathbb{R}^2]} dW^\varepsilon(s,y)\, p_{t-s}(x,y)\,.$$

Remark 5.2 a) Equation (5.6) has a unique solution with paths in $\mathcal{C}(\mathbb{R}_+;\mathcal{S}'(\mathbb{R}^2))$.
b) The map $\lambda \to U^\varepsilon(\lambda,t,x)$, the solution to (5.7), is analytic in λ. It admits an expansion of the type

$$U^\varepsilon(t,x) = X_\varepsilon(t,x) + \sum_{k=1}^{\infty} \frac{\lambda^k}{k!}\frac{\partial^k U^\varepsilon}{\partial\lambda^k}(0,t,x)\,.$$

Consequently, the expansion of order N is $X_\varepsilon(t,x) + Z_N^\varepsilon(\lambda,t,x)$ where

$$Z_N^\varepsilon(\lambda,t,x) = \sum_{k=1}^{N} \frac{\lambda^k}{k!}\frac{\partial^k U^\varepsilon}{\partial\lambda^k}(0,t,x)\,.$$

c) As $\varepsilon \to 0, X_\varepsilon(t,\varphi) \to X(t,\varphi)$ in $L^2(\Omega)$, for every $\varphi \in \mathcal{S}(\mathbb{R}^2)$, where $X = X^{0,0}$ is the solution to the free equation, see Section 2.

Theorem 5.3 If $A > 0$ is small enough, then $Z_N^\varepsilon(t,x) \to Z_N(t,x)$ in $L^2(\Omega)$, for all $t \geq 0, x \in \mathbb{R}^2$, where $Z_N(t,x)$ is a field which is a linear combination of iterated integrals of Wick non-linearities of the free solution of the type

$$(5.8)\qquad \int_{[0,t]\times\mathbb{R}^2} ds_n dy_n\, p_{t-s_n}(x,y_n)\ :F_n:(AX)(s_n,y_n)\,\chi(y_n)$$

$$\int_{[0,s_{j(n-1)}]\times\mathbb{R}^2} ds_{n-1}dy_{n-1}p_{s_{j(n-1)}-s_{n-1}}(y_{j(n-1)},y_{n-1})\ :F_{n-1}:(AX)(s_{n-1},y_{n-1})\,\chi(y_{n-1})$$

$$\vdots$$

$$\int_{[0,s_{j(1)}]\times\mathbb{R}^2} ds_1 dy_1\, p_{s_{j(1)}-s_1}(y_{j(1)},y_1)\ :F_1:(AX)(s_1,y_1)\,\chi(y_1)$$

where $t = s_{j(n)}, x = y_{j(n)}$ and $j(\ell) \in \{\ell+1,\ldots,n\}, \ell \in \{1,\ldots,n\}$ for some positive integer n and $F_\ell(y) = \int d\mu_\ell(a)\exp(iay)$, μ_ℓ being a complex measure with compact support such that $\mu_\ell(\{0\}) = 0$.

Remark 5.4 a) Iterated integrals of the form (5.9) can be defined as $L^2(\Omega)$-limits of the same quantities where $:F_j:(X)$ is replaced with $:F_j:(X_\varepsilon)_\varepsilon$.

b) The proof is technically complicated and it makes use of the logarithmic behavior of the covariance of the free solution, see Proposition 2.4 of [2].

Acknowledgments: The first author (FR) is grateful to M. Oberguggenberger and to the Erwin Schrödinger Institut in Wien for the opportunity of spending there one month in a very stimulating atmosphere. FR is also grateful to Professors S. Albeverio, Ph. Blanchard, Yu. Kondratiev, M. Röckner and L. Streit for the kind invitation to the BiBoS Research Center in Bielefeld.

References

[1] S. ALBEVERIO, Z. HABA, F. RUSSO: Trivial solution for a non-linear two-space dimensional wave equation perturbed by space-time white noise. Stochastics and Stochastic Reports **56**, 127-160 (1996).

[2] S. ALBEVERIO, Z. HABA, F. RUSSO: A two-space dimensional heat equation perturbed by a (Gaussian) white noise. BiBoS Preprint, Univ. Bielefeld 1997.

[3] S. ALBEVERIO, M. RÖCKNER: Stochastic differential equation in infinite dimensions: solutions via Dirichlet forms. J. Funct. Anal. **88**, 395-436 (1990).

[4] F. E. BENTH, TH. DECK, J. POTTHOFF: A white noise approach to a class of non-linear stochastic heat equations. J. Funct. Anal. **146**, 382-415 (1997).

[5] H.A. BIAGIONI: A Non-linear Theory of Generalized Functions. Lect. Notes Math. Vol. 1421, Springer-Verlag, Berlin (1990).

[6] J.F. COLOMBEAU: Elementary Introduction to New Generalized Functions. North-Holland Math. Studies Vol. 113, North-Holland, Amsterdam (1985).

[7] J.F COLOMBEAU: Multiplication of Distributions. Lect. Notes Math. Vol. 1532, Springer-Verlag, Berlin (1992).

[8] R. C. DALANG: Extending the martingale measure stochastic integral with applications to spatially homogeneous SPDE's. Preprint 1998.

[9] R. C. DALANG, N. FRANGOS: The stochastic wave equation in two spatial dimensions. Annals of Probab. **26-1**, 187-212 (1998).

[10] G. DA PRATO, L. TUBARO: Introduction to stochastic quantization. Preprint, Scuola Normale Superiore di Pisa 1996.

[11] G. DA PRATO, J. ZABCZYK: Stochastic Equations in Infinite Dimensions. Encyclopedia Math. Appl. Vol. 44, Cambridge University Press, Cambridge 1992.

[12] A. DERMOUNE: Around the stochastic Burgers equation. Preprint 55-2, Universités du Maine et D'Angers (1996).

[13] D. GATAREK, B. GOLDYS: On existence and uniqueness for the stochastic quantization equations in finite volume. Preprint (1995).

[14] J. GLIMM, A. JAFFE: Quantum Physics: A Functional Integral Point of View. Springer-Verlag, Berlin (1981).

[15] Y. HU, G. KALLIANPUR: Exponential integrability and application to stochastic quantization. Preprint 1997.

[16] T. HIDA, H.H. KUO, J. POTTHOFF, L. STREIT: White Noise – An Infinite Dimensional Calculus. Kluwer, Dordrecht (1993).

[17] H. HOLDEN, B. ØKSENDAL, J. UBOE, T.S. ZHANG: Stochastic Partial Differential Equations. Birkhäuser, Boston (1996).

[18] P. JONA-LASINIO, P.K. MITTER: On the stochastic quantization of field theory. Comm. Math. Phys. **101**, 409-436 (1985).

[19] A. KARCZEVSKA, J. ZABCZYK: A note on the stochastic wave equation. Preprint 574, Polish Academy of Sciences, Warsaw (1997).

[20] C. MARTIAS: Extension du calcul d'Itô aux distributions de Schwartz. C.R. Acad. Sci. Paris, Sér. I **321**, 331-334 (1995).

[21] C. MARTIAS: Stochastic integration on generalized function spaces and its applications. Stochastics and Stochastic Reports **57**, 289-301 (1996).

[22] R. MIKULEVICIUS, B. L. ROZOVSKI: Martingale problems for stochastic PDE's. Preprint 1998.

[23] A. Millet, M. Sanz: A stochastic wave equation in two dimensions: Smoothness of the law, Preprint 410, Laboratoire de Probabiliteés de Paris 6 (1997).

[24] M. Oberguggenberger: Multiplication of Distributions and Application to Partial Differential Equations, Pitman Research Notes Math. Vol. 259, Longman, Harlow (1992).

[25] M. Oberguggenberger: Generalized functions and stochastic processes. In: E. Bolthausen, M. Dozzi, F. Russo (Eds.), Seminar on Stochastic Analysis, Random Fields and Applications, Birkhäuser, Basel (1995), 215-229.

[26] M. Oberguggenberger, F. Russo: Singular limits in nonlinear stochastic wave equations. Proceedings of the Workshop on Microlocal Analysis, Pisa (1996), to appear.

[27] M. Oberguggenberger, F. Russo: Nonlinear SPDEs: Colombeau solutions and pathwise limits. In: L. Decreusefonds, J. Gjerde, B. Øksendal, A.S. Üstünel (Eds.), Stochastic Analysis and Related Topics VI, Birkäuser, Boston (1998), 319-332.

[28] M. Oberguggenberger, F. Russo: Nonlinear stochastic wave equations. Proc. Conf. Generalized Functions Novi Sad 1996. Integral Transforms and Special Functions **6**, 71-83 (1998).

[29] S. Peszat, J. Zabczyk: Nonlinear stochastic wave and heat equations, Preprint 584, Polish Academy of Sciences, Warsaw (1998).

[30] S. Pilipović: Generalized functions and pseudo-differential operators. Lect. Math. Sciences Vol. 4, Univ. Tokyo (1994).

[31] E.E. Rosinger: Generalized solutions of nonlinear partial differential equations, North-Holland Math. Studies Vol. 146, North-Holland, Amsterdam (1987).

[32] F. Russo: Colombeau generalized functions and stochastic analysis. In: A.I. Cardoso, M. De Faria, J. Potthoff, R. Sénéor, L. Streit, Stochastic Analysis and Applications in Physics, NATO ASI Conference Series C: Mathematical and physical sciences Vol. 449, Kluwer, Dordrecht (1994), 329-350.

[33] B. SIMON: The $P(\phi)_2$ Euclidean (Quantum) Field Theory. Princeton Univ. Press, Princeton (1974).

[34] J. B. WALSH: An Introduction to stochastic partial differential equations. In: R. CARMONA, H. KESTEN, J.B. WALSH (Eds.), Ecole d'Eté de Probabilité de Saint-Flour XIV-1984, Lect. Notes Math. Vol. 1180, Springer-Verlag, New York (1986), 265-439.

M. OBERGUGGENBERGER
Universität Innsbruck, Institut für Mathematik und Geometrie, A-6020 Innsbruck, Austria, E-mail: michael@mat1.uibk.ac.at

F. RUSSO
Université Paris 13, Institut Galilée, Département de Mathématiques, F-93430 Villetaneuse, France, E-mail: russo@math.univ-paris13.fr

Measurement Methods Related to Differential Equations

JAN UBØE

1 Introduction

In this paper we give a survey on some results with relevance to singular ordinary and partial stochastic differential equations. We approach such equations smoothing both the coefficients and the candidates for a solution. Passing to the limit, we formulate what we call a measurement method for the equation. The measurement method was introduced in [6] and extended in various directions in [12]. The basic mechanism is to smoothen distributions by an approximate identity and to study limits of such expressions. Of course this is nothing but a variation over a classical theme in functional analysis and as such not very original. The approach originates from ideas related to the theory of Colombeau distributions, see [1] and [7]. In this theory, generalized functions F are identified by mappings $(x, \phi) \mapsto F_\phi(x)$, where $x \mapsto F_\phi(x)$ are smooth functions. Functional objects of the above kind have found many applications. In particular they have been used extensively to study properties of stochastic partial differential equations, see, e.g., [2], [4], [5], and [8].

2 The measurement method

Let D be a domain in $\mathbb{R}^d$, let $\mathcal{D}$ denote the space of test functions on D, i.e., the set of all $\phi \in C_c^\infty(D)$ with the usual test function topology, and let $\mathcal{D}'$ denote the dual space of distributions. For each multi-index $\alpha = (\alpha_1, \alpha_2, \ldots, \alpha_d)$, we let D^α denote the differential operator

$$D^\alpha = \left(\frac{\partial}{\partial x_1}\right)^{\alpha_1} \cdots \left(\frac{\partial}{\partial x_d}\right)^{\alpha_d} \tag{1}$$

We consider a formal differential operator L on the form

$$Lu = \sum_{|\alpha| \le N} f_\alpha \cdot D^\alpha u \tag{2}$$

We will assume that $u \in \mathcal{D}'$ and also that all $f_\alpha \in \mathcal{D}'$. Note that the products $f_\alpha \cdot D_\alpha u$ do not necessarily make sense as distributions, so we will need to make a more refined definition. The interpretation goes like this; choose and fix $\delta_0 \in C_c^\infty(\mathbb{R}^d)$ with the properties $\delta_0 \ge 0$ and $\int_{\mathbb{R}^d} \delta_0(x)dx = 1$. Then consider the family $\delta_n(x) := n^d \delta_0(nx), n \in \mathbb{N}$. This we will call a family of measurements, and δ_0 is called the core of the measurement. In general we may want to perform different measurements on the different terms in (2). To this end, we will for each multi-index α consider two families of measurements $\delta_{\alpha,1,n}, \delta_{\alpha,2,n}$ each constructed from cores $\delta_{\alpha,1,0}, \delta_{\alpha,2,0}$, respectively. With the formal expression (2), we now associate what we call an errorfunction

$$E_n(u)(x) = \sum_{|\alpha| \le N} (f_\alpha * \delta_{\alpha,1,n}(x)) \cdot (D^\alpha u * \delta_{\alpha,2,n}(x)) \tag{3}$$

where the $*$ denotes a convolution product. We now want to consider differential equations which formally can be written in the form

$$Lu = 0 \tag{4}$$

If τ is some fixed topology, we say that $u \in \mathcal{D}'$ is a measurement solution to (4) in the τ-topology if

$$\lim_{n \to \infty} E_n(u) = 0 \qquad \text{in the } \tau\text{-topology} \tag{5}$$

When all the coefficients $f_\alpha \in C^\infty(\mathbb{R}^d)$, then (4) also makes sense in the space of distributions. In this case, however, we have the theorem below.

Theorem 2.1 *If all $f_\alpha \in C^\infty(\mathbb{R}^d)$, $D = \mathbb{R}^d$, and τ is the topology of convergence in the sense of distributions, then the following statements are equivalent:*
A: u solves $Lu = 0$ in the sense of distributions
B: u is a measurement solution to $Lu = 0$ in the τ-topology.

Proof. Choose and fix any multi-index α. Since each f_α is smooth, $f_\alpha * \delta_{\alpha,1,n} \to f_\alpha$ in $C^\infty(\mathbb{R}^d)$ with its usual Fréchet space topology. Clearly every $D^\alpha u * \delta_{\alpha,2,n} \to D^\alpha u$ in $\mathcal{D}'$. Using Theorem 6.18 in [9], we get

$$\lim_{n \to \infty} E_n(u) = Lu \tag{6}$$

and the theorem follows trivially from this. □

If the coefficients f_α are not smooth, however, the measurement solutions are more flexible than classical solutions in the sense of distributions. Note that with the definition above, we are not required to make sense to the product of the distributions $f_\alpha \cdot u$. If only we can prove that $E_n(u) \to 0$ in some appropriate topology, we have a solution.

3 Random distributions

It is well known that noise in SPDEs leads to very singular problems. Even simple equations like the Poisson equation:

$$\Delta u = W \qquad\qquad W \text{ White noise in } \mathbb{R}^d \tag{7}$$

have solutions which are pure Schwartz distributions (i.e., not functions) when the space dimension $d > 3$. See e.g., [13]. The expression:

$$u(x,\omega) = \int G(x,y)dB(y,\omega) \tag{8}$$

where $G(x,y)$ is the Green function, is our obvious candidate for the solution. In higher dimensions, however, the Green function is not square integrable w.r.t. y and it is not clear how to make sense out of the integral. To bypass this difficulty, the strategy is to smooth the expression by a test function $\phi \in \mathcal{D}$. More precisely we consider:

$$u(\phi,\omega) = \int\int G(x,y)\phi(x)dxdB(y,\omega) \tag{9}$$

The Green function will in general satisfy an estimate of the form

$$\left|\left| \int G(x,y)\phi(x)dx \right|\right|_{L^2(D)} \leq C||\phi||_{L^2(D)} \tag{10}$$

Hence the expression in (9) is well defined as an Ito-integral. We are, however, still left with some rather delicate technical problems. The Ito-integral is only defined as an L^2-object. Hence we only know that

$$u(\phi+\psi,\omega) = u(\phi,\omega) + u(\psi,\omega) \text{ a.s.} \tag{11}$$

The problem with this is that for each different pair of testfunctions (ϕ, ψ), there might be a different exceptional set. Hence we cannot conclude that

$$u(\phi + \psi, \omega) = u(\phi, \omega) + u(\psi, \omega) \text{ for all } \phi, \psi \text{ a.s.} \tag{12}$$

A functional object satisfying (11) for each pair ϕ, ψ is called a random linear functional, and the following nontrivial results, originally due to Ito, see [13] p. 322, comes to our support.

Theorem 3.1 [Ito] *Let u be a random linear functional which is continuous in probability on a nuclear space Θ. Then u has a version with values in Θ'.*

Since $\mathcal{D}'$ is a nuclear space, the previous result applies if only we can prove that u is continuous in probability. Here we must assume that D is a bounded domain. It suffices to prove that u satisfies the convergence properties at 0, so let $\phi \to 0$ in $\mathcal{D}$. Then by Sobolev's theorem, $\sup_{x \in D} |\phi(x)| \to 0$. The Ito-isometry together with (10) gives

$$\begin{aligned} E[|u(\phi, \omega)|^2] &= \left|\left| \textstyle\int G(x,y)\phi(x)dx \right|\right|^2_{L^2(D)} \\ &\leq C^2 ||\phi||^2_{L^2(D)} \leq C' \sup_{x \in D} |\phi(x)|^2 \to 0 \end{aligned} \tag{13}$$

Hence u is continuous in L^2 which in turn implies continuity in probability. By the theorem above, we can then find a version s.t. $u(\cdot, \omega) \in \mathcal{D}'$ for a.a. ω. Actually it is possible to prove the slightly stronger result that $u(\cdot, \omega) \in H_{-N}$ (Sobolev space of negative index) for a.a. ω where N does not depend on ω. See [13] for more details on this. Once we know that $u \in \mathcal{D}'$, then both sides of (7) make sense as distributions. It is then straightforward to verify that u solves (7) in the sense of distributions. In the next section, however, the operator does not extend to $\mathcal{D}'$ and we will apply the extended framework of the measurement method to make sense of the equation.

4 The Poisson equation

In this section we formulate a result from [6] on the boundary value problem

$$LV = W \qquad\qquad V|_{\partial D} = 0 \tag{14}$$

We assume that $L = \nabla[a\,\nabla(\cdot)]$ is a strictly elliptic operator on a bounded domain D in $\mathbb{R}^d$ where the coefficients $a = [a_{ij}]$ are bounded measurable functions. If we let $G(x, y)$ denote the Green function, then the expression:

$$V(\phi) = \int_D \int_D G(x, y)\phi(x)dxdB(y) \tag{15}$$

is the obvious candidate for the solution. If $\psi \in L^2(D)$, then (see [3] p. 173) the problem:

$$Lu = \psi \qquad\qquad u|_{\partial D} = 0 \tag{16}$$

is weakly solvable with solution $u_\psi(x) = \int_D G(x, y)\psi(y)dy$, and we have:

$$||u_\psi||_{W^{1,2}(D)} \leq C||\psi||_{L^2(D)} \tag{17}$$

i.e., u_ψ is in the Sobolev space $W^{1,2}(D)$, and:

$$\int_D |u_\psi(x)|^2 + |\nabla u_\psi(x)|^2 dx \leq C^2||\psi||^2_{L^2(D)} \tag{18}$$

By this estimate and the Ito isometry we see that:

$$E[|V(\phi)|^2] = \int_D |u_\phi(x)|^2 dx \leq C^2||\phi||^2_{L^2(D)} \leq C' \sup_{x\in D} |\phi(x)|^2 \tag{19}$$

This shows that the mapping $\phi \mapsto V(\phi)$ is well-defined, almost linear and continuous in probability on $\mathcal{D}$. By Theorem 3.1, it then has a version in $\mathcal{D}'$. It is not clear, however, in what sense this distribution is a solution of the original equation. The operator involves a singular product, and it is necessary to examine the product of the non-smooth function a and the pure distribution ∇V. We phrase this in terms of the following error function:

$$E_n(V)(x) = \nabla[a * \delta_n(x) \cdot \nabla[V * \delta_n(x)]] - W * \delta_n(x) \tag{20}$$

In [6] we were able to prove the following result:

Theorem 4.1 *The distribution*

$$V(\phi) = \int_D \int_D G(x, y)\phi(x)dxdB(y) \tag{21}$$

solves the problem

$$LV = W \qquad\qquad V|_{\partial D} = 0 \tag{22}$$

in the sense that

$$E_n(V)(x) \to 0 \qquad \textit{in the sense of distributions.} \tag{23}$$

Remark In the language of Colombeau distributions, the result says that we have equality between LV and W in the sense of association. It is known, see [1], that two representations of Schwartz distributions are strictly equal if and only if they are associated. Since the objects in Theorem 4.1 just slightly extend beyond the class of distributions, it seems reasonable to conjecture that the equality above is strict.

5 The heat equation

In [6] we were also able to modify the construction in the previous section to verify the same type of result for the stochastic heat equation. Again we assume that $L = \nabla[a\,\nabla(\cdot)]$ is a strictly elliptic operator on $\mathbb{R}^d$ where the coefficients $a = [a_{ij}]$ are bounded measurable functions. It is then well-known, see [10], that there exists a unique Feller continuous Markov semigroup P_t with the properties that:
a) $t \in [0,\infty) \to P_t\phi \in W^{1,2}(\mathbb{R}^d)$ is a weakly continuous map for each $\phi \in C_0^\infty(\mathbb{R}^d)$.
b) For all $\phi, \psi \in C_0^\infty(\mathbb{R}^d)$:

$$< P_t\phi, \psi > = < \phi, \psi > - \int_0^t < \nabla P_s\phi, a\,\nabla\psi > ds, \;\; t \in [0,\infty)$$

Moreover there is a $p \in C((0,\infty) \times \mathbb{R}^d \times \mathbb{R}^d)$ such that:

$$P_t\phi(x) = \int_{\mathbb{R}^d} p(t,x,y)\phi(y)dy \tag{24}$$

$$||\nabla P_t\phi||_{L^2} \leq \frac{1}{\lambda^{\frac{1}{2}}\, t^{\frac{1}{2}}}||\phi||_{L^2}, \; t > 0 \tag{25}$$

$$\frac{1}{M\, t^{\frac{d}{2}}} e^{-M\frac{|y-x|^2}{t}} \leq p(t,x,y) \leq \frac{M}{t^{\frac{d}{2}}} e^{-\frac{|y-x|^2}{Mt}} \tag{26}$$

This time we consider the boundary value problem:

$$\frac{\partial u}{\partial t} = Lu + W_{t,x} \qquad\qquad u(0,x) = 0 \tag{27}$$

Our formal candidate for the solution can then be expressed as follows:

$$u(t,x) = \int_0^t \int_{\mathbb{R}^d} p(t-s,x,y)W(ds,dy) \tag{28}$$

We make sense of this through the distribution:

$$u_t(\phi) = \int_0^t \int_{\mathbb{R}^d} \left(\int_{\mathbb{R}^d} p(t-s,x,y)\phi(x)dx \right) W(ds,dy) \tag{29}$$

Using the Ito-isometry, one can verify that:

$$\begin{aligned} E[|u_t(\phi)|^2] &= \int_0^t \int_{\mathbb{R}^d} \left(\int_{\mathbb{R}^d} p(t-s,x,y)\phi(x)dx\right)^2 dsdy \\ &\leq \int_0^t \left(\int_{\mathbb{R}^d} \phi^2(x)dx\right) ds = t\,||\phi||_{L^2}^2 \end{aligned} \tag{30}$$

This shows that the mapping $\phi \mapsto u_t(\phi)$ is well-defined, almost linear and continuous in probability on $\mathcal{S}(\mathbb{R}^d)$. Thus by Theorem 2.1 again, $u_t \in \mathcal{S}'(\mathbb{R}^d)$. We now consider the errorfunction:

$$E_n(u) = \frac{\partial}{\partial t}(u * \delta_n(t,x)) - \nabla[a * \delta_n(t,x) \cdot \nabla[u * \delta_n(t,x)]] - W * \delta_n(t,x) \tag{31}$$

where the ∇ is acting on the x variable only. The main result, see [6], can then be phrased as follows:

Theorem 5.1 *The distribution*

$$u_t(\phi) = \int_0^t \int_{\mathbb{R}^d} \left(\int_{\mathbb{R}^d} p(t-s,x,y)\phi(x)dx \right) W(ds,dy) \tag{32}$$

solves the problem

$$\frac{\partial u}{\partial t} = Lu + W_{t,x} \qquad\qquad u(0,x) = 0 \tag{33}$$

in the sense that

$$E_n(u)(t,x) \to 0 \qquad \textit{in the sense of distributions.} \tag{34}$$

6 Itô/Stratonovich solutions

The initial value problem

$$F' + \mu \cdot F = 0 \qquad F(0) = F_0 \tag{35}$$

where μ is a measure on $\mathbb{R}$, is discussed in detail in [12]. It turns out that even when there is no random dependence in μ, this problem has several different solutions depending on the interpretation of the equation. If the core of the measurement and the topology is fixed, however, one can sometimes prove existence and uniqueness results for the equation, see [12]. Here we will limit the discussion to the random measure $\mu = \alpha + \beta W$, $\alpha, \beta \in \mathbb{R}$. Formally this version of (35) corresponds to the linear stochastic differential equation

$$dF_t = -\alpha F_t dt - \beta F_t dB_t \tag{36}$$

An important property of stochastic differential equations is that the solution is heavily dependent on the interpretation of the equation; we have, e.g., Itô and Stratonovich solutions. It is well known that if we consider the limit of a sequence

$$F_n' + (\mu * \delta_n) \cdot F_n = 0 \qquad F_n(0) = F_0 \tag{37}$$

then the sequence F_n converges to the Stratonovich solution of (36), see, e.g., [14] and [11]. As it is demonstrated in [12], however, the measurement method can be expected to offer a different interpretation. With a carefully selected collection of measurements, it turns out that the Itô solution of (36) is a solution to (35) in the measurement sense.

One cannot expect that the Itô solution solves (35) for any collection of measurements. The measurements must somehow reflect the underlying structure in the Itô integral. The construction goes like this: We call a measurement δ_0 anticipating if it is supported at $(-\infty, 0]$, and non-anticipating if it is supported at $[0, \infty)$. We let $\hat{\delta}_0(x) = \delta_0(-x)$. So if δ_0 is anticipating, then $\hat{\delta}_0$ is non-anticipating and vice versa. The errorfunction we want to consider is as follows

$$E_n(F)(t, \omega) = \frac{dF}{dt} * \hat{\delta}_n(t) + \left(\mu * \hat{\delta}_n(t)\right)(F * \delta_n(t)) \tag{38}$$

Within this framework one can prove the following theorem, see [12]:

Theorem 6.1 *If $\mu = \alpha + \beta W$, $\alpha, \beta \in \mathbb{R}$, δ_0 is a non-anticipating measurement, and the errorfunction is specified through (38), then the Geometric Brownian Motion*

$$F(t, \omega) = F_0 e^{-(\alpha + \frac{1}{2}\beta^2)t - \beta B_t} \tag{39}$$

is a solution to the initial value problem (35) in the measurement sense, pointwise in t, in the weak-topology on $L^2(P)$.*

Remark Note that while $F(t, \omega)$ given by (39) is indeed the Itô-solution of (36), the Itô-calculus plays no part in the formulation of the above theorem.

7 Some concluding remarks

It is common to all the examples in this paper that there is some kind of natural candidate for the solution which we can insert in the errorfunction and proceed to prove the desired limiting property. Once we are faced with a new type of singular equation, however, it will be far from clear how to produce new candidates. Typically one must expect that solutions are depending on the choice of cores and also on the topology of convergence. Once these are fixed, one can then have some hope of proving existence/uniqueness results. The first question to emphasize will be whether or not it is possible to determine cores and topologies that single out physically acceptable solutions. Seen from this viewpoint, the theory has much in common with basic viewpoints in the theory of, e.g., nonlinear conservation laws. Indeed such conservation laws can easily be phrased within the suggested framework. So far, however, we have not been able to provide any new results for nonlinear problems.

Acknowledgment: This work is supported by VISTA, a research cooperation between The Norwegian Academy of Science and Letters and Den Norske Stats Oljeselskap (Statoil). I also wish to thank M. Oberguggenberger and the ESI-institute for their support and hospitality during my visit to Vienna.

References

[1] COLOMBEAU, J. F.: Multiplication of distributions, Bull. Am. Math. Soc. 23, 251–268, 1990.

[2] COLOMBEAU, J. F., A. HEIBIG AND M. OBERGUGGENBERGER: Generalized solutions to partial differential equations of evolution type, Acta Applicandae Mathematicae 45(2), 115–142, 1996.

[3] GILBARG, D. AND N. S. TRUDINGER: Elliptic Partial Differential Equations of Second Order, Springer Verlag 1977.

[4] HOLDEN, H., T. LINDSTRØM, B. ØKSENDAL, J. UBØE AND T.-S.ZHANG: Stochastic boundary value problems. A white noise functional approach, Probability Theory and Related Fields 95, 391–419, 1993.

[5] Holden, H., T. Lindstrøm, B. Øksendal, J. Ubøe and T.-S.Zhang: The Burgers equation with a noisy force and the stochastic heat equation, Communications of Partial Differential Equations 19, 119–141, 1994.

[6] Lindstrøm, T., B. Øksendal, J. Ubøe and T.-S. Zhang:, Stability properties of stochastic partial differential equations, Stochastic Analysis and Applications 13(2), 177–204, 1995.

[7] Oberguggenberger, M.: Multiplication of Distributions and Applications to Partial Differential Equations, Longman, Wiley, 1992.

[8] Oberguggenberger, M. and F. Russo: Colombeau Solutions and Pathwise Limits. In L. Decreusefond, J. Gjerde, A. S. Üstunel and B. Øksendal (editors), Stochastic Analysis and Related Topics, pp. 325–340, Birkhäuser 1997.

[9] Rudin, W.: Functional Analysis, Springer, 1980.

[10] Stroock, D.W.: Diffusion semigroups corresponding to uniformly elliptic divergence form operators. In T.Azéma, P.A.Meyer and M.Yor (editors): Séminaire de Probabiletés XXII. Springer LNM 1321, 316-347, 1988.

[11] Sussmann, H. J.: On the gap between deterministic and stochastic ordinary differential equations, Annals of Probability 60, 19–41, 1978.

[12] Ubøe, J.: Measurements of ordinary and stochastic differential equations, manuscript, 1997.

[13] Walsh, J. B.: An introduction to stochastic partial differential equations. In R. Carmona, H. Kesten and J. B. Walsh (editors), École d'Été de Probabiletés de Saint-Flour XIV-1984, pp. 265–437, Springer LNM 1180 1986.

[14] Wong, E. and M. Zakai: Riemann-Stieltjes approximations of stochastic integrals, Z. Wahr. verw. Geb. 120, 87–97, 1969.

Jan Ubøe

Stord/Haugesund College, Skåregaten 103, 5 500 Haugesund, Norway, E-mail: jan.uboe@hsh.no

On the Small-Time Asymptotics of Solutions of Linear and Nonlinear Stochastic Differential Equations

TUSHENG ZHANG

1 Introduction

In this paper we give a survey of some results of the small-time asymptotics of solutions of stochastic differential equations on Hilbert spaces, which include solutions of some simple stochastic partial differential equations. Let us start by recalling the basic results in finite dimensions. Let $L = \frac{1}{2}\Delta$ be one half of the Laplacian operator on R^d. Then we know that the heat kernel is the transition density of the Brownian motion which is given by

$$P_t(x,y) = (2\pi t)^{-\frac{d}{2}} \exp\{-\frac{d^2(x,y)}{2t}\} \tag{1}$$

where $d(x,y)$ stands for the usual distance on R^d.

It is clear that the following small time asymptotics holds

$$\lim_{t\to 0} 2t \log P_t(x,y) = -d^2(x,y) \tag{2}$$

Much work has been done to extend the above asymptotics to more general situations where the Laplacian is replaced by general elliptic operators, R^d is replaced by some finite-dimensional Riemannian manifolds and $d(x,y)$ is the corresponding Riemannian distance. The results so far are quite satisfactory, see [3], [9], [11].

Now we are concerned with the above asymptotics in infinite-dimensional cases where L will be the generator of a diffusion process $X_t, t \geq 0$ on some Hilbert space E. Because of the lack of the transition density, the natural replacement for $P_t(x,y)$ in the equation (2) is $P(X_0 \in A, X_t \in B)$, where A,B are two Borel subsets. The distance $d(x,y)$ between two points x, y is replaced by the distance of the two sets A and B. More precisely, we are interested in the following small time asymptotics

$$\lim_{t\to 0} 2t \log P(X_0 \in A, X_t \in B) = -d^2(A,B), \tag{3}$$

where d is the appropriate Riemannian distance associated with the diffusion. The aim of this paper is to present some results (without proofs) obtained in this direction. The main references are [4], [5] and [12]. The paper is organized as follows. In Section 2, we discuss the Gaussian cases. The results of this section come from a joint work with S. Fang [5]. Section 3 is devoted to the general situations.

2 The Gaussian cases

The Ornstein-Uhlenbeck process $X_t, t \geq 0$ on Hilbert space E formally solves the following infinite-dimensional stochastic differential equation:

$$dX_t = dw_t - \frac{1}{2}AX_t\, dt \tag{4}$$

where w_t is an E-valued Brownian motion and A in general is an unbounded linear operator. The simplest example is the solution of the following SPDE

$$\frac{\partial u(t,x)}{\partial t} = \dot{w}_t(x) + \frac{1}{2}\frac{\partial^2 u(t,x)}{\partial x^2} \tag{5}$$

The precise formulation of the process is given as follows. Let μ be a mean zero Gaussian measure on $(E, \mathcal{B}(E))$ with the reproducing kernel space H, where $\mathcal{B}(E)$ denotes the Borel σ-field. Then (H, E, μ) is an abstract Wiener space in the sense of Gross.

For $h \in H$, define a Gaussian random variable $<< h, x >> \in L^2(E, \mu)$ by

$$<< h, x >> = \lim_{n\to\infty} < k_n, \cdot > \quad \text{in} \quad L^2(E, \mu) \tag{6}$$

where $(k_n)_{n\in N}$ is any sequence in E' such that $k_n \to h$ in H as $n \to \infty$. Let $P^{(n)}$ be the closed linear span of the monomials $\prod_{j=1}^{m} << h_j, x >>$, $m \leq n$, $h_1, \cdots, h_m \in H$, in $L^2(E, \mu)$. Define

$$\mathcal{H}_0 = R, \mathcal{H}_n = P^{(n)} \ominus P^{(n-1)}\ , \text{ for } \quad n \geq 1. \tag{7}$$

The following well-known (Wiener-Ito) chaos decomposition holds,

$$L^2(E, \mu) = \oplus_{n\geq 0}\mathcal{H}_n. \tag{8}$$

We recall that for any linear operator S on H with operator norm $||S|| \leq 1$, there exists a unique bounded linear operator $\Gamma(S)$ on $L^2(E, \mu)$ with $||\Gamma(S)|| \leq 1$ defined by

$$\Gamma(S)1 = 1,\ \Gamma(S)(:\prod_{j=1}^{n} << h_j, \cdot >>:_n) =: \prod_{j=1}^{n} << Sh_j, \cdot >>:_n, \quad h_1, \cdots, h_n \in H$$

and by linearity. Here $:\ :_n$ means orthogonal projection onto $\mathcal{H}_n$.

Now let A be a self-adjoint operator on H satisfying $A \geq cI_H$, where $c > 0$, and I_H stands for the identity operator on H. The associated semigroup is denoted by $T_t = e^{-tA}$. It is not difficult to show that $\Gamma(e^{-tA}), t \geq 0$, forms a strongly continuous semigroup of symmetric contractions on $L^2(E, \mu)$. So there exists a unique positive definite form $(\mathcal{E}, D(\mathcal{E}))$ on $L^2(E, \mu)$ associated with $(\Gamma(e^{-tA}))_{t \geq 0}$ such that $\mathcal{E}(u, v) = (-d\Gamma(A)u, v)$, for all $u \in D(d\Gamma(A))$, $v \in D(\mathcal{E})$, $d\Gamma(A)$ being the generator of $\Gamma(e^{-tA})$, $(,)$ being the scalar product in $L^2(E, \mu)$. $d\Gamma(A)$ is called the second quantization of A. Define $H_0 = D(\sqrt{A})$ with inner product $< h_1, h_2 >_{H_0} = < \sqrt{A}h_1, \sqrt{A}h_2 >_{H_0}$.

Then

$$E' \subset H_0 \subset H \subset E \quad \text{densely and continuously.}$$

The inclusion $H_o \subset E$ is also Hilbert-Schmidt.

Introduce

$$\mathcal{F}C_b^\infty = \{u(x) = f(l_1(x), \cdots, l_m(x)); f \in C_b^\infty(\mathbf{R}^m), l_1, \cdots, l_m \in E',\ m \geq 1\}.$$

Given $u \in \mathcal{F}C_b^\infty$, denote by $\nabla u(x) \in H_o$ such that for all $k \in H_o$,

$$< \nabla u(x), k >_{H_o} = \lim_{\varepsilon \to 0} \frac{u(x + \varepsilon k) - u(x)}{\varepsilon} \tag{9}$$

The following theorem is taken from [2]:

Theorem 2.1. (i) $(\mathcal{E}, D(\mathcal{E}))$ is the closure of the form

$$\mathcal{E}(u, v) = \int_E < \nabla u, \nabla v >_{H_o} d\mu, \quad u, v \in \mathcal{F}C_b^\infty; \tag{10}$$

(ii) there exists a diffusion process $M = ((X_t)_{t \geq 0},\ P_x,\ x \in E)$ on E associated with $(\mathcal{E}, D(\mathcal{E}))$, which solves (4) with a Brownian motion w having the reproducing space H_o (see [2, theorem 5.7]).

Now we are ready to formulate the main results. For $u, v \in \mathcal{F}C_b^\infty$, let

$$\Gamma(u, v)(z) =< \nabla u(z), \nabla v(z) >_{H_0} . \tag{11}$$

The intrinsic metric $d(x,y)$ on $E\times E$ determined by the Dirichlet form (10) is defined as

$$d(x,y) = \sup_{u\in\mathcal{F}C_b^\infty, \Gamma(u,u)\leq 1} (u(x) - u(y)) \tag{12}$$

From the definition of $\Gamma(u,v)$, it is easy to see that $d(x,y)$ has the following expression

$$d(x,y) = \begin{cases} |x-y|_{H_0} & \text{if } x-y\in H_0, \\ +\infty & \text{otherwise.} \end{cases} \tag{13}$$

Let B be a Borel subset of E, we define: $d(x,B) = \inf\{d(x,y);\ y\in B\}$. Now for two Borel subsets B, C with $\mu(B) > 0$ and $\mu(C) > 0$, define

$$d(B,C) = \sup(\text{ess}\inf_{z\in\tilde{B}} d(z,\tilde{C}),\ \text{ess}\inf_{z\in\tilde{C}} d(z,\tilde{B})) \tag{14}$$

where $\tilde{B}$ denotes the support of the measure $\chi_B d\mu$.

Define

$$P_\mu(\cdot) = \int_E P_x(\cdot)d\mu. \tag{15}$$

Theorem 2.2. Let B, C be two Borel subsets with $\mu(B) > 0$, $\mu(C) > 0$. Then

$$\limsup_{t\to 0} 2t\log P_\mu(X_0\in B,\ X_t\in C) \leq -d^2(B,C). \tag{16}$$

To give the lower bound, we need the following:

Assumption 2.3: The semigroup $T_t = e^{-tA}$, $t\geq 0$, generated by $-A$, extends to a strongly continuous semigroup of bounded linear operators on E.

Remark 2.4. This assumption is not as strong as it looks. According to [2], for a given operator A and Hilbert space H one can always properly choose E so that the above assumption holds.

Theorem 2.5. Assume the assumption 2.3 holds. Let B, C be two Borel subsets with $\mu(B) > 0$, $\mu(C) > 0$. If B or C is open, then

$$\liminf_{t\to 0} 2t\log P_\mu(X_o\in B,\ X_t\in C) \geq -d^2(B,C). \tag{17}$$

3 The general case

In this section, we treat more general diffusions than Ornstein-Uhlenbeck processes. Keep the same notations as before. Let $\mathcal{L}_s(H_0)$ denote the set of all symmetric non-negative definite bounded linear operators on H_0. Let $A(z)$ be a strongly continuous map from E into $\mathcal{L}_s(H_0)$ such that

$$c^{-1}I_{H_0} \leq A(z) \leq cI_{H_0} \tag{18}$$

where c is a positive constant, and I_{H_0} denotes the identity operator on H_0. Introduce the quadratic form

$$\mathcal{E}_A^0(u,v) = \int_E < A(z)\nabla u(z), \nabla v(z) >_{H_0} d\mu, \qquad u,v \in \mathcal{F}C_b^\infty \tag{19}$$

This form is closable on $L^2(E,\mu)$ due to Theorem 2.1. Its closure, denoted by $(\mathcal{E}_A, D(\mathcal{E}_A))$, is a Dirichlet form. Let $M_A = (X_t, t \geq 0, Q_x, x \in E)$ be the diffusion process associated with $(\mathcal{E}_A, D(\mathcal{E}_A))$, see [6]. The process M_A in general is very different from the Ornstein-Uhlenbeck process discussed in Section 2. It is not Gaussian any more. To get similar results as in the previous section, a different approach has to be adopted. Set, for $u, v \in \mathcal{F}C_b^\infty$,

$$\Gamma(u,v)(z) =< A(z)\nabla u(z), \nabla v(z) >_{H_0} \tag{20}$$

Then the intrinsic metric $d(x,y)$ can be defined exactly as in Section 2 with Γ there replaced by the Γ here.

Let B be Borel subset of E with positive measure. Put $u(x) = d(x,B)$.

Lemma 3.1. For any constant $c > 0$, $u_c(x) = u(x) \wedge c \in D(\mathcal{E}_A)$ and $\Gamma(u_c, u_c)(z) \leq 1$.

Set

$$Q(\cdot) = \int_E Q_x(\cdot) d\mu \tag{21}$$

Theorem 3.2. Let B, C be two Borel subsets with $\mu(B) > 0$, $\mu(C) > 0$. Then

$$\limsup_{t\to 0} 2t \log Q(X_0 \in B, \ X_t \in C) \leq -d^2(B,C). \tag{22}$$

For the lower bound, a much stronger assumption on $A(z)$ is needed. We do not want to go into details here. Basically the assumption can be formulated as follows:

Assumption 3.3: The operator function $A(z)$ is smooth enough so that the diffusion M_A is a solution of a stochastic differential equation on the Hilbert space E.

Theorem 3.4 Assume that Assumption 3.3 holds. Let B, C be two Borel subsets with $\mu(B) > 0$, $\mu(C) > 0$. If B or C is open, then

$$\liminf_{t\to 0} 2t \log Q(X_o \in B, \ X_t \in C) \geq -d^2(B,C). \tag{23}$$

Acknowledgment: This work was initiated when I visited the Schrödinger Institute at Vienna. I would like to thank M. Oberguggenberger and the ESI-institute for their hospitality and financial support during my pleasant stay.

References

[1] ALBEVERIO, S. AND RÖCKNER, M.: Classical Dirichlet forms on topological vector spaces: construction of an associated diffusion process, Probab. Theor. Relat. Fields 83, 405–434, 1989.

[2] ALBEVERIO, S. AND RÖCKNER, M.: Stochastic differential equations in infinite dimensions: solutions via Dirichlet forms, Probab. Theor. Relat. Fields 89, 347–386, 1991.

[3] DAVIES, E.B.: Heat Kernels and Spectral Theory, Cambridge University Press, 1989.

[4] FANG, S.: On the Ornstein-Uhlenbeck process, Stochastics and Stochastic Reports 46, 141–159, 1994.

[5] FANG, S. AND ZHANG, T.S.: On the small time behavior of Ornstein-Uhlenbeck processes with unbounded linear drifts, preprint, 1997.

[6] LYONS, T.J. AND AND ZHANG, T.S.: Decomposition of Dirichlet processes and its applications, Annals of Prob. 22:1, 494–524, 1994.

[7] MA, Z. AND RÖCKNER, M.: Introduction to the Theory of (Non-Symmetric) Dirichlet Forms, Springer-Verlag, Berlin, Heidelberg, 1992.

[8] RÖCKNER, M. AND ZHANG, T.S.: Uniqueness of generalized Schrödinger operators and applications, Journal of Functional Analysis 105, 187-231, 1992.

[9] STROOCK, D.W.: Diffusion semigroups corresponding to uniformly elliptic divergence form operators. In T. Azéma, P.A. Meyer and M. Yor (Editors): Séminaire de Probabiletés XXII. Springer LNM 1321, 316-347, 1988.

[10] TAKEDA, M.: On a martingale method for symmetric diffusion processes and its applications, Osaka J. Math. 26, 605-625, 1989.

[11] VARADHAN, S.R.S.: Diffusion processes in small time intervals, Comm. Pure Appl. Math. 20, 659-685, 1967.

[12] ZHANG. T.S.: On the small time asymptotics of diffusion processes on Hilbert spaces, preprint, 1998.

TUSHENG ZHANG
Faculty of Engineering, HSH, Skåregaten 103, 5 500 Haugesund, Norway, E-mail: tz@hsh.no

NONSTANDARD METHODS

The Global Control of Shock Waves

JULIO E. RUBIO

1 Introduction

We have developed (see [2, 3]) an approach to the study of global optimization of nonlinear optimal control problems based on the consideration of measure spaces. In this paper we deal with nonlinear wave equations and their solutions in the sense of Colombeau; see [13, 4]. The solutions of these equations may exhibit discontinuities—shock waves—so that there is difficulty dealing with the derivatives of the main variables as well as with their products with the variables themselves; this is the main problem solved by Colombeau by having recourse to particular kinds of quotient algebras.

From the standpoint of the optimization problem, we need to consider unbounded sets for the slopes of the solutions. The actual discontinuities happen, if at all, at infinitesimal values $\lambda = [< \lambda_p >]$ of a parameter, λ, appearing in the equations chosen; we will consider the optimization problem for each of the values λ_p, and put the results together at the end. Thus the need to consider unbounded sets for the values taken by the slopes; the parameters λ_p tend to 0, so that we get closer and closer to the discontinuities, the shocks.

The problem still remains, how do we enlarge our spaces? It may appear natural to simply include 'delta functions', impulses, there are many ways of doing this. Alas, it is very difficult to work with impulses; in particular, it is very hard to define functions of impulses; see [2], Chapter 6, for a discussion of this point. In this paper, our idealized elements lack the familiarity of impulses and such; they are *nonstandard elements*, not easily visualized but easily handled mathematically.

Thus, our path is as follows. The control problem will be written in a manner involving the solution of a set of integral equalities; these are mapped then into a nonstandard framework, in which the use of Loeb measures gives rise to an important result, that a near-minimizer for the nonstandard optimization problem always exists. The standard part map provided us with a *global minimizer* for the original problem, as well as with a measure-theoretical framework in the standard world in which a linear program is obtained with the minimizer as a solution.

Approximation tools developed in our previous work [3] are then used to develop a finite-dimensional approximation of the linear program, and construct nearly optimal solutions of the variational problem. A numerical example is given. A more complete set of references, as well as all the proofs, appear in [1].

2 The problem

Consider a system of partial differential equations:

$$\begin{aligned} -\lambda u_{xx} + u_t + u u_x &= F \\ -\lambda v_{xx} + v_t + u v_x &= G, \\ u(0,x) = u_0(x),\ v(0,x) &= v_0(x). \end{aligned} \tag{1}$$

where $u, v : [0,\infty) \times \mathbb{R} \to \mathbb{R}$, $u_0, v_0 : \mathbb{R} \to \mathbb{R}$ and $\lambda > 0$. The functions $F, G : (t,x) \to \mathbb{R}$ can be considered as *distributed control functions*; we shall assume that their support is in $J \times J$, with $J := [0,1]$. Bounds will be put on some of these variables later on. Equations like these—without the control functions—have been much studied; see [15, 14]. Shock waves may happen for *infinitesimal* λ, so that they will be approximated for small values of this parameter. Our program consists then in choosing an appropriate superstructure, not necessarily the same as in Section 3—$\aleph_0$-saturation would suffice here—and write

$$\lambda = [< \lambda_p >], \tag{2}$$

with $p \in \Pi$, an appropriate index set. We shall study below an optimization problem associated with (1) for standard, fixed values of λ, and we shall come back to our nonstandard setting later on, once this problem is solved.

Our interest now then is to study a control problem associated with (1) for small—as small as we wish—values of the *standard* parameter λ, so that it is convenient to consider as unbounded the sets in which the derivatives u_x, v_x take values. As we shall see below, the second derivative u_{xx} will play no role.

Our objective is to choose the control functions so as to improve the behaviour of the system, by for instance minimizing a performance criterion such as the one defined in (3) below. However, it may happen—it most likely will happen—that no minimizer exists for this functional in the 'nice' class of control-trajectories quadruples $\mathcal{F}$ defined below; as indicated above, these equations may have solutions approximating shock waves arbitrarily.

Let us consider then the class $\mathcal{F}$ of *admissible quadruples* (u, v, F, G), in which F, G are continuously differentiable and (u, v) are the corresponding *classical solutions* of (1). We assume that this class is nonempty and seek to minimize the functional $I : \mathcal{F} \to \mathbb{R}$

$$I(u, v, F, G) = \int_{J\times J} f_0(t, x, u, v, u_x, v_x, F, G)\, dt\, dx, \tag{3}$$

for $(u, v, F, G) \in \mathcal{F}$. Of course, we are looking for an optimal control function. Here f_0 is a continuous function defined on $\Omega := J^2 \times \mathbb{R}^6$.

We will be interested in the solutions of (1) on the square $J \times J$; thus the supports of f_0 and F, G. Note we are putting no boundary conditions for u, v or their derivatives on the boundary of this set.

We develop some equalities. Let $D_0 := \{t = 0\} \times J$, $D_1 := \{x = 1\} \times J, \Gamma_1 := J \times \{x = 0\} \cup J \times \{x = 1\}$, and $\mathcal{K}$ be the class of C_1-functions ϕ on $J \times J$ which are zero on D_1 and Γ_1. If we multiply pointwise the equations in (1) by ϕ and do some simple algebra, we obtain

$$\begin{aligned} &\int_{J\times J} [-u\phi_t + \lambda u_x\phi_x - \frac{1}{2}u^2\phi_x - F\phi]\, dtdx = \int_{D_0} u\phi\, dx \\ &\int_{J\times J} [-v\phi_t + +\lambda v_x\phi_x + uv_x\phi - G\phi]\, dtdx = \int_{D_0} v\phi\, dx, \forall\phi \in \mathcal{K}. \end{aligned} \tag{4}$$

It will be necessary to explicitly use integral relationships between u and u_x, v and v_x. Let $\mathcal{B}$ be the class of C_1 functions ψ on $J \times J$ which have zero normal derivative ψ_x on Γ_1. Then,

$$\begin{aligned} &\int_{J\times J} [-u\psi_{xx} + u_x\psi_x]\, dtdx = 0 \\ &\int_{J\times J} [-v\psi_{xx} + v_x\psi_x]\, dtdx = 0, \forall\psi \in \mathcal{B}. \end{aligned} \tag{5}$$

All integrals above can and will be considered as the action of an admissible quadruple (u, v, F, G) on a continuous function defined on Ω. It will be necessary to consider in the same light integrable functions $\xi : \Omega \to \mathbb{R}$ which depend only on (x,t). Then, trivially,

$$\int_{J\times J} \xi(t,x,u,v,u_x,v_x,F,G)\, dtdx = a_\xi, \tag{6}$$

with a_ξ the Lebesgue integral of ξ.
As explained in [3, 8] we shall take for each of these spaces countable sets of functions whose linear combinations are dense in the corresponding spaces in appropriate topologies, $\{\phi_i\}$ for $\mathcal{K}$, $\{\psi_j\}$ for $\mathcal{B}$, $\{\xi_h\}$ for the space associated with equation (6). Further, we shall consider a finite number of the resulting infinite number of equalities:

$$\begin{aligned}
\int_{J\times J} [-u\phi_{it} + \lambda u_x \phi_{ix} - \frac{1}{2}u^2\phi_{ix} - F\phi_i]\, dtdx &= \int_{D_0} u\phi_i\, dx, i = 1,\dots, M_1;\\
\int_{J\times J} [-v\phi_{jt} + \lambda v_x \phi_{jx} + uv_x\phi_j - G\phi_j]\, dtdx &= \int_{D_0} v\phi_j\, dx, j = 1,\dots, M_2;\\
\int_{J\times J} [-u\psi_{kxx} + u_x\psi_{kx}]\, dtdx &= 0, k = 1,\dots, M_3\\
\int_{J\times J} [-v\psi_{kxx} + v_x\psi_{kx}]\, dtdx &= 0, k = 1,\dots, M_3\\
\int_{J\times J} \xi_h(u,v,u_x,v_x,F,t,x)\, dtdx &= a_{\xi_h}, h = 1,\dots, M_4.
\end{aligned} \tag{7}$$

Eventually we shall take limits as $M_i \to \infty, i = 1,2,3,4$.

3 The nonstandard way

We shall change our framework here—in a manner that appears minor. Let $\overline{\mathbb{R}}$ be the extended real line. Instead of assuming that the derivatives u_x, v_x corresponding to functions in admissible quadruples $(u,v,F,G) \in \mathcal{F}$ take values in $\mathbb{R}$, we shall take $\overline{\mathbb{R}}$ as a place of abode for these values. There will be no apparent change—the derivatives do take values in $\mathbb{R}$ and $\mathbb{R} \subset \overline{\mathbb{R}}$. But, as we shall see below, the introduction of $\overline{\mathbb{R}}$ is fundamental to our development.
In what follows we shall write equations (7) in a more economical way as in (9) below; we put $M := M_1 + M_2 + 2M_3 + M_4$. We consider therefore the problem of

minimizing the functional

$$I(u,v,F,G) = \int_{J\times J} f_0(t,x,u(t,x),v(t,x),u_x(t,x),v_x(t,x),F(t,x),G(t,x))\,dt\,dx, \tag{8}$$

on the class $\mathcal{F}_M$ of quadruples (u,v,F,G) satisfying

$$\int_{J\times J} f_i(t,x,u(t,x),v(t,x),u_x(t,x),v_x(t,x),F(t,x),G(t,x))\,dtdx = b_i,\ i=1,\dots,M, \tag{9}$$

where $u(\cdot),v(\cdot) \in C_1(J)$ and take values in closed bounded sets $A,B \subset \mathbb{R}$ respectively, and F,G take values in a closed bounded set $U \subset \mathbb{R}$. Here $f_0, f_i, i=1,\dots,M$, are in $C(\Omega')$, with $\Omega' := J^2 \times A \times B \times \overline{\mathbb{R}}^2 \times U^2$. The integer $M \geq 1$ is fixed, and so are the constants $b_i, i=1,\dots,M$. We assume that the class $\mathcal{F}_M$ is nonempty. We shall develop in this section a procedure to enlarge the set $\mathcal{F}_M$, while at the same time extending the functional (8) to the whole of the new, larger set of admissible elements. This procedure will be based on nonstandard techniques incorporating $\overline{\mathbb{R}}$, the extended real line.

For general treatments of this topic, see [2]. We will work in a nonstandard framework given by a superstructure $V(W)$, $\overline{\mathbb{R}} \subset W$. The superstructure $V({}^*V)$ is also an enlargement, and $\aleph_1$-saturated.

We study integrals of the form (8,9), that is,

$$\int_{J\times J} f(t,x,u(t,x),v(t,x),u_x(t,x),v_x(t,x),F(t,x),G(t,x))\,dtdx, \tag{10}$$

with $(u,v,F,G) \in \mathcal{F}_M$ and $f \in C(\Omega')$. Then,

$$(\forall (u,v,F,G) \in \mathcal{F}_M)\ (\int_{J\times J} f(t,x,u(t,x),v(t,x),u_x(t,x),v_x(t,x),F(t,x),G(t,x))\,dtdx \in \overline{\mathbb{R}}, \tag{11}$$

by transfer,

$$(\forall (u,v,F,G) \in {}^*\mathcal{F}_M)\ ({}^*\int_{{}^*J\times{}^*J} {}^*f(t,x,u(t,x),v(t,x),u_x(t,x),v_x(t,x),F(t,x),G(t,x))\,dtdx \in {}^*\overline{\mathbb{R}}, \tag{12}$$

where here and below we write $u_x(\cdot)$ for $(^*\frac{\partial}{\partial x}u(\cdot)$, and similarly for $v(\cdot)$. Thus, the nonstandard version of the optimization problem (8,9) consists in minimizing

$$\begin{aligned} &{}^*I(u,v,F,G) \\ &= {}^*\int_{^*J\times{}^*J} {}^*f_0(t,x,u(t,x),v(t,x),u_x(t,x),v_x(t,x),F(t,x),G(t,x))\,dt\,dx, \end{aligned} \tag{13}$$

on the class ${}^*\mathcal{F}_M$ of quadruples (u,v,F,G) satisfying

$$\begin{aligned} &{}^*\int_{^*J\times{}^*J} {}^*f_i(t,x,u(t,x),v(t,x),u_x(t,x),v_x(t,x),F(t,x),G(t,x))\,dtdx \\ &= b_i, i=1,\dots,M, \end{aligned} \tag{14}$$

For instance,

$${}^*f_i(t,x,u,v,F,G) := -u{}^*\phi_{it} + {}^*\lambda u_x + -(1/2)u_2{}^*\phi_{ix} - F{}^*\phi, i=1,\dots,M_1.$$

Note that the standard λ becomes of course ${}^*\lambda$.
Consider now the map suggested by (11). If $(u,v,F,G)\in\mathcal{F}_M$ is fixed, the map

$$\begin{aligned} &\nu_{(u,v,F,G)} : f \rightarrow \\ &\int_{J\times J} f(t,x,u(t,x),v(t,x),u_x(t,x),v_x(t,x),F(t,x),G(t,x))\,dxdt \in \overline{\mathbb{R}}, \end{aligned} \tag{15}$$

$f\in C(\Omega')$, is linear and positive. By Riesz' Theorem, there is a measure, to be denoted also by $\nu_{(u,v,F,G)}$, on the Borel sets $\mathcal{B}$ of Ω', that represents this map; remember that Ω' is compact. Then $({}^*\Omega', {}^*\mathcal{B}, {}^*\nu_{(u,v,F,G)})$ is a nonstandard measure space and then [2],

Lemma 1 *There is a measure space* $({}^*\Omega', \mathcal{A}, \mu_L^{(u,v,F,G)})$ *so that* $\mu_L^{(u,v,F,G)}$ *is the Loeb measure associated with* $\nu_{(u,v,F,G)}$*; then,*

$$\begin{aligned} &{}^*\int_{^*J\times{}^*J} f(t,x,u(t,x),v(t,x),u_x(t,x),v_x(t,x),F(t,x),G(t,x))\,dtdx = \\ &\mu_L^{(u,v,F,G)}(f) := \int_{^*\Omega'} f d\,\mu_L^{(u,v,F,G)}, f\in C({}^*\Omega'). \end{aligned} \tag{16}$$

The algebra $\mathcal{A}$ *is an extension of the algebra* ${}^*\mathcal{B}$.

Thus, one can write the optimization problem (13,14) as the problem of minimizing

$$J(\mu_L^{(u,v,F,G)}) := \mu_L^{(u,v,F,G)}({}^*f_0), \tag{17}$$

over the set $\mathcal{M}_M^L$ of measures of the form $\mu_L^{(u,v,F,G)}$ defined by

$$\mu_L^{(u,v,F,G)}(^*f_i) = b_i, i = 1, \ldots, M. \tag{18}$$

The following two propositions show that the solution of our problem is a *global optimizer.*

Proposition 1 *(i) The infima associated with the problems* (13 – 14) *and* (17 – 18) *are equal.*
(ii) For any positive infinitesimal $s \in {}^*\overline{\mathbb{R}}$, we can find a near-minimizer $\mu_s \in \mathcal{M}_M^L$ *for the functional J in (17) in the set $\mathcal{M}_M^L$, so that*

$$J(\mu_s) = \inf_{\mathcal{M}_M^L} J + s. \tag{19}$$

Let, then, s be a fixed positive infinitesimal in ${}^*\overline{\mathbb{R}}$, and μ_s the corresponding near-minimizer for J on $\mathcal{M}_M^L$. We can proceed to map back this measure to the standard world, by means of the *standard part map*, see [2].

Proposition 2 *There is a Baire measure μ_{opt} on Ω' so that:*
(i) If S is a Baire set in Ω',

$$\mu_{opt}(S) = {}^\circ\mu_s(\mathrm{st}_{\Omega'}^{-1}(S)),$$

where $\mathrm{st}_{\Omega'}^{-1}(S)$ is the union of the monads of the elements of S.
(ii)

$$\mu_{opt}(f_0) := \int_{\Omega'} f_0 d\,\mu_{opt} \leq \inf_{\mathcal{F}_M} \int_{J\times J} f_0(t, x, u, v, u_x, v_x, F, G)\, dtdx.$$

(iii) The measure μ_{opt} is a solution of the following optimization problem. Minimize

$$\mu(f_0) \tag{20}$$

over the set $\mathcal{M}_M^+(\Omega')$ of positive Baire measures on Ω' satisfying

$$\mu(f_i) = b_i, i = 1, \ldots, M. \tag{21}$$

(iv) If the support of μ_{opt} contains subsets of Ω' in which at least one of the variables $u_x, v_x \in \overline{\mathbb{R}}$ is either $-\infty$ or ∞, the measure μ_{opt} is defined by a Baire measure on

$$J \times J \times A \times B \times \mathbb{R} \times \mathbb{R} \times U \times U$$

plus atomic measures on those subsets.

Note that the infimum for the problem (20-21) can be strictly less than the classical infimum, as indicated in (ii) above; this is discussed in detail in [3], Chapters 4-5. In fact, values $\pm\infty$ do not actually happen for λ standard. Even so, in problems of interest, in which the function f_0 tends to infinity at infinity, and in which the infimum is finite, elements in Ω' with value ∞ or $-\infty$ do not really occur anyhow in the support of μ_{opt}; note that expressions such as $\infty - \infty$ are not defined for the extended real line. Thus,

Proposition 3 *Suppose that*

$$|f_0(t, x, u, v, u_x, v_x, F, G)| = \infty$$

whenever u_x or v_x are either ∞ or $-\infty$, and that the minimum associated with the linear program (20)-(21) is finite. Then such elements are not present in the support of μ_{opt}.

We are now in a strong position to solve our original problem— the optimization problem (8) and (9) in the standard world. Note that we have been able to construct an extension of the original space $\mathcal{F}_M$, made up of elements which are not quadruples; however, the action of μ_{opt}—a global optimizer—can be approximated by members of $\mathcal{F}_M$.

4 Existence and approximation

From the results of the optimization problem (20-21) we can build a near-optimal control pair (F, G); these are piecewise continuous functions on $\mathbb{R}^2$ with support in $J \times J$. We consider then the equations in $G_{s,g}(\mathbb{R} \times [0, \infty))$:

$$\begin{aligned} \lambda U_{xx} + U_t + UU_x &= \hat{F} \\ \lambda V_{xx} + V_t + UV_x &= \hat{G} \\ U(0, \cdot) = U_0, V(), \cdot) &= V_0. \end{aligned} \tag{22}$$

Here $U_0, V_0 \in G_{s,g}(\mathbb{R}), \hat{F}, \hat{G} \in G_{s,g}(\mathbb{R})$; of course, $\hat{F}, \hat{G}, U_0, V_0$ are sequences of smooth functions obtained by mollifying F, G, u_0, v_0, respectively.
It is not difficult to prove the existence of the solutions of (22) in $G_{s,g}(\mathbb{R} \times [0, \infty))$; see [4, 14]. Then

Proposition 4 *Suppose that the minimum in the linear program (20)-(21) is finite, that the conditions of Proposition 3 are satisfied, and that the function f_0 is Lipschitz, that is, that there is a constant h so that*

$$\begin{aligned}
&|f_0(t',x',u',v',z',r',F',G') - f_0(t,x,u,v,z,r,F,G)| \\
&\leq h(|t'-t| + |x'-x|| + |u'-u| + \\
&|v'-v| + |z'-z| + |r'-r| + |F'-F| + |G'-G|)
\end{aligned} \tag{23}$$

for all $(t',x',u',v',z',r',F',G'),(t,x,u,v,z,r,F,G)$ in Ω. Then it is possible to construct a quadruple in $\mathcal{F}_M$ so that as $M_1, M_2, M_3, M_4 \to \infty$, the corresponding value of the performance index tends to

$$\inf_S \mu(f_0),$$

with $S := \cap_{(M_1,M_2,M_3,M_4 \in \mathrm{N})} S(M_1,M_2,M_3,M_4)$, where $S(M_1,M_2,M_3,M_4)$ is the set of positive Baire measures on Ω' satisfying (21); recall that $M = M_1+M_2+2M_3+M_4$.

5 Further approximation

We consider again the optimization problem (20),(21). By means of a result of Rosenbloom, and since Ω' is compact, we can state that the minimizer μ_{opt} for this problem is a linear combination of atomic measures; thus, we wish to minimize

$$\sum_{\ell=1}^{M} \alpha_\ell f_0(w_\ell), \tag{24}$$

on the set defined by the elements $\alpha_\ell \geq 0, w_\ell \in \Omega', \ell = 1, \ldots, M$, which satisfy

$$\sum_{\ell=1}^{M} \alpha_\ell f_i(w_\ell) = b_i, i = 1, \ldots, M, \tag{25}$$

Note that we have in (24)-(25) a *nonlinear* optimization problem, in which the unknowns are the coefficients α_ℓ and supports $w_\ell, \ell = 1, \ldots, M$. In order to find a linear approximation to this problem, we consider ω, a countable dense subset of Ω'. Taking $N >> M$ elements from ω, including all elements of the form introduced in (iv) above in which some variables take values either $-\infty$ or ∞, we can write (24)-(25) as follows. We wish to minimize

$$\sum_{\ell=1}^{N} \alpha_\ell f_0(w_\ell), \tag{26}$$

on the set defined by the elements $\alpha_\ell \geq 0, \ell = 1, \dots, M$, which satisfy, further,

$$\sum_{\ell=1}^{N} \alpha_\ell f_i(w_\ell) = b_i, i = 1, \dots, M. \tag{27}$$

Here, then, the supports w_ℓ are fixed, in ω; the coefficients $\alpha_\ell, \ell = 1, \dots, M$, are the only unknowns; this is an $M \times N$ (finite-dimensional) linear program. Of course as $N \to \infty$, the support of the optimal measure μ_{opt} in (24)-(25) can be approximated closer and closer by that of μ_{opt}^N, the solution of (26)-(27). The approximation process has been studied in detail in [3], and a quadruple in $\mathcal{F}_M$ can be constructed approximating the action of μ_{opt} on f_0. It is necessary to modify the set ω into a set ω_Q, in which the coordinates of the u_x, v_x directions take values in a portion of the dense set ω defined by a number Q; if Q is large enough, all the elements in the support of μ_{opt} will be approximated adequately. Then,

Proposition 5 *Suppose that the minimum in the linear program (20)-(21) is finite, that the conditions of Proposition 3 are satisfied, and that the function f_0 is Lipschitz. Then it is possible to construct suboptimal admissible control pairs (F, G) so that:*
(i) As $Q \to \infty$, the corresponding values of the performance criterion tend to $\mu_{opt}^N(f_0)$.
(ii) As $N \to \infty$,

$$\mu_{opt}^N(f_0) \to \mu_{opt}(f_0).$$

The proof is much as that of Proposition 4 in [8]. The actual construction is explained in detail in [11] and [10]; see also the next Section.

6 An example

We have carried out the numerical computations associated with the estimation of a nearly-optimal control for the system (1) introduced in the last two sections. We took $F \equiv 0$, $v_0 \equiv 0$, while $u_0(x) = 0.5, x \in (0.5, 1]$, zero otherwise. The variables u, v, G are constrained to take values in the same set, $[0, 10]$; we took $\lambda = 0.01$ and aimed to approximate the high values of u_x, v_x by making their domain the set $[0, 1000]$, that is $Q = 1000$. Our aim was to control the extent of the shocks in the variable v, so we took

$$f_0(t, x, u, v, u_x, v_x, F, G) := v^2 + v_x^2.$$

We set a total of 35 equations ($M_1 = M_2 = 6, M_3 = 3, M_4 = 17$). The domains corresponding to the variables (u, v, F, v_x, u_x) were divided into 4 subintervals, those corresponding to (x, y) into 5, so that we had a total of 36864 variables. The suboptimal control is shown in Figure 1, constructed by means explained in detail in [11], [12]; the variable v_x in Figure 2 constructed by the same means.

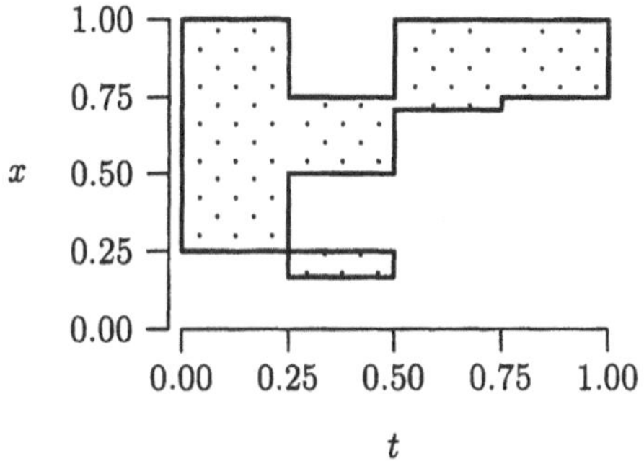

FIG. 1. *Graph of the suboptimal control, taking a value 10 at the shaded area of the (t, x)-plane, zero otherwise.*

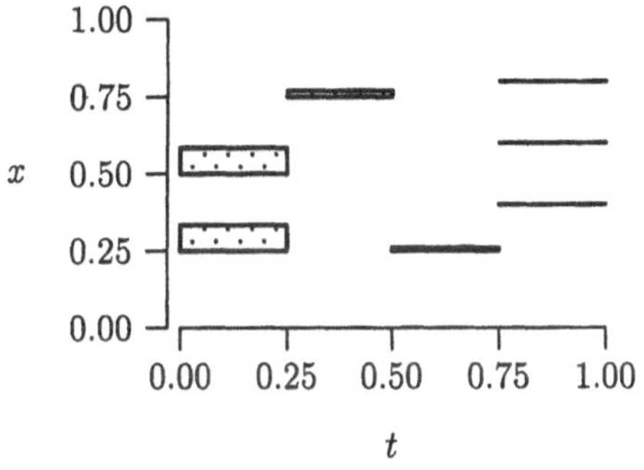

FIG. 2. *Graph of the variable v_x, taking a value 1000 at the shaded area of the (t, x)-plane, zero otherwise.*

Several points can be made concerning these results:

- In both figures we show the supports of the functions F and v_x in the (t, x)-plane, taking values 10 and 1000, respectively. Figure 2 shows therefore the regions in this plane where the function v_x has large positive values, a precursor of the shocks most likely to be exhibited when λ is taken as an infinitesimal, as explained in the next section.

- Note then the action of the control in minimizing these areas: as times goes by, from 0 to 1, these 'shocks' get weaker and weaker, the initial ones caused by the initial conditions disappearing at the end into very weak regions; the last three 'shocks' are not to scale, their width being about 1/100 the width of the preceding one.

- These results are tentative; a thorough numerical investigation is needed, with much larger matrices and many values of λ; also, the actual numerical solution of (1) corresponding to the control F should be obtained for each value of λ.

7 A final step

Finally, we go back to our original nonstandard setting, associated with equation (2). If we solve the optimization problem for each value $\lambda_p, p \in \Pi$, obtaining for instance each time a suboptimal admissible control pair $(F, G)_p$, we can say that the nonstandard object

$$[< (F, g)_p >]$$

is a suboptimal pair for the infinitesimal λ defined in (2). Presumably, as explained above, the 'shocks' of the example would become actual shocks, with infinite slopes.

References

[1] RUBIO, J. E.: The global control of shock waves, Preprint No. 532, Erwin-Schrödinger-Institute (ESI), Vienna, 1998.

[2] RUBIO, J. E. : Optimization and nonstandard analysis, Marcel Dekker 1994.

[3] RUBIO, J. E.: Control and optimization: The linear treatment of nonlinear problems, MUP-Wiley 1986.

[4] OBERGUGGENBERGER, M.: Multiplication of distributions and applications to partial differential equations, Longman 1992.

[5] LAWDEN, D. F.: Discontinuous solutions of variational problems, *Journal of the Australian Mathematical Society, 1, 27*, 1959.

[6] BRESSAN, A. AND RAMPAZZO, F.: On differential systems with quadratic impulses and their application to Lagrangian mechanics, *SIAM Journal on Control and Optimization, 31,1205*, 1993.

[7] RENDER, H.: Pushing down Loeb measures, *Mathematica Scandinavica,72,61*, 1993.

[8] RUBIO, J. E.: The global optimization of variational problems with discontinuous solutions, *Journal of Global Optimization, 7, 1*, 1997.

[9] RUBIO, J.E.: Semi-infinite programming and control Problems, Chapter in Encyclopedia of Optimization (P. M. Pardalos and C. A. Floudas, Eds.), Kluwer 1998.

[10] FARAHI, M. H., RUBIO, J. E. AND WILSON, D. A.: The global control of a nonlinear wave equation, *International Journal on Control, 65,1*, 1996.

[11] FAKHARZADEH, A.: Shapes, measures and elliptic equations, PhD Thesis, University of Leeds, 1997.

[12] FARAHI, M. H.: The boundary control of the wave equation, PhD Thesis, University of Leeds, 1996.

[13] ARAGONA, J. AND BIAGIONI, H. A.: Intrinsic definition of the Colombeau algebra of generalized functions, *Analysis Mathematica, 17, 75*, 1991.

[14] BIAGIONI, H. A. AND OBERGUGGENBERGER, M.: Generalized solutions to Burger's equation, *Journal of Differential Equations, 97,263*, 1992.

[15] HOPF, H.: The partial differential equation $u_t + uu_x = \mu u_{xx}$, *Communications on Pure and Applied Mathematics, 3, 201*, 1950.

[16] COLOMBEAU, J. F, HEIBIG, A. AND OBERGUGGENBERGER, M.: Generalized solutions to partial differential equations of evolution type, *Acta Applicandae Mathematicae, 45, 115*, 1996.

J. E. RUBIO
University of Leeds, School of Mathematics, Leeds LS2 9JT, England, U.K., E-mail: amt6jer@leeds.ac.uk

Pointwise Values and Fundamental Theorem in the Algebra of Asymptotic Functions

TODOR D. TODOROV

Introduction: The algebra of asymptotic functions ${}^{\rho}\mathcal{E}(\Omega)$ on an open set $\Omega \subset \mathbb{R}^d$ was introduced by M. Oberguggenberger [8] and the author of this paper [14] in the framework of A. Robinson's nonstandard analysis. It can be described as a differential associative and commutative ring (of generalized functions) which is an algebra over the field of A. Robinson's asymptotic numbers ${}^{\rho}\mathbb{C}$ (A. Robinson [13] and A. H. Lightstone and A. Robinson [6]). Moreover, ${}^{\rho}\mathcal{E}((\Omega)$ is supplied with the chain of imbeddings:

$$\mathcal{E}(\Omega) \subset \mathcal{D}'(\Omega) \subset {}^{\rho}\mathcal{E}(\Omega), \tag{0.1}$$

where $\mathcal{E}(\Omega)$ denotes the differential ring of the C^∞-functions (complex valued) on $\Omega, \mathcal{D}(\Omega)$ denotes the differential ring of the functions in $\mathcal{E}(\Omega)$ with compact support in Ω and $\mathcal{D}'(\Omega)$ denotes the differential linear space of Schwartz distributions on Ω. Here $\mathcal{E}(\Omega) \subset \mathcal{D}'(\Omega)$ is the usual imbedding in the sense of distribution theory (V. Vladimirov [16]). The imbedding $\mathcal{D}'(\Omega) \subset {}^{\rho}\mathcal{E}(\Omega)$, constructed in (M. Oberguggenberger and T. Todorov [10]), preserves all linear operations, including partial differentiation of any order, and the pairing between $\mathcal{D}'(\Omega)$ and $\mathcal{D}(\Omega)$. Finally, the imbedding $\mathcal{E}(\Omega) \subset {}^{\rho}\mathcal{E}(\Omega)$ preserves all differential ring operations. In addition, if T^d denotes the usual topology on $\mathbb{R}^d$, then the family $\{{}^{\rho}\mathcal{E}(\Omega)\}_{\Omega \in T^d}$ is a sheaf of differential rings and the imbeddings in (0.1) are sheaf preserving. That means, in particular, that the restriction $F|\Omega'$ is a well-defined element in ${}^{\rho}\mathcal{E}(\Omega')$ for any $F \in {}^{\rho}\mathcal{E}(\Omega)$ and any open $\Omega' \subseteq \Omega$. On these grounds we consider the asymptotic functions in ${}^{\rho}\mathcal{E}(\Omega)$ as "generalized functions on Ω": they are "generalized" functions (rather than "classical") because they are not mappings from Ω into $\mathbb{C}$. On the other hand, they are still "functions" (although, "generalized" ones) : **a)** because of the imbeddings (0.1) and **b)** because of the sheaf properties mentioned above. As a result, ${}^{\rho}\mathcal{E}(\Omega)$, supplied with the imbeddings (0.1), offers a solution to the problem for multiplication of Schwartz distributions - similar to the solution to the same problem presented not long ago by J. F. Colombeau (and other authors) in the framework of standard analysis (J. F. Colombeau [1]-[2]). Our interest in the algebra ${}^{\rho}\mathcal{E}(\Omega)$ is due to its importance for the theory of partial differential equations — linear (T. Todorov [15]) and nonlinear (M. Oberguggenberger [9]).

The purpose of this paper is to show that ${}^{\rho}\mathcal{E}(\Omega)$ is isomorphic to a subring of the ring $({}^{\rho}\mathbb{C})^{{}^{\rho}\tilde{\Omega}}$ of the pointwise functions from ${}^{\rho}\tilde{\Omega}$ into ${}^{\rho}\mathbb{C}$. Here ${}^{\rho}\mathbb{R}$ and ${}^{\rho}\mathbb{C}$ are the fields of the A. Robinson's asymptotic numbers (A. Robinson [13]), real and complex, respectively, ${}^{\rho}\mathbb{R}^d$ is the corresponding vector space, and ${}^{\rho}\tilde{\Omega} = \Omega + \mathcal{I}({}^{\rho}\mathbb{R}^d)$ denotes the set of the points in ${}^{\rho}\mathbb{R}^d$ of the form $x + h$, where $x \in \Omega$ and h is an infinitesimal point in ${}^{\rho}\mathbb{R}^d$. As a result, every asymptotic function in ${}^{\rho}\mathcal{E}(\Omega)$ can be identified with a unique pointwise function defined on ${}^{\rho}\tilde{\Omega}$ taking values in ${}^{\rho}\mathbb{C}$. In the particular case of classical smooth functions our evaluation coincides with the usual evaluation. Moreover, since ${}^{\rho}\mathcal{E}(\Omega)$ contains a copy of the space of Schwartz distributions, it follows that the distributions are also pointwise functions taking values in ${}^{\rho}\mathbb{C}$. We also prove the Fundamental Theorem of Integration Calculus in ${}^{\rho}\mathcal{E}(\Omega)$ and show that the set of the scalars of ${}^{\rho}\mathcal{E}(\Omega)$ coincides exactly with the field ${}^{\rho}\mathbb{C}$ and the set of the real scalars of ${}^{\rho}\mathcal{E}(\Omega)$ coincides exactly with the field ${}^{\rho}\mathbb{R}$.

Recall that if L is a differential linear space, then the set of the scalars $\mathcal{S}$ in L consists of the elements F in L with zero partial first derivatives $\partial^{\alpha} F = 0, |\alpha| = 1$, in symbols, $\mathcal{S} = \{f \in L : \nabla F = 0\}$. If, in addition, S is supplied with complex conjugation, then the set of the real scalars in L is given by $\mathrm{Re}(\mathcal{S}) = \{\pm|F| : F \in \mathcal{S}\}$.

The theory of asymptotic functions is similar to, but somewhat different from, J. F. Colombeau's nonlinear theory of generalized functions, where we also have a chain of imbeddings $\mathcal{E}(\Omega) \subset \mathcal{D}'(\Omega) \subset \mathcal{G}(\Omega)$ with similar properties. Here $\mathcal{G}(\Omega)$ denotes a typical algebra of J. F. Colombeau's generalized functions (J. F. Colombeau [1]-[2]). The set of the scalars $\bar{\mathbb{C}}$, and the real scalars $\bar{\mathbb{R}}$ of $\mathcal{G}(\Omega)$ are called J. F. Colombeau's generalized numbers (J. F. Colombeau [2], 2.1). Recently, M. Oberguggenberger and M. Kunzinger [11] proved that the algebras of the type $\mathcal{G}(\Omega)$ can be canonically imbedded in the ring $\bar{\mathbb{C}}^{\tilde{\Omega}_c}$ of pointwise functions from $\tilde{\Omega}_c$ into $\bar{\mathbb{C}}$, where $\tilde{\Omega}_c$ is a particular set of generalized points such that $\Omega \subset \tilde{\Omega}_c \subset \bar{\mathbb{R}}^d$. Thus, the generalized functions in $\mathcal{G}(\Omega)$ can also be viewed as pointwise functions.

The difference between the algebra of asymptotic functions ${}^{\rho}\mathcal{E}(\Omega)$ and J. F. Colombeau's algebras of generalized functions of the type $\mathcal{G}(\Omega)$ can be best expressed by comparing their scalars:

a) The set of the scalars $\bar{\mathbb{C}}$ of $\mathcal{G}(\Omega)$ is a non-Archimedean ring with zero divisors, containing a copy of $\mathbb{C}$. In contrast, the set of the scalars ${}^\rho\mathbb{C}$ of ${}^\rho\mathcal{E}(\Omega)$ is an algebraically closed non-Archimedean field, containing a copy of $\mathbb{C}$. In addition, ${}^\rho\mathbb{C}$ is Cantor complete in the sense that every sequence of closed circles in ${}^\rho\mathbb{C}$ with the finite intersection property has a non-empty intersection in ${}^\rho\mathbb{C}$. Finally, ${}^\rho\mathbb{C}$ contains a canonical copy of the T. Levi-Cività field of asymptotic series with complex coefficients (that property gives the name "asymptotic numbers" for the elements of ${}^\rho\mathbb{C}$).

b) The set of the real scalars $\bar{\mathbb{R}}$ of $\mathcal{G}(\Omega)$ is a partially ordered non-Archimedean ring with zero divisors containing a copy of $\mathbb{R}$. In contrast, the set of the real scalars ${}^\rho\mathbb{R}$ of ${}^\rho\mathcal{E}(\Omega)$ is a real closed non-archimedean field extension of $\mathbb{R}$ which is Cantor complete in the sense that every sequence of closed intervals in ${}^\rho\mathbb{R}$ with the finite intersection property has a non-empty intersection in ${}^\rho\mathbb{R}$. Moreover, ${}^\rho\mathbb{R}$ contains a canonical copy of the T. Levi-Cività field of asymptotic series with real coefficients (A. H. Lightstone and A. Robinson [6], p.93).

The improvement of the properties of the scalars ${}^\rho\mathbb{R}$ and ${}^\rho\mathbb{R}$ of ${}^\rho\mathcal{E}(\Omega)$ compared with the properties of the scalars $\bar{\mathbb{C}}$ and $\bar{\mathbb{R}}$ in J. F. Colombeau's theory is the main reason to involve the methods of nonstandard analysis in our approach.

1 Preliminaries: asymptotic numbers, asymptotic vectors and asymptotic functions

Our framework is a nonstandard model of the real numbers $\mathbb{R}$, with degree of saturation larger than card($\mathbb{N}$). Throughout this paper Ω denotes an open subset of $\mathbb{R}^d$. We denote by ${}^*\mathbb{R}$, ${}^*\mathbb{R}_+$, ${}^*\mathbb{C}$, ${}^*\mathcal{E}(\Omega)$ and ${}^*\mathcal{D}(\Omega)$ the nonstandard extensions of $\mathbb{R}$, $\mathbb{R}_+$, $\mathbb{C}$, $\mathcal{E}(\Omega)$ and $\mathcal{D}(\Omega)$, respectively. In particular, ${}^*\mathcal{E}(\Omega)$ consists of pointwise functions of the form f: ${}^*\Omega \to {}^*\mathbb{C}$. If X is a set of complex numbers or a set of (standard) functions, then ${}^*X$ will be its nonstandard extension and if $f : X \to Y$ is a (standard) mapping, then ${}^*f : {}^*X \to {}^*Y$ will be its nonstandard extension. For integration in ${}^*\mathbb{R}^d$ we use the $*$-Lebesgue integral. For a very short introduction to nonstandard analysis we refer to the Appendix in T. Todorov [15]. For a more detailed exposition we recommend T. Lindstrøm [5], where the reader will find many references to the subject.

The purpose of this section is to present a summary of those definitions and results of *Nonstandard Asymptotic Analysis* which we need for our exposition. For more details we refer to A. Robinson [13], A.H. Lightstone and A. Robinson [6], W.A.J. Luxemburg [7], Li Bang-He [4] and V. Pestov [12]. For the algebra of asymptotic functions and its applications we refer to M. Oberguggenberger [8], T. Todorov [14], R. F. Hoskins and J. Sousa Pinto [3], M. Oberguggenberger and T. Todorov [10], M. Oberguggenberger [9] and T. Todorov [15].

Part 1. *Asymptotic Numbers:* Let $\rho \in {}^*\mathbb{R}$ be a positive infinitesimal. We shall keep ρ fixed in what follows. Following A. Robinson [13] we define the field of A. Robinson's complex *ρ-asymptotic numbers* as the factor space ${}^\rho\mathbb{C} = \mathcal{M}_\rho({}^*\mathbb{C})/\mathcal{N}_\rho({}^*\mathbb{C})$, where

$$\mathcal{M}_\rho({}^*\mathbb{C}) = \{\xi \in {}^*\mathbb{C} : |\xi| < \rho^{-n} \quad \text{for some} \quad n \in \mathbb{N}\}, \tag{1.1}$$

$$\mathcal{N}_\rho({}^*\mathbb{C}) = \{\xi \in {}^*\mathbb{C} : |\xi| < \rho^{n} \quad \text{for all} \quad n \in \mathbb{N}\}, \tag{1.2}$$

are the sets of the r-moderate and ρ-null numbers in ${}^*\mathbb{C}$, respectively. We define the embedding $\mathbb{C} \subset {}^\rho\mathbb{C}$ by $c \to q(c)$, where $q : \mathcal{M}_\rho({}^*\mathbb{C}) \to {}^\rho\mathbb{C}$ is the quotient mapping. If $\mathcal{A} \subseteq {}^*\mathbb{C}$, we let

$$\mathcal{M}_\rho(\mathcal{A}) = \{\xi \in \mathcal{A} : |\xi| < \rho^{-n} \quad \text{for some} \quad n \in \mathbb{N}\}, \tag{1.3}$$

and ${}^\rho\mathcal{A} = q[\mathcal{M}_\rho(\mathcal{A})]$. The set ${}^\rho\mathcal{A} \subseteq {}^\rho\mathbb{R}$ is called the ρ-extension of $\mathcal{A}$ in ${}^\rho\mathbb{R}$. In the particular case $\mathcal{A} = {}^*A$, where $A \subseteq \mathbb{C}$, we shall write simply ${}^\rho A$ instead of the more precise ${}^\rho({}^*A)$. The field of *A. Robinson's real ρ-asymptotic numbers* is defined by ${}^\rho\mathbb{R} = q[\mathcal{M}_\rho({}^*\mathbb{R})]$. The imbedding $\mathbb{R} \subset {}^\rho\mathbb{R}$ is defined by $r \to q(r)$. The asymptotic number $s = q(\rho)$ is the scale of ${}^\rho\mathbb{R}$; it is a positive infinitesimal, i.e., $s \in {}^\rho\mathbb{R}$, $s > 0$ and $s \approx 0$. Finally, we observe that if $\alpha \in {}^*\mathbb{C}$, then $\alpha \in \mathcal{M}_\rho({}^*\mathbb{C})$ iff $|\alpha| \in \mathcal{M}_\rho({}^*\mathbb{R})$ and, similarly, $\alpha \in \mathcal{N}_\rho({}^*\mathbb{C})$ iff $|\alpha| \in \mathcal{N}_\rho({}^*\mathbb{R})$.

It is easy to check that $\mathcal{N}_\rho({}^*\mathbb{C})$ is a maximal ideal in $\mathcal{M}_\rho({}^*\mathbb{C})$ and hence ${}^\rho\mathbb{C}$ is a field. Also ${}^\rho\mathbb{R}$ is a real closed totally ordered non-Archimedean field (since ${}^*\mathbb{R}$ is a real closed totally ordered field) containing $\mathbb{R}$ as a totally ordered subfield. Thus, it follows that ${}^{\rho}\mathbb{C} = {}^{\rho}\mathbb{R}(i)$ is an algebraically closed field, where $i = \sqrt{-1}$. Finally ${}^{\rho}\mathbb{R}$ is Cantor complete in the sense that every sequence of closed intervals in ${}^{\rho}\mathbb{R}$ with the finite intersection property has a non-empty intersection (since ${}^{*}\mathbb{R}$ has this property). Unless it is said otherwise, we shall always topologize ${}^{\rho}\mathbb{R}$ and ${}^{\rho}\mathbb{C}$ by the order topology in ${}^{\rho}\mathbb{R}$. The sequence of intervals:

$$I_n = \{x \in {}^{\rho}\mathbb{R} : -s^n < x < s^n\}, \tag{1.4}$$

forms a base for the neighborhoods of the zero in the *interval topology* in ${}^{\rho}\mathbb{R}$.

Part 2. *Asymptotic Vectors*: Let ${}^{*}\mathbb{R}^d$ and ${}^{\rho}\mathbb{R}^d$ be the corresponding d-dimensional vector spaces over the fields ${}^{*}\mathbb{R}$ and ${}^{\rho}\mathbb{R}$, repsectively. We shall write always $\|\ \|$ for the norms of $\mathbb{R}^d$, ${}^{*}\mathbb{R}^d$ and ${}^{\rho}\mathbb{R}^d$, leaving to the reader to decide from the context which of these norms we have in mind. Notice that $\|\ \|$ takes non-negative values in $\mathbb{R}$, ${}^{*}\mathbb{R}$ and ${}^{\rho}\mathbb{R}$, respectively. In the case $d = 1$ this norm reduces to the absolute value in $\mathbb{R}$, ${}^{*}\mathbb{R}$ and ${}^{\rho}\mathbb{R}$, respectively. It is clear that $\mathbb{R}^d$ is a vector subspace of both ${}^{*}\mathbb{R}^d$ and ${}^{\rho}\mathbb{R}^d$ over $\mathbb{R}$, in symbols, $\mathbb{R}^d \subset {}^{*}\mathbb{R}^d$ and $\mathbb{R}^d \subset {}^{\rho}\mathbb{R}^d$. We call the elements in ${}^{\rho}\mathbb{R}^d$ *asymptotic points* or *asymptotic vectors*. Unless it is said otherwise, we shall always topologize ${}^{\rho}\mathbb{R}^d$ by the *product-interval topology*, i.e., by the product topology in ${}^{\rho}\mathbb{R}^d$, generated by the order topology in ${}^{\rho}\mathbb{R}$. The sequence of balls:

$$B_n = \{x \in {}^{\rho}\mathbb{R}^d : \|x\| < s^n\}, \quad n \in \mathbb{N}, \tag{1.5}$$

forms a base for the neighborhoods of the zero in ${}^{\rho}\mathbb{R}^d$.

Let $x \in {}^{\rho}\mathbb{R}^d$. Then x is called *infinitesimal, finite or infinitely large* if its ρ-norm $\|x\|$ is infinitesimal, finite or infinitely large in ${}^{\rho}\mathbb{R}$, respectively. We denote by $\mathcal{F}({}^{*}\mathbb{R}^d)$ and $\mathcal{F}({}^{\rho}\mathbb{R}^d)$ the sets of the finite elements of ${}^{*}\mathbb{R}^d$ and ${}^{\rho}\mathbb{R}^d$, respectively, and by $\mathcal{I}({}^{*}\mathbb{R}^d)$ and $\mathcal{I}({}^{\rho}\mathbb{R}^d)$ we denote the sets of the infinitesimals points in ${}^{*}\mathbb{R}^d$ and ${}^{\rho}\mathbb{R}^d$, respectively. If Ω is an open set of $\mathbb{R}^d$, then $\tilde{\Omega} = \Omega + \mathcal{I}({}^{*}\mathbb{R}^d)$ denotes the set of the *nearstandard points of* ${}^{*}\Omega$ in ${}^{*}\mathbb{R}^d$ and ${}^{\rho}\tilde{\Omega} = \Omega + \mathcal{I}({}^{\rho}\mathbb{R}^d)$ denotes the set of the *nearstandard points of* ${}^{*}\Omega$ in ${}^{\rho}\mathbb{R}^d$, in symbols,

$$\tilde{\Omega} = \Omega + \mathcal{I}({}^{*}\mathbb{R}^d) = \{x + \xi \,:\, x \in \Omega,\ \xi \in \mathcal{I}({}^{*}\mathbb{R}^d)\}, \tag{1.6}$$

$${}^{\rho}\tilde{\Omega} = \Omega + \mathcal{I}({}^{*}\mathbb{R}^d) = \{x + h \,:\, x \in \Omega,\ h \in \mathcal{I}({}^{*}\mathbb{R}^d)\}, \tag{1.7}$$

where the sum in $x+\xi$ is in ${}^*\mathbb{R}^d$, while the sum in $x+h$ is in ${}^\rho\mathbb{R}^d$. We observe that $\Omega \subset \tilde{\Omega} \subseteq {}^*\Omega$ and $\Omega \subset {}^\rho\tilde{\Omega} \subseteq {}^\rho\Omega$. In the particular case $\Omega = \mathbb{R}^d$ we have

$$\mathbb{R}^d + \mathcal{I}({}^*\mathbb{R}^d) = \mathcal{F}({}^*\mathbb{R}^d) \text{ and } \mathbb{R}^d + \mathcal{I}({}^\rho\mathbb{R}^d) = \mathcal{F}({}^\rho\mathbb{R}^d). \tag{1.8}$$

We denote by $({}^\rho\mathbb{C})^{{}^\rho\tilde{\Omega}}$ or $({}^\rho\mathbb{C})^{\Omega+\mathcal{I}({}^\rho\mathbb{R}^d)}$ the ring (under the pointwise operations) of the functions from ${}^\rho\tilde{\Omega}$ into ${}^\rho\mathbb{C}$.

The purpose of the next several lines is to connect the sets $\tilde{\Omega}$ and ${}^\rho\tilde{\Omega}$. First, we observe that the vector space ${}^\rho\mathbb{R}^d$ can, alternatively, be constructed as the factor space ${}^\rho\mathbb{R}^d = \mathcal{M}_\rho({}^*\mathbb{R}^d)/\mathcal{N}_\rho({}^*\mathbb{R}^d)$, where

$$\mathcal{M}_\rho({}^*\mathbb{R}^d) = \{\xi \in {}^*\mathbb{R}^d : \|\xi\| < \rho^{-n} \quad \text{for some} \quad n \in \mathbb{N}\}, \tag{1.9}$$

$$\mathcal{N}_\rho({}^*\mathbb{R}^d) = \{x \in {}^*\mathbb{R}^d : \|x\| < \rho^{-n} \quad \text{for all} \quad n \in \mathbb{N}\}, \tag{1.10}$$

are the sets of the ρ-moderate and ρ-null points in ${}^*\mathbb{R}^d$, respectively. We denote by $q^d : \mathcal{M}_\rho({}^*\mathbb{R}^d) \to {}^\rho\mathbb{R}^d$ the corresponding quotient mapping. We let $q^1 = q$ in the case $d = 1$. The imbedding $\mathbb{R}^d \subset {}^\rho\mathbb{R}^d$ is reproduced by $x \to q^d(x)$. If $\mathcal{A} \subseteq {}^*\mathbb{R}^d$, we let

$$\mathcal{M}_\rho(\mathcal{A}) = \{\xi \in \mathcal{A} : \|\xi\| < \rho^{-n} \quad \text{for some} \quad n \in \mathbb{N}\}, \tag{1.11}$$

and ${}^\rho\mathcal{A} = q[\mathcal{M}_\rho(\mathcal{A})]$. The set ${}^\rho\mathcal{A} \subseteq {}^\rho\mathbb{R}^d$ is called the ρ-extension of $\mathcal{A}$ in ${}^\rho\mathbb{R}^d$. In the particular case $\mathcal{A} = {}^*A$, where $A \subseteq \mathbb{R}^d$, we shall write simply ${}^\rho A$ instead of the more precise ${}^\rho({}^*A)$. We observe that ${}^\rho\tilde{\Omega}$ is the ρ-extension of $\tilde{\Omega}$ and also we have:

$$\xi \in \tilde{\Omega} \Leftrightarrow q^d(\xi) \in {}^\rho\tilde{\Omega}. \tag{1.12}$$

For more details about the vector spaces over the field ${}^\rho\mathbb{R}$ we refer to W.A.J. Luxemburg [7].

Part 3. *Asymptotic Functions*: The algebra of *asymptotic functions* ${}^\rho\mathcal{E}(\Omega)$ on Ω is the factor space ${}^\rho\mathcal{E}(\Omega) = \mathcal{M}_\rho({}^*\mathcal{E}(\Omega))/\mathcal{N}\rho({}^*\mathcal{E}(\Omega))$, where

$$\mathcal{M}_\rho({}^*\mathcal{E}(\Omega)) = \{f \in {}^*\mathcal{E}(\Omega) : (\forall\alpha \in \mathbb{N}_0^d)(\forall\xi \in \tilde{\Omega})(\partial^\alpha f(\xi) \in \mathcal{M}_\rho({}^*\mathbb{C}))\}, \tag{1.13}$$

$$\mathcal{N}_\rho({}^*\mathcal{E}(\Omega)) = \{f \in {}^*\mathcal{E}(\Omega) : (\forall\alpha \in \mathbb{N}_0^d)(\forall\xi \in \tilde{\Omega})(\partial^\alpha f(\xi) \in \mathcal{N}_\rho({}^*\mathbb{C}))\}. \tag{1.14}$$

The functions in $\mathcal{M}_\rho({}^*\mathcal{E}(\Omega))$ are called ρ-moderate (or, simply, *moderate)* and those in $\mathcal{N}_\rho({}^*\mathcal{E}(\Omega))$ are called *ρ-null functions* (or, simply, *null functions*). The *canonical (differential ring) imbedding* $\mathcal{E}(\Omega) \subset {}^\rho\mathcal{E}(\Omega)$ is defined by $f \to \sigma(f)$, where $\sigma(f) = Q_\Omega({}^*f)$ and $Q_\Omega : \mathcal{M}_\rho({}^*\mathcal{E}(\Omega)) \to {}^\rho\mathcal{E}(\Omega)$ is the quotient mapping. For the construction of an imbedding of the Schwartz distributions: $\mathcal{E}(\Omega) \subset \mathcal{D}'(\Omega) \subset {}^\rho\mathcal{E}(\Omega)$, discussed in the Introduction of this paper, we refer to (M. Oberguggenberger and T. Todorov [10]).

2 Asymptotic functions as pointwise functions

We define an evaluation of the asymptotic functions in ${}^\rho\mathcal{E}(\Omega)$ and show that this evaluation coincides with the usual evaluation in the particular case of classical smooth functions. Moreover, we construct a ring imbedding of ${}^\rho\mathcal{E}(\Omega)$ into the ring of pointwise functions $({}^\rho\mathbb{C})^{{}^\rho\tilde{\Omega}}$, thus showing that the asymptotic functions can be characterized as pointwise functions. In particular, the Schwartz distributions also can be characterized by their values. In what follows we shall use the notations in Section 1.

Recall that every asymptotic function is, by definition, an equivalence class of nonstandard internal functions in ${}^*\mathcal{E}(\Omega)$ (Section 1, Part 3). On the other hand, the functions in ${}^*\mathcal{E}(\Omega)$ are pointwise functions from ${}^*\Omega$ into ${}^*\mathbb{C}$. We should notice, however, that the asymptotic functions $Q_\Omega(f)$ cannot inherit literally the values of its representatives f because the factorization in $Q_\Omega(f)$ destroys this evaluation.

(2.1) Definition *(Values)*: Let $F = Q(f)$ be an asymptotic function in ${}^\rho\mathcal{E}(\Omega)$ with a representative $f \in \mathcal{M}_\rho({}^*\mathcal{E}(\Omega))$. Let $x = q(\xi)$ be an asymptotic point in ${}^\rho\tilde{\Omega}$ with a representative $\xi \in \tilde{\Omega}$. The asymptotic number $q(f(\xi)) \in {}^\rho\mathbb{C}$ is called the *value* of F at x, in symbols,

$$F(x) = q(f(\xi)) \quad or \quad Q(f)(q(\xi)) = q(f(\xi)).$$

We should mention that the evaluation of f at ξ is in the framework of ${}^*\mathbb{C}$.

(2.2) Lemma*(Correctness)*: Let $\alpha, \beta \in \tilde{\Omega}$, $\|\alpha-\beta\| \in \mathcal{N}_\rho({}^*\mathbb{R})$ and $f, g \in \mathcal{M}_\rho({}^*\mathcal{E}(\Omega))$, $f - g \in \mathcal{N}_\rho({}^*\mathcal{E}(\Omega))$. Then $f(\alpha) - g(\beta) \in \mathcal{N}_\rho({}^*\mathbb{C})$ (or, equivalently, $|f(\alpha) - g(\beta)| \in \mathcal{N}_\rho({}^*\mathbb{R})$).

Proof. By the Mean Value Theorem, applied by Transfer Principle (T. Todorov [15], p. 686), there exists $\gamma \in {}^*\mathbb{R}^d$ between α and β (in the sense that $\gamma = \alpha + \varepsilon(\beta - \alpha)$ for some $\varepsilon \in {}^*\mathbb{R}, 0 \leq \epsilon \leq 1$) such that

$$|f(\alpha) - f(\beta)| = |\nabla f(\gamma) \cdot (\alpha - \beta)| \leq \|\nabla f(\gamma)\| \, \|\alpha - \beta\| \, .$$

The point γ is in $\tilde{\Omega}$, since $\alpha \approx \beta$, by assumption, thus, $\|\nabla f(\gamma)\|$ is a moderate number in $\mathcal{M}_\rho({}^*\mathbb{R})$, by the definition of $\mathcal{M}_\rho({}^*\mathcal{E}(\Omega))$. It follows that the right-hand side of the above inequality is in $\mathcal{N}_\rho({}^*\mathbb{R})$, since $\mathcal{N}_\rho({}^*\mathbb{R})$ is an ideal in $\mathcal{M}_\rho({}^*\mathbb{R})$. On the other hand, we have

$$\|f(\alpha) - g(\beta)\| = \| f(\alpha) - f(\beta) + f(\beta) - g(\beta) \| \leq \|f(\alpha) - f(\beta)\| +$$

$$+ \|f(\beta) - g(\beta)\| \leq \|\nabla f(\gamma)\| \, \|\alpha - \beta\| + \|f(\beta) - g(\beta)\| \in \mathcal{N}_\rho({}^*\mathbb{R}),$$

as required, since $\|f(\beta) - g(\beta)\| \in \mathcal{N}_\rho({}^*\mathbb{R})$, by assumption. □

Recall that the algebra of standard smooth functions $\mathcal{E}(\Omega)$ is imbedded in ${}^\rho\mathcal{E}(\Omega)$ in a canonical way, in symbols, $\mathcal{E}(\Omega) \subset {}^\rho\mathcal{E}(\Omega)$, by the mapping $f \to \sigma(f)$, where $\sigma(f) = Q({}^*f)$. Recall as well that $\mathbb{C} \subset {}^\rho\mathbb{C}$ by the mapping $c \to q(c)$. In what follows we shall identify c with its image $q(c)$, writing simply $c = q(c)$.

The next result shows that the concept of *value* for the asymptotic functions, introduced above, is a generalization of the usual evaluation of the classical functions.

(2.3) Lemma *(Standard Functions at Standard Points)*: If $f \in \mathcal{E}(\Omega)$ and $x \in \Omega$, then $\sigma(f)(x) = f(x)$.

Proof. $\sigma(f)(x) = Q({}^*f)(q(x)) = q({}^*f(x)) = q(f(x)) = f(x)$ since ${}^*f$ is an extension of f and $f(x)$ is a standard (complex) number. □

The next two lemmas are in the framework of ${}^*\mathbb{R}^d$ and ${}^*\mathbb{C}$.

(2.4) Lemma: Let $f \in \mathcal{M}_\rho({}^*\mathcal{E}(\Omega))$ and $\xi \in \tilde{\Omega}$ and $\chi \in \mathcal{N}_\rho({}^*\mathbb{R}^d)$. Then:

(i) $\|f(\xi + \chi) - f(\xi)\| \in \mathcal{N}_\rho({}^*\mathbb{C})$.

(ii) $\dfrac{\|f(\xi + \chi) - f(\xi) - \nabla f(\xi) \cdot \chi\|}{\|\chi\|} \in \mathcal{N}_\rho({}^*\mathbb{C}), \quad \chi \neq 0.$

Proof. (i) By the Mean Value Theorem applied by Transfer Principle (T. Todorov [15], p. 686), there exists $\varepsilon \in {}^*\mathbb{R}$, $0 \leq \varepsilon \leq 1$, such that

$$\|f(\xi + \chi) - f(\xi)\| = \|\nabla f(\xi + \varepsilon\chi) \cdot \chi\|$$

It follows $\|f(\xi+\chi)-f(\xi)\| \leq \|\nabla f(\xi+\varepsilon\chi)\| \, \|\chi\|$. Notice that $\xi+\varepsilon\chi$ is also in $\tilde{\Omega}$ (since $\varepsilon\chi$ is an infinitesimal point) which implies $\|\nabla f(\xi+\varepsilon\chi)\| \in \mathcal{M}_\rho({}^*\mathbb{R})$ since $f \in \mathcal{M}_\rho({}^*\mathcal{E}(\Omega))$, by assumption. Now, the result follows since $\mathcal{N}_\rho({}^*\mathbb{R})$ is an ideal in $\mathcal{M}_\rho({}^*\mathbb{R})$.

(ii) By the Taylor Theorem applied by Transfer Principle, there exists $\varepsilon \in {}^*\mathbb{R}$, $0 \leq \varepsilon \leq 1$, such that

$$\frac{\|f(\xi+\chi)-f(\xi)-\nabla f(\xi)\cdot\chi\|}{\|\chi\|} = \frac{1}{2}\left\| \sum_{|\alpha|=2} \partial^\alpha f(\xi+\varepsilon\chi)\frac{\chi^\alpha}{\|\chi\|} \right\|$$

As before, $\|\partial^\alpha f(\xi+\varepsilon\chi)\| \in \mathcal{M}_\rho({}^*\mathbb{R})$ and also we have $\chi^\alpha/\|\chi\| \in \mathcal{N}_\rho({}^*\mathbb{R})$ due to $|\alpha| = 2$. Thus, it follows that the right-hand side of the above equality is in $\mathcal{N}_\rho({}^*\mathbb{R})$, since $\mathcal{N}_\rho({}^*\mathbb{R})$ is an ideal in $\mathcal{M}_\rho({}^*\mathbb{R})$. □

In what follows $K \subset\subset \Omega$ means that K is a compact subset of Ω.

(2.5) Lemma: Let $f \in \mathcal{M}_\rho({}^*\mathcal{E}(\Omega))$. Then:

(i) $(\forall K \subset\subset \Omega)(\forall n \in \mathbb{N})(\exists m \in \mathbb{N})(\forall \xi \in {}^*K)(\forall \chi \in {}^*\mathbb{R})$
$[(\|\chi\| < \rho^m) \Rightarrow (\|f(\xi+\chi)-f(\xi)\| < \rho^n)]$.

(ii) $(\forall K \subset\subset \Omega)(\forall n \in \mathbb{N})(\exists m \in \mathbb{N})(\forall \alpha \in {}^*K)(\forall \chi \in {}^*\mathbb{R})$
$$\left[(0 < \|\chi\| < \rho^m) \Rightarrow \quad \frac{\|f(\xi+\chi)-f(\xi)-\nabla f(\xi)\cdot\chi\|}{\|\chi\|} < \rho^n\right].$$

Proof. (i) Suppose (for contradiction) that (i) fails, i.e.,

$$(\exists K \subset\subset \Omega)(\exists n \in \mathbb{N})(\forall m \in \mathbb{N})(S_m \neq \emptyset),$$

where

$$S_m = \left\{(\xi,\chi) \in {}^*K \times {}^*\mathbb{R}^d : (\|\chi\| < \rho^m) \quad \textit{and} \quad (\|f(\xi+\chi)-f(\xi)\| \geq \rho^n)\right\}.$$

It is clear that $S_m \supset S_{m+1} \supset \ldots$, hence, by the Saturation Principle (T. Todorov [15], p. 687), there exists a pair $(\xi_0, \chi_0) \in S_m$ for all $m \in \mathbb{N}$. The latter means $\|\chi_0\| \in \mathcal{N}_\rho({}^*\mathbb{R})$ and $\|f(\xi_0+\chi_0)-f(\chi_0)\| \notin \mathcal{N}_\rho({}^*\mathbb{R})$, contradicting the assumption $f \in \mathcal{M}_\rho({}^*\mathcal{E}(\Omega))$ in view of Lemma 2.4.

(ii) As above, suppose (for contradiction) that (ii) fails, i.e., $(\exists K \subset\subset \Omega)(\exists n \in$

$\mathbb{N})(\forall m \in \mathbb{N})(T_m \neq \emptyset)$, where

$$T_m = \Big\{(\xi, \chi) \in {}^*K \times {}^*\mathbb{R}^d : \\ 0 < \|\chi\| < \rho^m \quad \text{and} \quad \frac{\|f(\xi + \chi) - f(\xi) - \nabla f(\xi) \cdot \chi\|}{\|\chi\|} \geq \rho^n \Big\}.$$

As before, we observe that $T_m \subset T_{m+1} \dots$, hence, by the Saturation Principle, there exists a pair $(\xi_0, \chi_0) \in T_m$ for all $m \in \mathbb{N}$. The latter means $\|\chi_0\| \in \mathcal{N}_\rho({}^*\mathbb{R})$ and

$$\frac{\|f(\xi_0 + \chi_0) - f(\xi_0) - \nabla f(\xi_0) \cdot \chi_0\|}{\|\chi_0\|} \notin \mathcal{N}_\rho({}^*\mathbb{R}),$$

contradicting the assumption $f \in \mathcal{M}_\rho({}^*\mathcal{E}(\Omega))$ in view of Lemma 2.4. □

(2.6) Theorem: Let $F \in {}^\rho\mathcal{E}(\Omega)$, $\partial^\alpha F$ be its partial derivatives and ∇F be the corresponding gradient of F in ${}^\rho\mathcal{E}(\Omega)$. Let $x \in {}^\rho\tilde{\Omega}$ be a finite asymptotic number and $F(x)$, $(\partial^\alpha F)(x)$ and $(\nabla F)(x)$ be the values of F, $\partial^\alpha F$ and ∇F at the point x, respectively. Then:

(i) $\lim_{{}^\rho\mathbb{R}^d \ni h \to 0} F(x + h) = F(x)$.

(ii) $\lim_{{}^\rho\mathbb{R}^d \ni h \to 0} \dfrac{\|F(x + h) - F(x) - (\nabla F)(x) \cdot h\|}{\|h\|} = 0$

where both limits are in the product-interval topology of ${}^\rho\mathbb{R}^d$ (1.5).

Proof. We have $F = Q(f), x = q(\xi)$ and $h = q(\chi)$ for some $f \in \mathcal{M}_\rho({}^*\mathcal{E}(\Omega))$ some $\xi \in \tilde{\Omega}$ and some $\chi \in \mathcal{M}_\rho({}^*\mathbb{R}^d)$. Now, both (i) and (ii) follow immediately from the above lemma after taking quotient mappings Q_Ω and q^d, taking into account that the sequence of balls $\{B_n\}$ (1.5) forms a base for the neighborhoods of the zero in the product-interval topology in ${}^\rho\mathbb{R}^d$. □

(2.7) Corollary (*The Case* $\Omega = \mathbb{R}$): Let $F \in {}^\rho\mathcal{E}(\mathbb{R})$ and F' be its derivative in the algebra ${}^\rho\mathcal{E}(\mathbb{R})$. Let $x \in \mathcal{F}({}^\rho\mathbb{R})$ be a finite asymptotic number and $F(x)$ and $F'(x)$ be the values of F and F' at the point x, respectively. Then:

(i) $\lim_{{}^\rho\mathbb{R}^d \ni h \to 0} F(x + h) = F(x)$.

(ii) $\lim_{{}^\rho\mathbb{R}^d \ni h \to 0} \dfrac{F(x + h) - F(x)}{h} = F'(x)$

where both limits are in the interval topology of ${}^\rho\mathbb{R}$ (1.4) (or, equivalently, in the valuation-metric topology of ${}^\rho\mathbb{R}$).

Proof. The result follows as a particular case from the above theorem for $d = 1$ and $\Omega = R$, taking into account that in this case we have $\widetilde{\mathbb{R}} = \mathbb{R} + \mathcal{I}({}^*\mathbb{R}) = \mathcal{F}({}^*\mathbb{R})$ and ${}^\rho\widetilde{\mathbb{R}} = \mathbb{R} + \mathcal{I}({}^\rho\mathbb{R}) = \mathcal{F}({}^\rho\mathbb{R})$. □

(2.8) Theorem *(The Imbedding)*: Define $V : {}^\rho\mathcal{E}(\mathbb{R}) \to ({}^\rho\mathbb{C})^{{}^\rho\tilde{\Omega}}$, by $F \to V(F)$, where $V(F)(x) = F(x)$, $x \in {}^\rho\tilde{\Omega}$. Then V is a ring imbedding in the sense that it is injective and it preserves the addition and multiplication.

Proof. Suppose that $V(F) = 0$ in $({}^\rho\mathbb{C})^{{}^\rho\tilde{\Omega}}$, i.e., $F(x) = 0$ in ${}^\rho\mathbb{C}$ for all $x \in {}^\rho\tilde{\Omega}$. We have to show that $F = 0$ in the algebra ${}^\rho\mathcal{E}(\Omega)$, i.e., to show that $F = Q_\Omega(f)$ for some $f \in {}^*\mathcal{E}(\mathbb{R})$ such that $\partial^\alpha f(\xi) \in \mathcal{N}_\rho({}^*\mathbb{C})$ for all $\xi \in \tilde{\Omega}$ and all $\alpha \in \mathbb{N}_0^d$ (1.14)). By our assumption we have $F(x+h) - F(x) = 0$ for all $x \in {}^\rho\tilde{\Omega}$ and all $h \in \mathcal{I}({}^\rho\mathbb{R}^d)$ (since $x + h \in {}^\rho\tilde{\Omega}$ which implies $F(x+h) = 0$ along with $F(x) = 0$). It follows that $\|\nabla F(x)\| = 0$ for all $x \in^\rho \tilde{\Omega}$, by the second part of Theorem 2.6. In other words, $\partial^\alpha F(x) = 0$ for all $x \in {}^\rho\tilde{\Omega}$ and all $\alpha \in \mathbb{N}_0^d$. Next, observe that each partial derivative $\partial^\alpha F$ with $|\alpha| = 1$, also belongs to ${}^\rho\mathcal{E}(\Omega)$, so, we can repeat the above arguments by induction in $|\alpha|$; the result is $(\partial^\alpha F)(x) = 0$ in ${}^\rho\mathbb{C}$ for all $x \in {}^\rho\tilde{\Omega}$ and all $\alpha \in \mathbb{N}_0^d$. On the other hand, we have $F = Q_\Omega(f)$ for some $f \in \mathcal{M}_\rho({}^*\mathcal{E}(\mathbb{R}))$ and $\partial^\alpha F = Q_\Omega(\partial^\alpha f)$. Also, $x \in {}^\rho\tilde{\Omega}$ means $x = q^d(\xi)$ for some $\xi \in \tilde{\Omega}$ (1.12). Hence, we conclude that $q(\partial^\alpha f(\xi)) = 0$ for all $\xi \in \tilde{\Omega}$ and all $\alpha \in \mathbb{N}_0^d$. The latter means $\partial^\alpha f(\xi) \in \mathcal{N}_\rho({}^*\mathbb{C})$ for all $\xi \in \tilde{\Omega}$ and all $\alpha \in \mathbb{N}_0^d$, as required. To show the preservation of the ring operations, let us take another asymptotic function $G = Q(g)$ with a representative $g \in \mathcal{M}_\rho({}^*\mathcal{E}(\mathbb{R}))$. For the addition we have

$$(F+G)(x) = Q(f+g)(q(\xi)) = q(f(\xi) + g(\xi)) = q(f(\xi)) + q(g(\xi)) =$$

$$= Q(f)(q(\xi)) + Q(g)(q(\xi)) = F(x) + G(x),$$

and similar for the multiplication. □

(2.9) Corollary *(C^∞-Functions)*: Let $F \in {}^\rho\mathcal{E}(\Omega)$ and $V(F)$ be its image in $({}^\rho\mathbb{C})^{{}^\rho\tilde{\Omega}}$. Then $V(F)$ is a pointwise C^∞-function from ${}^\rho\tilde{\Omega}$ to ${}^\rho\mathbb{C}$ in the sense that for any $x \in {}^\rho\tilde{\Omega}$ and any $\alpha \in \mathbb{N}_0^d$ the partial derivative $\partial^\alpha V(F)$ exists and it is continuous at x in the product-order topology in ${}^\rho\mathbb{R}^d$ (1.5).

Proof. The result follows directly from Theorem 2.6, by induction in $|\alpha|$. □

3 Fundamental theorem of Calculus in ${}^{\rho}\mathcal{E}(\Omega)$

We prove the Fundamental Theorem of Integration Calculus in ${}^{\rho}\mathcal{E}(\Omega)$ and show that the set of the scalars of ${}^{\rho}\mathcal{E}(\Omega)$ coincides exactly with A. Robinson's field ${}^{\rho}\mathbb{C}$ and the set of the real scalars of ${}^{\rho}\mathcal{E}(\Omega)$ coincides exactly with A. Robinson's field ${}^{\rho}\mathbb{R}$.

(3.1) Definition (*Scalars in* ${}^{\rho}\mathcal{E}(\Omega)$): We call $F \in {}^{\rho}\mathcal{E}(\Omega)$ a scalar in ${}^{\rho}\mathcal{E}(\Omega)$ if $\partial^{\alpha} F = 0$ in ${}^{\rho}\mathcal{E}(\Omega)$ for all $\alpha \in \mathbb{N}_0^d$, $|a| = 1$, in short, if $\nabla F = 0$ in ${}^{\rho}\mathcal{E}(\Omega)$.

(3.2) Theorem (*Fundamental Theorem in* ${}^{\rho}\mathcal{E}(\Omega)$): Let Ω be an open subset of $\mathbb{R}^d$ which in addition is arcwise connected. Let $F \in^{\rho} \mathcal{E}(\Omega)$ be an asymptotic function and ∇F be its gradient in ${}^{\rho}\mathcal{E}(\Omega)$. Then the following are equivalent:

(i) F is a scalar in ${}^{\rho}\mathcal{E}(\Omega)$, i.e., $\nabla F = 0$ in ${}^{\rho}\mathcal{E}(\Omega)$.

(ii) $(\nabla F)(x) = 0$ in ${}^{\rho}\mathbb{C}^d$ for all $x \in {}^{\rho}\tilde{\Omega}$ (pointwisely) in the sense that $(\partial^{\alpha} F)(x) = 0$ in ${}^{\rho}\mathbb{C}$ for all $x \in {}^{\rho}\mathcal{E}(\Omega)$ and all $\alpha \in \mathbb{N}_0^d$, $|\alpha| = 1$.

(iii) F is a constant asymptotic function in the sense that $F(x) = C$ in ${}^{\rho}\mathbb{C}$ for some $C \in {}^{\rho}\mathbb{C}$ and all $x \in {}^{\rho}\tilde{\Omega}$.

(iv) $F = Q_{\Omega}(c)$ in ${}^{\rho}\mathcal{E}(\Omega)$ for some $c \in \mathcal{M}_{\rho}({}^{*}\mathbb{C})$, more precisely, $F = Q_{\Omega}(f_c)$, defined by $f_c(\xi) = c$ for all $\xi \in {}^{*}\Omega$.

The connection between (iii) and (iv) is given by $C = q(c)$, where $q : \mathcal{M}_{\rho}({}^{*}\mathbb{C}) \to {}^{\rho}\mathbb{C}$ is the quotient mapping (Section 1, Part 1).

Proof. (i) $\Leftrightarrow$ (ii) follows immediately from Theorem 2.8.
(ii) $\Rightarrow$ (iii): Observe that $V(\|\nabla F\|) = 0$ in $({}^{\rho}\mathbb{C})^{{}^{\rho}\tilde{\Omega}}$ implies $\|\nabla F\| = 0$ in ${}^{\rho}\mathcal{E}(\Omega)$, by Theorem 2.8. Let $x_1, x_2 \in {}^{\rho}\tilde{\Omega}$. We have to show that $F(x_1) = F(x_2)$ in ${}^{\rho}\mathbb{C}$. We need the representatives: we have $F = Q_{\Omega}(f)$ for some $f \in \mathcal{M}_{\rho}((\mathcal{E}(\Omega))$ and $x_i = q^d(\xi_i)$, for some $\xi_i \in \tilde{\Omega}$, $i = 1, 2$. Since Ω is arcwise connected, it follows that ${}^{*}\Omega$ is arcwise connected, by Transfer Principle (T. Todorov [15], p. 686). Hence, ξ_1 and ξ_2 can be connected with a $*$-continuous curve $L \subset {}^{\rho}\tilde{\Omega}$. It follows

$$f(\xi_2) - f(\xi_1) = \int_L \nabla f(\xi) \cdot d\xi,$$

by Transfer Principle. Continuing the arguments, $\|\nabla F\| = 0$ in ${}^\rho\mathcal{E}(\Omega)$ implies $\|\nabla f\| \in \mathcal{N}_\rho({}^*\mathcal{E}(\Omega))$. Thus,

$$|f(\xi_2) - f(\xi_1)| \leq \|\nabla f(\xi_0)\| \|\xi_2 - \xi_1\| \in \mathcal{N}_\rho({}^*\mathbb{R})$$

where ξ_0 is some point on L. It follows $f(\xi_2) - f(\xi_1) \in \mathcal{N}_\rho({}^*\mathbb{C})$. Hence, $F(x_2) - F(x_1) = q(f(\xi_2) - f(\xi_1)) = 0$, as required.

(ii) $\Leftarrow$ (iii): We have $F(x + h) - F(x) = 0$ for all $x \in {}^\rho\tilde{\Omega}$ and all $h \in \mathcal{I}({}^\rho\mathbb{R}^d)$, by assumption (since $x + h$ is also in ${}^\rho\tilde{\Omega}$). It follows $\nabla F(x) = 0$ for all $x \in {}^\rho\tilde{\Omega}$, by the (ii) - part of Theorem 2.6.

(iii) $\Rightarrow$ (iv): Suppose that $F = Q_\Omega(f)$ for some $f \in \mathcal{M}_\rho({}^*\mathcal{E}(\Omega))$ and $C = q(c)$ for some $c \in \mathcal{M}_\rho({}^*\mathbb{C})$. Now, $F(x) = C$ for all $x \in {}^\rho\tilde{\Omega}$ implies $q(f(\xi) - c) = 0$ or, equivalently, $f(\xi) - c \in \mathcal{N}_\rho({}^*\mathbb{C})$, for all $\xi \in \tilde{\Omega}$ (1.14). Hence, $F - Q_\Omega(f_c) = Q_\Omega(f - f_c) = Q_\Omega(f - c) = 0$, as required.

(iii) $\Leftarrow$ (iv): We have $F(x) = q(f_c(\xi)) = q(c) = C$, where $x = q(\xi)$. □

(3.3) Corollary:

(i) The set of the scalars in ${}^\rho\mathcal{E}(\Omega)$:

$$\{F \in^\rho \mathcal{E}(\Omega) : \nabla F = 0 \quad in \quad {}^\rho\mathcal{E}(\Omega)\}$$

is an algebraically closed non-Archimedean field which is isomorphic to the field of A. Robinson's complex asymptotic numbers ${}^\rho\mathbb{C}$ (Section 1, Part 1) under the imbedding $\mathcal{C} : F \to F(x)$, where x is an arbitrary point in ${}^\rho\tilde{\Omega}$. Consequently, ${}^\rho\mathbb{C}$ is imbedded in ${}^\rho\mathcal{E}(\Omega)$, in symbols, ${}^\rho\mathbb{C} \subset {}^\rho\mathcal{E}(\Omega)$, by $\mathcal{C} : C \to Q_\Omega(f_c)$, where $f_c \in {}^*\mathcal{E}(\Omega)$ is defined by $f_c(\xi) = c$ for all $\xi \in {}^*\Omega$, where $C = q(c)$. The imbedding $\mathcal{C}^{-1}$ preserves the addition and multiplication.

(ii) The set of the real scalars in ${}^\rho\mathcal{E}(\Omega)$:

$$\{\pm|F| : F \in {}^\rho\mathcal{E}(\Omega) \quad \text{such that} \quad \nabla F = 0 \quad \text{in} \quad {}^\rho\mathcal{E}(\Omega)\}$$

is a totally ordered real closed Cantor complete non-Archimedean field which is isomorphic to the field of A. Robinson's real asymptotic numbers ${}^\rho\mathbb{R}$ (Section 1, Part 1) under the imbedding $\mathcal{C} : \pm|F| \to \pm|F(x)|$, where x is an arbitrary point in ${}^\rho\tilde{\Omega}$.

Proof. (i) For every scalar F in ${}^{\rho}\mathcal{E}(\Omega)$ there exists $C \in {}^{\rho}\mathbb{C}$ such that $F(x) = C$ for all $x \in {}^{\rho}\tilde{\Omega}$, by Theorem 3.2. So, the mapping $F \to F(x)$ from the constant functions in ${}^{\rho}\mathcal{E}(\Omega)$ into ${}^{\rho}\mathbb{C}$ is well-defined. The preservation of the addition and multiplication follows from Theorem 2.8. Conversely, if $C = q(c)$ is in ${}^{\rho}\mathbb{C}$, then the function $F = Q_{\Omega}(f_c)$ is the preimage of C.

(ii) follows immediately from (i) taking into account that ${}^{\rho}\mathbb{R} = \{\pm|a| : \alpha \in {}^{\rho}\mathbb{C}\}$. □

Acknowledgment: The author thanks Michael Oberguggenberger for the discussion of the preliminary version of this paper.

References

[1] J. F. Colombeau, New Generalized Functions and Multiplication of Distributions, North-Holland Math. Studies 84, North-Holland, Amsterdam 1984.

[2] J. F. Colombeau, Elementary Introduction to New Generalized Functions, North-Holland Math. Studies 113, North-Holland, Amsterdam 1985.

[3] R. F. Hoskins, J. Sousa Pinto, Nonstandard Treatments of New Generalised Functions, In: R.S. Pathak (Ed.), Generalized Functions and Their Applications, Plenum Press, New York 1993, 95-108.

[4] Li Bang-He, Non-standard analysis and multiplication of distributions, Sci. Sinica **21**(1978), 561-585.

[5] T. Lindstrøm, An Invitation to Nonstandard Analysis. In: N. Cutland (Ed.), Nonstandard Analysis and its Applications, London Mathematical Society Student Texts 10, Cambridge University Press, Cambridge 1988, 1-105.

[6] A. H. Lightstone, A. Robinson, Nonarchimedean Fields and Asymptotic Expansions, North-Holland Math. Library 13, North-Holland, Amsterdam; American Elsevier, New York 1975.

[7] W. A. J. Luxemburg, On a class of valuation fields introduced by A. Robinson, Israel J. Math. **25**(1976), 189-201.

[8] M. OBERGUGGENBERGER, Products of distributions: nonstandard methods, Z. Anal. Anwendungen **7**(1988), 347-365. Corrections: ibid. **10**(1991), 263-264.

[9] M. OBERGUGGENBERGER, Contributions of Nonstandard Analysis to Partial Differential Equations. In: N. J. CUTLAND, V. NEVES, F. OLIVEIRA, J. SOUSA-PINTO (Eds.), Developments in Nonstandard Mathematics, Longman, Harlow 1995, 130-150.

[10] M. OBERGUGGENBERGER, T. TODOROV, An embedding of Schwartz distributions in the algebra of asymptotic functions, Int. J. Math. & Math. Sci. **21**(1998), 417-428.

[11] M. OBERGUGGENBERGER, M. KUNZINGER, Characterization of Colombeau generalized functions by their pointvalues, Math. Nachr., to appear.

[12] V. PESTOV, On a valuation field invented by A. Robinson and certain structures connected with it. Israel J. Math. **74**(1991), 65-79.

[13] A. ROBINSON, Function theory on some nonarchimedean fields, Amer. Math. Monthly **80**(6)(1973), Part II: Papers in the Foundations of Mathematics, 87-109.

[14] T. TODOROV, Colombeau's New Generalized Functions and Non-Standard Analysis, In: B. STANKOVIĆ, E. PAP, S. PILIPOVIĆ, V. S. VLADIMIROV (Eds.), Generalised Functions, Convergence Structures and Their Applications, Plenum Press, New York 1988, 327-339.

[15] T. TODOROV, An existence result for linear partial differential equations with $\mathcal{C}^{\infty}$-coefficients in an algebra of generalized functions, Trans. Am. Math. Soc. **348**(1996), 673-689.

[16] V. S. VLADIMIROV, Generalized Functions in Mathematical Physics, Mir-Publishers, Moscow, 1979.

T. D. TODOROV
California Polytechnic State University, Department of Mathematics, San Luis Obispo, California 93407, USA, E-mail: ttodorov@polymail.calpoly.edu